Understanding Telecommunications Business

Other volumes in this series:

Understanding Telecommunications Business

Andy Valdar and Ian Morfett

The Institution of Engineering and Technology

Published by The Institution of Engineering and Technology, London, United Kingdom

The Institution of Engineering and Technology is registered as a Charity in England & Wales (no. 211014) and Scotland (no. SC038698).

The Institution of Engineering and Technology
Michael Faraday House
Six Hills Way, Stevenage
Herts, SG1 2AY, United Kingdom

www.theiet.org

British Library Cataloguing in Publication Data
A catalogue record for this product is available from the British Library

ISBN 978-1-84919-745-8 (paperback)
ISBN 978-1-84919-746-5 (PDF)

Typeset in India by MPS Limited
Printed in the UK by CPI Group (UK) Ltd, Croydon

To our grandchildren, the next generation:

- *Finn Johnston*
- *Niamh Johnston*
- *Lara Marriott*
- *Abby Marriott*
- *Rhys Valdar*
- *Hayley Valdar*
- *Joseph Valdar*

ARV and IMM
April 2015

Contents

Foreword

Today life is truly dominated by readily available and all-pervasive information, communications and entertainment, delivered to wherever we are, usually to our personal device. People now expect ubiquitous access to the world wide web and many couldn't operate their lives without it. Whether booking tickets online, purchasing goods and services, studying, being entertained, working or simply keeping-up with friends through social networking, such technology is central to supporting people's lives.

Most of those enjoying the benefits of this 'information age' would not be aware of the advanced technology in the hand terminals (smart phones, tablets, etc.), but rather assume that communication is provided over some wireless connection, vaguely linked to 'clouds' and the Internet. However, not only is there a wide range of infrastructure – in the form of cell-sites, masts, cables, switches, routers, computer servers and mass storage – but also many companies are involved in the business of providing and operating such resources. This book attempts to explain the complex interplay between the companies and how their businesses operate.

Our focus in this book is on the telecommunications that underpin all Internet, cloud, broadband, mobile and fixed services. We consider how the companies tackle the challenging information and communication technology (ICT) market-place; how they make a case for investment; and how they operate tele-communications networks and computer server resources. In particular, we have tried to provide a comprehensive introduction to the tools for analysing markets, constructing business cases and providing customer service – all with specific reference to telecommunications.

It is intended that the book will act as a text for undergraduate and graduate degree students. However, we feel sure that many people already working in the industry, or considering joining it, whatever their discipline, will also find our wide-ranging coverage helpful in showing how all the elements of the tele-communications and ICT business fit together.

We have based much of the content on the material used to teach Masters degree students over the last 10 years or so, as well as drawing on our knowledge gained through working within the industry. Our combined experience totals over 80 years, covering planning, strategy, financial management, network development, regulation, working variously for BT, United Nations and the Cabinet Office – and more recently teaching at University College London.

This book forms a companion to Understanding Telecommunications Networks, also in the IET Telecommunications series. Although self-contained, our book examines and extends the various business and commercial aspects of the technologies and networks described in the companion book. We feel that the combination of the two books will give the reader a holistic view of the fascinating world of telecommunications.

ARV and IMM
April 2015

Acknowledgements

We have found the writing of this book hard work, although also enjoyable and stimulating. But in undertaking this task we have also been delighted by the willingness of many of our colleagues to review the draft chapters and provide helpful suggestions and advice. Their generosity, both with time and ideas, has been greatly appreciated.

Our thanks and acknowledgements go to: Prof. Cliff Bowman (Chapters 3 and 10), Jane Britton (Appendix), Dr Meryll Bushell (Chapter 3), Keith Carrington (Chapters 1, 3, 4, 5 and 10), Prof. Moira Clark (Chapters 7 and 9), Prof. Izzat Darwazeh (Chapter 6), Lucy Freidman (Chapter 7), Liam Johnston (Chapter 9), Susan Kay (Chapter 10), Prof. Roger Maull (Chapter 10), Peter McCarthy-Ward (Chapters 1, 2, 5 and 8), Dr John Mitchell (Chapter 6), Prof Joe Nellis (Chapter 3), Richard Pettinger (Chapter 10), Prof Sri Shrikanithan (Chapter 4), Steve Thomas (Chapter 8), Prof Keith Ward (Chapters 6 and 9), Peter Willmott (Chapter 10), Simon Wood (Chapter 9) and Lucy Woods (Chapter 7).

Our final acknowledgements must go to our families and friends for their interest and encouragement. We finish by expressing our utmost gratitude to our wives – Jean Morfett and Susan Valdar for their enduring love and support.

ARV and IMM
April 2015

Abbreviations

1G	1st Generation (mobile network system)
2G	2nd Generation (mobile network system)
3G	3rd Generation (mobile network system)
4G	4th Generation (mobile network system)
5G	5th Generation (mobile network system)
4P's	Product, Place, Price, Promotion
ADSL	Asymmetric Digital Subscriber Line
AON	Activity-On-Node
API	Application Program Interface
Ar	Aggregation ratio
ARPU	Average Revenue per User
ATM	Asynchronous Transfer Mode
AUF	Asset Utilisation Factor
BBC	British Broadcasting Company
BER	Bit Error Rate
BIS	Brought into Service
BORSCHT	Battery, Overload-protection, Ringing, Signalling, Codec, Hybrid, & Test
BSC	Base Station Controller
BSG	Boston Consulting Group
BSP	Burden of Spare Plant
BT	British Telecommunications Plc
BTC	Base Station Controller
CapEx	Capital Expenditure
Cs & Bs	Clicks and Bricks
CAPM	Capital Asset Pricing Model
CDMA	Code Division Multiple Access
CDN	Content Distribution Network
CEO	Chief Executive Officer

CER	Cell Error Ratio
CFO	Chief Financial Officer
CNN	Cable News Network
CP	Communication Provider
CPI	Consumer Price Index
CR	Corporate Responsibility
CRD	Customer-Required-by-Date
CRM	Customer Relationship Management
CSR	Corporate Social Responsibility
CTD	Cell Transfer Delay
CTV	Cable Television
D-side	Distribution-side
DCF	Discounted Cash Flow
DER	Digital Error Rate
DMSU	Digital Main Switching Unit
DNS	Domain Name System
DOS	Denial of Service
DP	Distribution Point
DSC	District Switching Centre
DSLAM	Digital Subscriber Line Access Multiplexer
DT	Deutsche Telecom
E	Erlang (unit of telephone traffic)
EBITDA	Earnings Before Interest, Tax, Depreciation and Amortisation
EFT	Earliest Finishing Time
ELF	Early Life Faults
EPS	Earnings per Share
E-side	Exchange-Side
EST	Earliest Start Times
EU	European Union
ETO	Economic, Technical or Organisational
eTOM	extended Telecommunications Operations Map
FAB	Fulfilment, Assurance and Billing
F&F	Friends and Family
FAQ	Frequently Asked Questions
FCA	The Financial Conduct Authority
FCC	Federal Communications Commission

FD	Finance Director
FDM	Frequency Division Multiplexing
FRIACO	Flat-Rate Internet Access Call Origination
FTTH	Fibre to the Home
FTTO	Fibre to the Office
G/	Gateway
GAAT	Generally Accepted Accounting Standard
GDP	Gross Domestic Product
GOS	Grade of Service
GPRS	General Packet Radio Service
GPS	Global Positioning System
GSC	Group Switching Centre
GSM	Global System for Mobile
HD	High Definition
HFC	Hybrid Fibre-Coax
HLR	Home Location Register
HP	Hewlett Packard
HQ	Head Quarters
HR	Human Resources
HRM	Human Resources Management
HSE	Health and Safety Executive
IaaS	Infrastructure as a Service
ICT	Information and Communication Technology
IDA	Integrated Digital Access
IDV	Degree of Individualism
IFRS	International Financial Reporting Standard
IM	Instant Messaging
IMS	Internet Protocol Multimedia Subsystem
IMSI	International Mobile Subscriber Identity
IN	Intelligent Network
IP	Intellectual Property
IP	Internet Protocol
IPv4	Internet Protocol Version 4
IPv6	Internet Protocol Version 6
IPTV	Internet Protocol Television
IRR	Internal Rate of Return

ISC	International Switching Centre
ISDN	Integrated Services Digital Network
ISP	Internet Service Provider
ITIL	The Information Technology Infrastructure Library
ITT	Invitation to Tender
ITU	International Telecommunications Union
ITU-T	International Telecommunication Union – Telecommunications section
IVR	Interactive Voice Response
JD	Job Description
LAN	Local Area Network
LC	Line Card
LDDS	Long Distant Discount Service
LE	Local Exchange
LFT	Latest Finish Time
LoP	Life of Plant
LSP	Label Switched Path
LST	Latest Start Time
LTE	Long-Term Evolution
M&A	Mergers and Acquisitions
MAS	Masculinity versus Femininity
MDF	Main Distribution Frame
MGC	Media Gateway Controller
MMC	The Monopolies and Mergers Commission
MMS	Multimedia Messaging Service
MNO	Mobile Network Operator
MPLS	Multi-Protocol Label Switching
MSC	Main Switching Centre
MSC	Mobile Switching Centre
MSISDN	Mobile Station International Subscriber Directory Number
MSP	Multi-Service Platform
MTTR	Mean Time To Repair
MVNO	Mobile Virtual Network Operator
NFC	Network Field Centres
NGN	Next Generation Network
NHS	National Health Service

NNI	Network-Network Interface
NP	Number Portability
NPV	Net Present Value
NOC	National Operations Centre
NTL	National Transcommunications Limited
NTTP	Network Test and Termination Point
NUF	Network Utilisation Factor
O&M	Operations and Maintenance
Ofcom	The Office of Communications
Oftel	The Office of Telecommunication
Ofgem	The Office of Energy Regulation
Ofwat	The Office of Water Regulation
OFTA	The Office of the Telecommunications Authority
OLO	Other Licenced Operator
OpEx	Operational Expenditure
OSI	Open Systems Interconnection
OSS	Operational Support Systems
OTT	Over-the-Top (Application Provider)
PaaS	Platform as a Service
PABX	Private Automatic Branch Exchange
P&L	Profit and Loss
PC	Personal Computer
PCM	Pulse Code Modulation
PCP	Primary Connection Point
PDA	Personal Digital Assistant
PDH	Plesiochronous Digital Hierarchy
PDI	Power Distance Index
PE	Price Earnings
PERT	Project (or programme) Evaluation and Review Technique
PEST	Political, Economic, Social and Technology
PESTLE	Political, Economic, Social, Technology, Legal and Environmental
PM	Product Manager
PM	Project Manager
POLO	Payments to Other Licenced Operators
PoI	Points of Interconnect

PON	Passive Optical Network
POP	Point of Presence
PR	Public Relations
PSN	Packet Service Node
PSTN	Public Switched Telecommunications Network
PUV	Perceived Use Value
QoE	Quality of Experience
QoS	Quality of Service
R&D	Research and Development
RAN	Radio Access Network
RFI	Request for Information
RPI	Retail Price Index
ROC	Regional Operations Centres
ROCE	Return on Capital Employed
ROLO	Receipts from Other Licenced Operators
ROI	Return on Investment
RONA	Return on Net Assets
RSI	Repetitive Strain Injuries
SaaS	Software as a Service
SCP	Secondary Connection Point
SDH	Synchronous Digital Hierarchy
SIP	Session Initiation Protocol
SLA	Service-level Agreements
SLC	Subscriber Line Card
SMART	Specific, Measureable, Achievable, Relevant, Time-bound
SME	Small and Medium Enterprise
SMS	Short Message Service
SNS	Social Network Service
SS7 or SSno7	Signalling System 7
SSAP	Statement of Standard Accounting Practice
STR	Strategic Telecommunications Review
SWOT	Strengths, Weaknesses, Opportunities, and Threats
T&C	Terms and Conditions
TDM	Time Division Multiplexing
TDR	Test Discount Rate
TE	Trunk Exchange

TE	Telephone Exchange
Telco	Telecommunications Company
TMF	TeleManagement Forum (now known as 'TM forum')
TS16	Time Slot 16
TUPE	Transfer of Undertakings (Protection of Employment)
UAE	United Arab Emirates
UAI	Uncertainty Avoidance Index
UNI	User-Network Interface
USO	Universal Service Obligation
VCA	Value Chain Analysis
VDSL	Very-high-bit-rate Digital Subscriber Line
VLSI	Very Large Scale Integrated
VOIP	Voice Over Internet Protocol
VOLTE	Voice over LTE
VPN	Virtual Private Network
VULA	Virtual Unbundled Local Access
WACC	Weighted Average Cost of Capital
WiMAX	World-wide Interoperability for Microwave Access
WLC	Whole-life Cost
WRULD	Work Related Upper Limb Disorders

Chapter 1

Introduction to the telecommunications business

1.1 Introduction

The dictionary definition of telecommunications is 'communication over long distance by cable, telegraph, telephone or broadcasting', but since its initiation over 100 years ago things have moved rapidly. Telecommunications is now a very complex industry with many different pressures, operating in a highly dynamic environment. It is best viewed as part of a wider industry known as information and communication technology (ICT). The purpose of this chapter is to explain where telecommunication fits in, to highlight some of the complexities – hopefully to simplify them – and to position the industry in today's dynamic business environment.

We aim to:

- Explain value chain analysis (VCA) and its relevance to telecommunications.
- Position telecommunications in the ICT value chain.
- Describe how telecommunication companies make money i.e. the business model.
- Describe the wider business environment and its relevance.
- Use this analysis to explain likely future trends and changes occurring in the industry.

We all are customers of telecommunications in one form or the other. We make phone calls, use our mobile, surf the net, maybe use it to buy books, tickets, holidays, flights. We go online to study, work, pursue our hobbies and find out what's happening in the world. We send email, communicate through instant messaging and make voice calls through the computer. We download music and films; we enjoy radio and TV online, either in *real-time* or *time-shifted*; we may play games, either individually or interactively with other players.

And much, much more.

We are individuals, but companies also use telecommunications in a wide variety of ways. They use telecommunications to advertise their products and services; to reach their customers; to sell to them and to service products. They use them to manage their supply chains, to buy raw materials and to manage distribution. In some cases the whole company is designed around telecommunication e.g. First Direct or Amazon. In other cases, key components of the business couldn't operate without telecommunication e.g. just-in-time stock systems for the high-street retail outlets.

Table 1.1   Example companies in ICT

Category	Example company
Equipment (e.g. user terminals)	Nokia; HP; RIM; Apple
Broadband	BT; Sky; TalkTalk; Virgin Media
Calls	BT; Tesco; Sky; Virgin Media; Skype; Facetime
Mobile	Zain; EE; Vodafone; Optus
Internet service providers	MNS; Yahoo; BTInternet; Tesco; Virgin Media; TalkTalk
Applications	Google; YouTube; Facebook

Without telecommunications many companies would fail. Even a short outage can cost millions of pounds.

Whether we are talking about individuals, small companies or major firms (or indeed public sector organisations like the National Health Service), tele-communications services are provided by a wide variety of companies.

A few of the categories and companies involved are shown in Table 1.1.

1.2   Value chain analysis (VCA)

To provide a full range of telecommunication services all the categories of products and services are involved. Incidentally, a *product* might strictly be defined as a physical item such as a mobile phone or laptop, while a *service* is intangible, such as a broadband connection. However, in this book we use the words somewhat interchangeably to refer to anything offered to the market (usually for a price) that aims to satisfy a customer need or want. To make sense of this very complex world we can employ VCA. This is a business analysis tool originally invented by Prof. Michael E. Porter, who is a professor at the Institute for Strategy and Competitiveness, Harvard Business School. Initially used to analyse a company's activities, it is now also a very useful tool to analyse a complete industry.

In essence the value chain starts with the customer (usually drawn on the right) and works back through the various activities required to provide the product or service to the customer, drawing each activity or component as a block on the diagram. A simple example using food retailing is shown in Figure 1.1 [1].

Food retail value chain

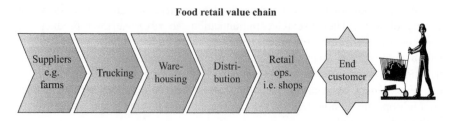

Figure 1.1   Food retailing value chain

ICT value chain

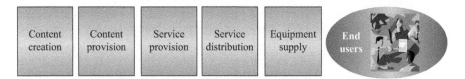

Figure 1.2   High-level ICT value chain

1.2.1   The ICT value chain

We have seen that telecommunications fits into a wider industry of ICT. This is a complex and crowded industry with many players, much competition, significant overlap and dynamic movement between firms. In these circumstances VCA is an ideal tool for making sense of the industry and getting to grips with what is going on. The diagram in Figure 1.2 represents a relatively simple view, but nonetheless it provides useful insight.

Within each of these categories we can identify some sub-categories and also some of the key players. For example:

Content creation: It consists of authors, musicians, sports clubs and even individuals who put their material on YouTube or Facebook. But it also includes the government (e.g. the content on public service websites e.g. www.direct.gov.uk) and advertising companies. Increasingly, content is coming to include the information that drives commerce, rather than just entertainment e.g. insurance policies, healthcare records, even road traffic data. Soon it will even cover digital money as value stored on a smartphone or tablet replaces coins, notes and credit card transactions.

Content provision: It includes companies that bring content together and package it in a way that people might want to view it. It includes book publishers such as Penguin, and music publishers such as Sony; not only personal content packagers such as Facebook but also Sky and the BBC who package a great deal of different types of content for presentation. It is important to note that the BBC, Channel 4 and ITV no longer rely solely on TV and radio broadcasting to distribute their content but use a variety of service distribution channels (Internet, digital, podcast, etc.).

Service provision: It is made up of internet service providers (ISPs) such as Yahoo, Tencent, Thus, PlusNet and Sky; it also includes application developers (such as Microsoft and Google) and specialised internet portals (such as eBay, iTunes and price comparison sites).

Service distribution: It includes many of the traditional telecommunications companies (Telco) because the activity is essentially to deliver the service, telephone call or information to the end user in a way they can make use of it. So BT provides telephone lines and broadband access. Mobile companies such as Vodafone are included here. So, strictly, are newsagents that deliver physical newspapers, CD retailers and book shops that sell goods on the High Street. They are all part of the ICT value chain.

Equipment supply: ICT services need to be delivered via a piece of kit, be it a personal computer (PC), laptop, tablet, games machine, TV or personal organiser. There is a wide range of companies providing either a generalised portfolio of electronic equipment e.g. Apple, Sony, or very specialised equipment such as Nintendo (a big player in the gaming market).

1.2.2   Creating value from the value chain

There are some obvious advantages for a company to operate in one specific part of the value chain. In particular, they aim to become expert, to be the lowest cost producer and to lead the market in their favoured activity. If this is their firm strategy, they will often demerge or outsource other activities; they may perhaps enter strategic partnerships with players in other parts of the value chain to deliver that part of the work.

However, many companies operate in more than one part of the value chain. This doesn't reduce the benefit of VCA. Use of the value chain enables companies to analyse where they have particular strengths and capabilities; where they might create new opportunities; which elements of the chain they might add to their capabilities or whether they should work with others. So VCA not only enables the companies to analyse the current situation; but it also enables them to develop new strategic moves for the future.

Let us look at some real-life examples. If we think about where IBM operate in the value chain, we find that they are certainly part of the service provision activity, supplying business computer services to major companies, but they also distribute services for many of their customers. They do this by renting, or in some cases by owning, telecommunications networks to deliver the applications. Similarly, if we consider Sky, they are certainly content providers (they were, at one time, considering buying content-generation capabilities), but they also distribute services over their own satellite broadcast network; finally they are also active in branding equipment and providing it to end users, though they may not actually manufacture it.

Overlap across the value chain is a normal business process. Companies use VCA as a strategic tool to identify new business opportunities. They might consider moving into adjacent parts of the value chain as a way of cutting cost, reducing reliance on others or providing a more complete service to their customers. Firms could do this by acquiring an existing company, by building their own capability or by making a strategic partnership with another player. A real life example, that of triple and quad play, is described in Box 1.1.

Therefore, in their hands, VCA can be a valuable tool to build future competitive advantage. We will look at the process of building competitive advantage in greater detail in Chapter 3.

Some examples:

- In food retailing, supermarket chains are buying farms – often abroad – both to reduce costs and to control the quality of the food that they sell.
- In car manufacture, Jaguar initially had a simple customer–supplier relationship with Ford to purchase basic components such as engines; they then

ICT value chain

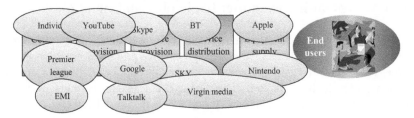

Figure 1.3   Example companies within the ICT value chain

developed the relationship further by entering a strategic partnership to jointly improve the product for customers; and eventually sold out to Ford to ensure greater manufacturing efficiency and lower costs.

- In ICT, Microsoft attempted to buy Google to overcome their own weakness in search capabilities; BT entered a strategic partnership with Yahoo to fast-track their Internet service provision; and Virgin Mobile, NTL and Telewest got together to create Virgin Media to gain greater leverage across the whole value chain.

In Figure 1.3 we show some of the key ICT players and their position on the Value Chain.

1.2.3   The expanded ICT value chain (Figure 1.5)

Within each of the elements of the high-level value chain a whole series of different contributing functions can be identified. In some cases these will be separate activities carried out by the same companies; in other cases, specialist organisations will be creating value and earning profit by focusing on narrow, but essential, elements. For example, within food retailing, distribution might cover bulk transport (e.g. Eastons plc), packaging (e.g. Tetra Pak UK plc) and companies specialising in movements of perishable goods, such as Perishable Movements plc.

This is equally true of the ICT marketplace, where each element of the value chain will cover a wide range of functions, either completed in-house or potentially outsourced to specialists.

Within the consumer equipment category, a company like Apple will potentially outsource component supply, assembly, even parts of the design process; while specialist companies will offer service and repair of one or a range of proprietary brands (with or without the brand's blessing), e.g. Geek Squad, a spin-off from Carphone Warehouse.

Similarly, within content provision category, some companies will be closely associated with the content creation and define their role in terms of preparing that material for digital distribution, e.g. Random House. Other companies will seek varying content to match customer needs and package it in a coherent and exciting way e.g. Fashion feed from Sina Weibo. There will even be backroom companies, far more focused on a single narrow role. For example, Digital Nirvana is a world leader in repurposing content (i.e. digitising it, or otherwise preparing it, to displays appropriately on mobile phones, tables, etc.).

Box 1.1   Real-life example: triple play and quad play

A specific example of how the value chain can be used in telecommunications to provide strategic competitive advantage is *triple play* (and now also *quad play*) (see Figure 1.4). In essence, the marketing thinking is that customers would prefer to buy all their communication needs in a single package (sometimes called a *one-stop-shop*). In *triple play* the provider, therefore, tries

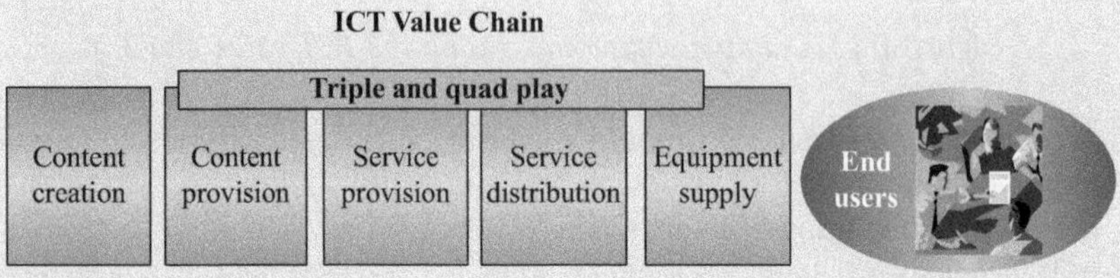

ICT Value Chain

Figure 1.4   Triple play and the ICT value chain

to provide broadband, television and telephone calls in a single package. Broadband incorporates high-speed access to the Internet, email, etc., while television may include *pay-as-you-go* entertainment, time-shifted programming and video on demand. The addition of mobile services makes this package into *quad play*.

The attraction of triple and quad plays is that they increase average revenue per user (ARPU); allow selective cross-subsidy between services (especially attractive for short-term promotions); and make customers more reluctant to leave one company for another supplier. In the colourful language of marketeers, they make customers *sticky*. However, a risk of multiple-service plays such as these is that, as the offer becomes more complex, so does the selling process required to capture the customer. If it goes beyond the capability of call centres, then the cost of sale will rise, and advantage may shift to those with retail distribution networks – such as Carphone Warehouse.

While in a technical sense triple and quad plays can be delivered over a single access line to the home (twisted copper pairs or co-axial cables [2]), several elements of the value chain are required to put the commercial offering together. For this reason companies that are developing a triple play strategy will need to determine which part of the value chain already represents a strength for them and where they need to acquire, partner or develop new strengths.

Virgin media: The best example of triple and quad plays in operation in the UK is the development of virgin media. Originally, two companies (NTL and Telewest) provided cable TV services to homes in the UK. They quickly found that they could increase margins by adding telephony to their package. Late in the 1990s, sales of home computers took off and it became important for the cable companies to offer access to the internet to attempt to stop their customers moving to the broadband services offered by BT and others. Finally, in an attempt to provide their customers with a full communications service, they added mobile services to the package. To do so they initially entered a strategic partnership with Virgin Mobile; finally all three companies merged to form virgin media in a hope of domdominating the value chain for provision of quad play [3].

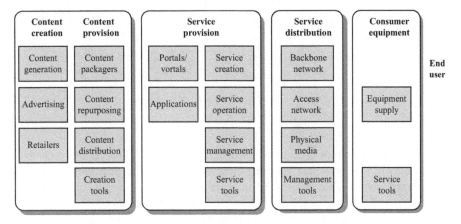

Figure 1.5   The expanded ICT value chain

So when ICT companies use VCA to develop strategy and to read the dynamics of their industry, they will be interested in the detailed set of activities covered by each of the broad elements.

1.2.4   The Internet value chain

Finally, we have spent significant time here looking at the ICT value chain, and the telecommunications element within it. However, we can broaden our viewpoint further and consider the extended value chain as it applies to the Internet and the content and services that it contains. A good way to look at this is represented in Figure 1.6.

We can see that all the activity associated with the Internet can be analysed into the standard sections of a value chain. It is interesting to note the wide range of online services based upon the Internet, ranging from the communications group covering Voice over Internet Protocol (VoIP) (e.g. Skype), content group, search group, entertainment group and transactions group. Furthermore the enabling technologies – web hosting, web design, etc. – have fuelled the massive expansion of the IT industry. On the other hand, the relatively constrained role of the Telco's in the connectivity space is clearly apparent, explaining their ambitions to move left within the Internet value chain.

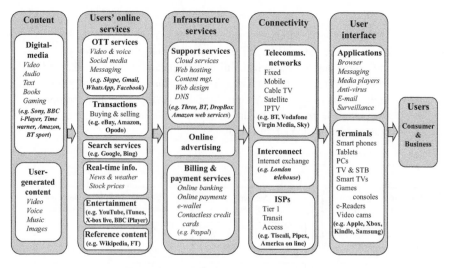

Figure 1.6   The Internet value chain

1.2.5   Consumer-generated content

Any view of the current activity progressing through the ICT value chain demonstrates clearly the force that has done more than anything to disrupt the world of content production. That force is the proliferation of what we might call consumer-produced media. We have moved from a world of media produced by a few major corporations for the consumption of the many; to a world where content is produced by the many for the entertainment of those who choose to opt-in. As John Naughton puts it, this is a world that is moving from the *push* of broadcast TV, blockbuster films and global pop music to a world where individual consumers *pull* content from the web. In the new world nothing comes to you that you haven't chosen, from the whole universe of consumer-produced media [4].

The prime example must be YouTube, but everything from blogs, digital photography, travel reviews and Twitter contribute to the trend. Indeed, every posting to any social network service (SNS) is a form of consumer production of media content.

Entertainment was the initial focus of this revolution, as we saw major disruption to the global major music studios and film publishers. However, consumer production has moved on to challenge journalism, book publishing, medicine and education. What is certain is that the relentless increase in innovation and creativity, fuelled by the technological innovation of the Internet, will continue to decimate the old guard of corporations reliant on the mass-production of media [5].

We expand on some of these issues when we look at the empowered consumer in section 7.8.

1.2.6   Ecommerce and Web 2.0

The digital world continues to evolve, and eCommerce is changing faster than any other aspect as newer and newer models develop to identify and extract value opportunities. Web 2.0 is the term (invented by Tim O'Reilly in 2004) to describe a world where users control and own their data; where they work collaboratively; where social networks play a central role in peoples' lives, affecting their buying decisions. In this world, companies must become customer-centred and pay attention to some very new aspects of business. We explore these issues in detail in Chapter 7, but this might also be a good moment to reflect on a life beyond online shopping.

Web 2.0 services are leading to a new level of, for example, interaction, uploading of self-generated material and applications that will revolutionise the creative arts, culture, choice and self-expression. Examples would include:

- Online museums and interactive museum guides.
- Web-casts of live opera and theatre performances.
- Art interacting with media (such as Julius Popp's sculpture called bit.fall that replicates words from real-time newsfeeds in a waterfall).
- Broadcast TV combined with online presentation and interactive services.
- Twitter and other social networks used as a tool to comment on TV shows, and TV shows reacting to tweets in real time.
- Individuals creating (as well as publishing) their own novels, music, and art using online applications and tools.
- The 2012 London Olympics dubbed the first digital games. One particular feature of the games, never previously observed, was that people in one venue were watching live events elsewhere on their smartphones or tablets e.g. watching the final of the men's 100 m race whilst attending the swimming.

These examples, and many others not even thought of, represent a huge burgeoning of creativity and personal choice driven by Web 2.0.

1.3   The high-level telecommunications commercial model

1.3.1   Introduction

In any understanding of the telecommunications industry it is vital to understand the commercial model that operates.

In essence, running a telecommunication business involves investing a significant sum of money in major infrastructure – the network and support systems. This is a long-term investment as most network technology lasts 10–15 years. This investment is recovered from products and services that run over the network e.g. voice calls, email traffic, surfing the net, etc. – often called *conveyance* or *traffic*. These products or services are sold on a short-term basis i.e. per call or message, often per minute or by unit. Business models are changing and we will look at the evolving business models later in the book, but, nonetheless, an important part of running a telecommunication company remains the need to recover major sunk costs by filling the network in the most efficient manner with as much profitable telecommunication traffic as possible.

Whilst we refer to telecommunications companies throughout this section, the analysis is equally applicable to other ICT companies that require significant investment and shared infrastructure, such as ISPs who run massive server farms, broadcasters or even e-commerce retailers such as Amazon.

We will look at three aspects: revenue, costs and investment. However, it is vital to realise that all these three are closely related and that they interact with each other. We will therefore also consider the whole model and what it tells us about how best to operate a telecommunication company to maximise shareholder returns.

1.3.2   Revenue

To maximise profit a Telco needs to sell as many products or services as possible (i.e. the volume of calls, the number of TV programmes viewed, etc.) at the highest price possible. Each of these products and services may be charged in a different way.

Some examples:

- Landlines are normally charged as a monthly or quarterly subscription (increasingly with minutes of use built into the bundle).
- Broadband may be a monthly subscription, but minute by minute dial-up remains an important product in some parts of the world.
- Telephone calls (fixed or mobile) are sometimes charged by the minute, but often with a package of *free* minutes.
- Entertainment is often charged by the value of the content (i.e. a tune, album or film).

To maximise revenue, a firm will wish to sell as many lines, calls, etc., as possible, but this also has a second consequence. In the same way that a car manufacturer needs to build as many cars as it hopes to sell, and a restaurant has to have enough food to make all the meals it sells, a Telco has to have the capacity to provide the lines, deliver the calls or show the films that customers want. In short, a Telco has to forecast demand, and the more demand it hopes to serve, the more investment it needs to make in the size and structure of its network (in this regard the term 'network' includes not only traditional cables, telephone exchanges and cellular masts but also the servers and systems required to provide the full range of products and services). This has been summarised in Figure 1.7.

We will return to the issue of investment in the network later but some examples of the consequences of increased demand would be as follows:

- More calls might require more capacity in telephone (fixed or mobile) exchanges (i.e. *switching capacity* in the jargon).
- More streaming of video clips from the Internet, or downloading of music might require more *backbone capacity*.
- Selling more lines to customers might require more investment in the *access network* (the lines that run through the streets to connect homes and businesses to the telephone exchange).

Not all of this investment will happen immediately because Telcos aim to have spare capacity in their networks; to achieve this they need good forward plans and excellent management of the lead-times (see more on the planning process in Chapter 5).

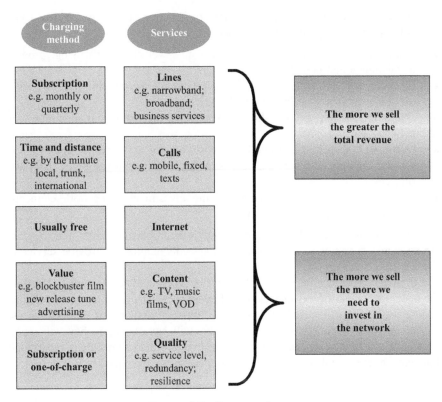

Figure 1.7   Revenue drivers

Before we leave revenue we can draw some clear conclusions. The more products and services you sell the more revenue you achieve, but the more you need to invest in the size and structure of your network. Finally, what you sell or plan to sell determines which part of the network you need to invest in.

1.3.3   Costs

Like all companies, Telcos incur a wide range of costs in delivering services to their customers. Clearly, to maximise profit a company will wish to minimise costs commensurate with maximising profitable sales. It does this through a focus on efficiency.

Costs can be categorised in a number of different ways. For example by:

- Cost element/nature (e.g. salaries and wages).
- Cost centre (e.g. Engine Plant or Regional Sales South).
- Function/activity/process (e.g. marketing or billing).
- Product/service (e.g. equipment or local call).
- Customer segment/market (e.g. retail, wholesale, small and medium enterprise [SME]).

If we look specifically at costs incurred by function or activity a Telco would certainly incur most, or all, of the following:

- *Marketing and selling*: such as sales teams, websites, advertising, product development and launch, etc.
- *Customer service*: such as maintaining customer records, billing, answering service questions, clearing line-faults, etc.
- *Operations*: such as installing lines, repairing faults, maintaining the network, etc.
- *Overheads*: such as accommodation costs of rent and rates, computing and systems support, and headquarters' staff such as human resources (HR) and finance.
- *Depreciation*: the annual charge for the investment in network elements, vehicles and support systems, etc.

Once again we find that there is an interrelation between costs and the size, structure and operation of the network and systems required to provide service to the company's customers.

First, the efficiency with which the network is operated has immediate effect on the operational costs of the business e.g. if telephone exchanges are very unreliable, excess costs will be incurred in fire-fighting repairs. Similarly, if the local access cables are poorly maintained, then customers will have outages; they will report frequent faults to the call centre; and they will require field engineers to repair their line, etc. Other examples of operational efficiencies are described in Box 1.2.

Box 1.2   Other examples of operational efficiency

- Providing new lines will be more effective if the local access network has a high percentage of spare unused pairs (cables).
- The number of faults dealt with per operator depends on investment in diagnostic systems.
- More modern equipment requires less routine maintenance.
- On-line help services (FAQs, etc.) will reduce customer calls to call centres.
  (There are many more examples.)

These operational efficacies are called *operational cost drivers*. We will return to this subject when we discuss network economics in Chapter 8.

Second, investment levels in the networks and systems also have an impact. If investment has been made to keep the equipment up-to-date, it will carry more traffic (calls, data, etc.) and fail less often. Similarly, if investment has been made to provide spare capacity in the network, the company will incur less cost in rapid actions to fulfil growth in service and will be able to handle unexpected demand more easily. And if investment has been made in state-of-the-art computer systems,

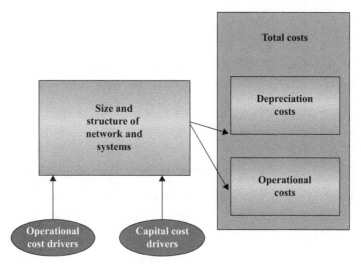

Figure 1.8   Cost drivers

then less manual processing (and therefore lower wage bills) will be incurred. These factors are described as *capital cost drivers*. This subject is addressed further when we discuss network economics in Chapter 8.

Finally, the greater the investment in network and systems, the more depreciation is required to reflect the costs incurred in running the business. The exact relationship between the levels of investment and depreciation is bound up in accountancy policy and therefore best left to the company finance department! Figure 1.8 shows these relationships.

Again, we can draw some clear conclusions. To maximise profit, operational costs need to be kept to a minimum, commensurate with providing a high-quality service to customers. Operational and capital cost drivers govern the relationship between size, structure and efficiency of the network and the level of operational costs incurred. And the greater the investment in network and systems, the greater the depreciation cost. (See Box 1.3 for a detailed explanation of how depreciation is calculated.)

1.3.4   Profit and profitability

Profit is simply total revenue (made up of all the subscriptions, rentals and usage – minutes and content) for all the relevant services, less the total costs (operational costs and depreciation).

Profitability is also important. Whilst we measure profit in £s or $s (£millions hopefully, or even £billions), we are also interested in profitability, i.e. the profit relative to the investment required to achieve it.

Clearly a company – all things being equal – aims to maximise profit with the lowest investment possible (commensurate with efficient operation and

Box 1.3   Depreciation

Depreciation is an accountancy technique used to spread the cost of an investment over its useful, revenue-earning life. Such an investment would be used to purchase an asset, typically a building, piece of machinery, or equipment; this type of expenditure is known as capital account expenditure (as opposed to current account expenditure) and can only be classed as such if the asset purchased has a revenue-earning life longer than one year. The cost of the investment is divided by the agreed life of plant (LoP) and the annual charge is shown in the expenditure statement to represent that year's cost for using the asset. Depreciation is therefore a non-cash transaction but does have a major effect on a company's finances because the more a company invests in assets, the more depreciation it incurs and the lower its profit.

LoPs are calculated by the Finance department, in conjunction with their engineering colleagues, and will be discussed and agreed with the auditors. They take into account both the long-term likely success of the products and services supported by the asset, the levels of wear and tear that may exhaust the asset, and the speed of obsolescence. In the 1970s and 1980s the key components of a telecommunications networks typically had a LoP of around 40 years. Today that might be only 15 years, perhaps less for some parts of a mobile network. (For a more complete discussion of network LoP's see section 5.2.2.)

Note: current account expenditure consists of all the running costs of a company that need to be spent to deliver today's services e.g. salaries of staff; rent and rates; raw materials.

safeguarding future revenue streams). Profitability is normally measured as return on investment (ROI) i.e. profit divided by value of assets employed.

Finally – like costs and revenue – profit also has an effect on investment levels, in two main ways:

- Greater profit provides more cash to reinvest without borrowing. This can be a major advantage if the cost of borrowing is high.
- High profitability will encourage potential investors to buy shares – everyone wants a piece of a successful operation. That will make the job of raising new funds both easier and cheaper.

So, a successful company will have a lower relative cost of capital than an unsuccessful one.

We can bring all these points together in a single diagram (see Figure 1.9) that effectively describes – at a very high level – the commercial model of any major Telco.

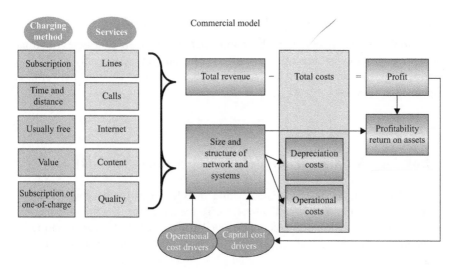

Figure 1.9   Telecommunications commercial model

1.4   The commercial model in practice

1.4.1   Introduction

We can take the concept of a commercial model and apply to a telecommunications service. Such a model will show how the service is delivered by the various service providers and how the money flows between them – sometimes this is captured by the expression of 'who's doing what to whom and who is paying for it'. In defining a service commercial model we need to define two distinct entities, namely: 'roles' and 'players'. There are many roles in the ICT and telecommunications industry e.g. access provider, core network provider, various level of ISP, mobile network operator (MNO), broadband service provider, cable TV provider and data centre provider. In addition, there are a range of so-called 'virtual operators' who provide a customer-facing function to offer mobile service as one of their branded offerings, in effect reselling capacity leased from network operators.

The players in the market are the companies involved, who may take just one, a few, or all of the roles – e.g. Virgin Media take the roles of cable TV, broadband service provider, virtual mobile operator and fixed telephony provider.

The commercial model for a service shows how the service flows from source to customer, in effect following the value chain, and how the money flows. Where third parties are involved in carrying the service or providing TV content, then further flows of money may be required. The final design of the service is often decided by the commercial ownerships of the various parts of its operation and conveyance. We look at the service commercial model as one of the four architectural perspectives involved in top-level network design in Chapter 6. In Chapter 6 we also introduce a simple pictorial model of the various pieces of telecoms and ICT infrastructure, and the way that services are delivered by the Telcos and the over-the-top (OTT) application providers e.g. Skype (see Figure 6.4).

1.4.2   The commercial model in practice – mobile

Let us look at virtual operators in the mobile market. There are four types of virtual mobile operators extending from a minimal sales-and-marketing only model to a nearly fully functional operator but without a wireless cellular network, as explained in Box 1.4.

Box 1.4   Mobile virtual network operators (MVNO)

A full set of functions associated with a mobile network are shown in the stack on the extreme right in Figure 1.10. To the left of the dividing line there are four functional stacks, each associated with a different class of virtual mobile operators – all of whom lease wireless call/session capacity from a full MNO [6–9].

Brand extender. This company has already a well-established brand and sells their branded mobile service as a way of adding value to their customer base. Such companies provide only marketing and sales distribution, and rely on other operators/service companies to provide the remaining functions. Examples within the United Kingdom include: RSPCA Mobile, Family Mobile and OwnFone.

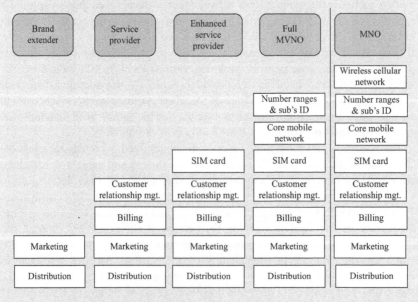

Brand extender	Service provider	Enhanced service provider	Full MVNO	MNO
				Wireless cellular network
			Number ranges & sub's ID	Number ranges & sub's ID
			Core mobile network	Core mobile network
		SIM card	SIM card	SIM card
	Customer relationship mgt.	Customer relationship mgt.	Customer relationship mgt.	Customer relationship mgt.
	Billing	Billing	Billing	Billing
Marketing	Marketing	Marketing	Marketing	Marketing
Distribution	Distribution	Distribution	Distribution	Distribution

Figure 1.10   The range of mobile virtual network operators

Service provider. This class of operator also provides full customer support in terms of sales, repairs, enquiries and billing. Examples in the United Kingdom include: Go mobile, DataWind and Highnet.

Enhanced service provider. The ESP class of operator also owns and provides to the customers the subscriber-information module (SIM), thereby gaining more control on the feature set of the service. Examples in the United Kingdom include: Afrimobile, Delight Mobile and MobiData.

Full MVNO. The class of virtual operators with the highest functionality is the full mobile virtual network operator (MVNO). Although it does not own the cellular wireless network or the switching nodes, it does own the telephone number range and the two subscriber identifiers – IMSI and MSISDN. Examples in the United Kingdom include: Stream Comms, Tesco Mobile and Virgin Mobile.

1.4.3   The commercial model in practice – telephony services

The commercial model for telephony services, as provided over conventional circuit-switched mobile or fixed networks, is shown in Figure 1.11. The revenue for a Telco comprises rental (or subscription) revenue from all connected customers plus call charges (related to the call duration and distance) paid by the calling subscriber (A). In the case where the call terminates on a connection of another network, the originating Telco-1 pays a conveyance charge (based also on call duration and distance covered) for traffic passed over the interconnect point to the

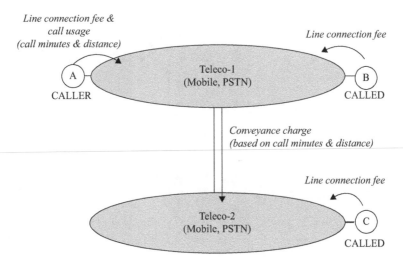

Figure 1.11   The commercial model for telephony service

receiving Telco-2, so the call can be terminated with subscriber C. (We examine interconnection charging in greater detail in section 2.2.4.) In principle, this pattern of interconnect charging continues with each additional network involved in the connection, including international connections, although there may be some variations in the exact charging arrangements. Also, the service providers may structure their tariffs differently for the various customer segments – particularly in highly competitive mobile markets.

1.4.4   The commercial model in practice – The Internet

The commercial model for the Internet is different to that for telephony for two reasons. The first is that the Internet provides inherently only data services, even if the use of these originates as voice. The second reason is that, unlike the Telco-based services, the design of the Internet is based on the philosophy of not charging for usage. All flows of money are, therefore, just subscription based. A simplified generic commercial model for the Internet is shown in Figure 1.12. This shows the basic structure of the Internet, in which subscribers gain access to the Internet via their serving ISP. Such ISPs rely on the Telcos to provide access (fixed or mobile), usually broadband, to the subscribers. They also rely on links over a Telco network to the transit ISPs, who provide interconnection to other ISPs in the region or elsewhere in the country. The transit ISPs in turn are connected over Telco links to one or more backbone ISPs, who provide international connectivity over Telco networks. As Figure 1.12 shows, the money flows from the subscriber through the various levels of ISPs. However, since there is no usage charging the interconnection between backbone ISPs has been historically on the basis of 'peering', whereby no money changes hands since it is assumed that such ISPs deliver as

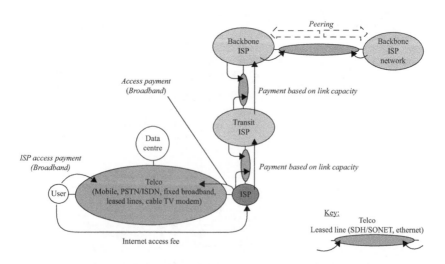

Figure 1.12   The commercial model for the Internet

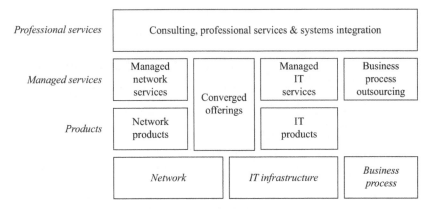

Figure 1.13   Typical ICT product range for a Telco

much data traffic as they receive. This means that the backbone ISPs have to be sufficiently big to warrant peering.

Figure 1.12 also shows how the role of data centres is included; although not part of the Internet, they are accessed through the Telco's network. Such data centres can, of course, provide a variety of cloud services.

1.4.5   The commercial model in practice – Telco

In today's market Telcos provide a full range of ICT services to their customers, using the resources of their networks and the support IT infrastructure, as well as their business processes. Figure 1.13 shows how the Telco's offerings may be grouped into plain products (managed by the customers), managed services (in which the Telco ensures continuity and provides full-life support) and professional services covering consultancy, facilities management, etc. Of course, as indicated earlier, this set of services may not all be provided by one Telco, and there may be several other players providing the various roles.

1.4.6   The commercial model in practice – over-the-top (OTT) services

An important role within today's ICT and telecoms markets is that of the OTT service. Figure 1.14 shows the range of such services and how they relate to the supporting infrastructure. The OTT service fall into the following categories:

- Communications – e.g. Skype (VOIP) and WhatsApp (messaging).
- Socialising – e.g. Facebook (social network).
- Entertainment – e.g. Spotify (music streaming).
- Information and data – e.g. Google (search engine).
- Commerce – e.g. Amazon (online retail).
- Productivity – e.g. iCloud (online backup and terminal synchronisation of content).

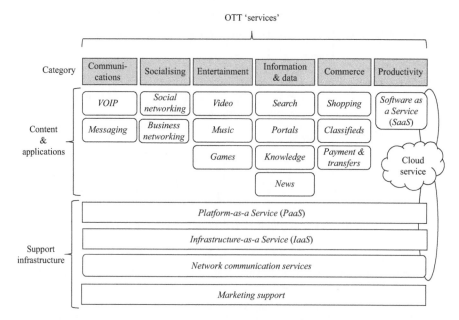

Figure 1.14   The range of OTT services

Also shown is the group of Cloud services: software as a service (SaaS), platform as a service (PaaS) and infrastructure as a service (IaaS) – with the latter two forming part of the supporting infrastructure, see also Box 1.5.

The key feature of the OTT services is that the supporting infrastructure is not directly involved in the commercial model, meaning that the Telcos do not receive any conveyance revenue from the OTT players, nor do they have any control on the quality of service (QoS) offered. The associated issue of network neutrality is discussed in Box 2.2.

1.5   Customers, products and the market

There is no business without customers.

It is clear that a key aspect of any profitable company is its revenue; indeed, success is often measured by revenue growth. In turn, revenue is the product of three aspects of customer management:

- winning customers
- retaining customers
- maximising average revenue per user (ARPU)

Therefore, understanding customer needs, aspirations and service requirements is a vital aspect of any marketing strategy and plan. Indeed, we can go further and say that successful marketing comes with a clear and focused understanding of the

Box 1.5 Cloud computing services

One of the most important additions to the portfolio of services provided by Telcos is that of cloud computing. This set of services relies on the availablity of reliable network communications and computing (processing and storage) facilities in the network. The Telcos have the advantage of a solid reliable brand, as well as the ability to gain economies of scale by managing the users computing requirements on large virtual machines in their data centres. For the customer cloud computing offers the attraction of not having to spend capital on building computer facilities, meaning their spend is a predictable current account annual charge (often incorporated into a long-term contract with the Telco or service provider). The customers, therefore, no longer have to worry about software updates and maintenance of the computing facilities within their company.

There are three distinct cloud computing products:

1. *Software as a service (SaaS)*. This service enables the running of software applications, macos, etc.
2. *Platform as a service (PaaS)*. Here the cloud service provider operates the computing platform (i.e. servers and storage) as a manged service which enables the customer to run their applications.
3. *Infrastructure as a service (IaaS)*. This service is concerned with the provision of individual component parts – virtual machines, servers, storage facilities, etc. – which customers can use as part of their computing activities.

target customer group (customer segment in the jargon) and how best to serve them. We will cover all of these subjects in depth in Chapter 7.

In addition, as we have seen, Telcos are in the business of developing, managing and selling products and services to customers. One of the best definitions of a market is *a place where products meet customers and customers make choices*. It is therefore essential for a successful company to have a clear understanding of the key characteristics of the market in which it operates, both at the global macro level and locally. Again, we will cover this in detail in Chapter 7.

A market can be defined as the place where products meet customers and customers make choices. It can be represented in a notional matrix where customer segments intercept with products. Most companies will have a clear idea of the market segments that they wish to address. Successful companies will build a strategy around a clear view of the characteristics of their ideal target customer. This may be relatively broad, e.g. Marks and Spencer – or quite narrow, e.g. Louis Vuitton, but the key to success will be clear focus. This focus will include a detailed understanding of customer needs, their demographics, propensity to buy, the likely

Market segments

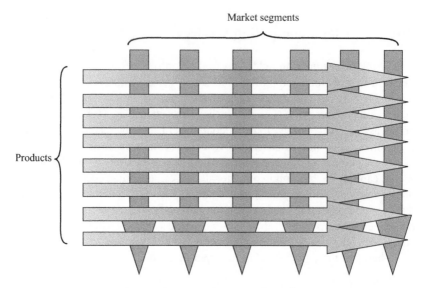

Figure 1.15   A theoretical market

changes in these factors, etc. Most companies will address more than one segment or have sub-segments within their target customer base.

Despite this, many ICT companies are far less concerned with specific target customer segments than with what they buy i.e. products. There are two reasons for this. One is that in ICT – as you might expect – value is often achieved through innovation and technology. The second reason is that many ICT companies are based on a single, multi-purpose network, whether it is a telecommunications network (e.g. BT, Vodafone) or an Internet protocol (IP) server-based network (e.g. an ISP or Sky). In these circumstances, a winning strategy will be to maximise utilisation by developing a wide range of products to operate across the shared investment, thereby spreading costs and providing a one-stop-shop to customers. This is covered in more detail in Chapter 8. These ideas have been shown diagrammatically in Figure 1.15.

Note: Not every intersection represents a market. In some cases certain market segments will buy all products; but in other cases, particular segments will buy only a limited range of products or particular groups of product.

1.6   The current business environment

1.6.1   Introduction

The telecommunications industry operates in a complex and dynamic environment. In particular, many of the external pressures on companies interact, so that a change in one will create a change in another. For example, growing competition will prompt a reduction in regulation; growing consumer expectations may require investment in new technology to satisfy them; similarly, advances in technology may prompt greater customer demand for more sophisticated services.

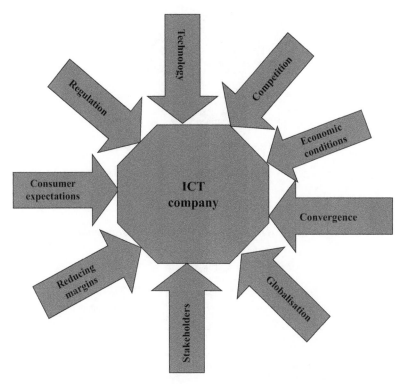

Figure 1.16   The current business environment

We might broadly express the most important pressures as shown in Figure 1.16. This list is certainly not exhaustive but it helps to illustrate the point.

It is intended that this book will explain, in some detail, each of these key influences. In this chapter we will explore four of the most important *viz* stakeholders, reducing margins, convergence and globalisation. (The other key influences will be covered in further chapters.)

1.6.2   Stakeholders

Every company has stakeholders and they form an important part of the business environment in which the company operates (Figure 1.17). They are the groups of people with an interest (or stake) in the company or its activities. Clearly, shareholders are an important stakeholder group but they are far from the only one. And stakeholder interests need not be purely financial or related to ownership as they are in the case of shareholders. If we take stakeholder in a wide sense, then you would certainly include customers and employees as stakeholders in any company.

The groups shown in Figure 1.18 may all have a stakeholder interest in a Telco.

Some of these groups might come as a surprise (e.g. government and the media), but it must be remembered that by any measure the telecommunications and ICT industries have national importance. They drive business efficiency, economic

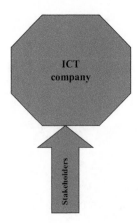

Figure 1.17   The business environment: stakeholders

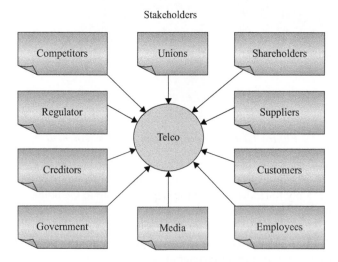

Figure 1.18   Key stakeholder groups

growth and enable individuals to organise and manage their lives. They touch everyone to a lesser or greater degree. They are central to national security. Certainly an attack – perhaps with Internet viruses or even with strategically placed bombs – could have devastating effects on the commercial success or security of a nation.

As we have seen, Telcos operate within the wider ICT value chain. Typically, they distribute services; they own and manage the physical telecommunications infrastructure such as local access cables, telephone exchanges, mobile cell sites, fibre rings and backbones networks; and they invest for the future. So, for similar reasons, they are sharply in the minds-eye of governments, regulators and the media.

In addition, ICT companies (and other firms utilising ICT services) now hold vast amounts of highly sensitive data on individuals. This is a significant worry to many people who fear anything and everything from identity fraud, government intrusion and Big Brother tactics from retail companies.

On the positive side, communication data is now a vital tool in the fight against terrorism and organised crime. And, again, government agencies have formal and informal interests in such data.

Apart from data protection and misuse, national security, and economic growth, the government is itself a major user of ICT (and pays big bills); it collects significant tax from the industry; and it is concerned about the *digital divide* which we will discuss in greater depth in Chapter 2. Finally, ICT is a major employer and investor in new equipment – both of which are of interest to any government.

1.6.2.1 Stakeholder analysis

As we have seen, it is important for managers of ICT companies to bear stakeholders in mind when making key decisions, whether those decisions are to do with strategy, markets, products, financing or employment. Indeed, the purpose of any stakeholder analysis is to influence the outcome to the company's advantage. To do this it is helpful to categorise stakeholders as direct or indirect. This categorisation will differ from company to company, and decision to decision, but broadly:

- Shareholders, customers, competitors and suppliers will be directly affected. The regulator will often be a direct stakeholder for an incumbent Telco but not for an alternative provider (we will cover regulation in greater depth in Chapter 2). Again, depending on the company in question, trade unions may or may not be directly involved in some decisions.
- The government, the media and creditors will be indirect stakeholders; this does not imply that they are less important, only that they may be harder to influence and require a different approach to gain their support or understanding.

1.6.2.2 Influencing stakeholders

Having identified and categorised relevant stakeholder groups, it is important to calculate the form of influence that is appropriate and most likely to be effective. A company may choose to:

- engage certain stakeholder groups, discussing the issue early, being open to options and happy to be guided, etc.
- consult groups (either formally or informally), allowing preferences to be expressed between different options
- invite comment on a single firm proposal
- inform stakeholders, effectively after the decision has been made

Which route is chosen will depend on many factors such as the relationship between groups and the company; the importance and complexity of the decision; and the strength of feeling and influence of the stakeholders. Whatever approach is taken the purpose must be to achieve the most successful outcome for the company in the long run. This may involve calculating the effects not only on profit and revenue but also on the company's reputation, national standing and competitive position. Given the importance of stakeholders and their potential impact on company decision-making, firms will allocate resources to achieve positive out-comes. Major shareholders will be regularly briefed by the company's investor relations manager. Human resource departments will include an industrial rela-tions department focused on working with trade unions, ideally in a mutually supportive atmosphere. Large ICT companies will certainly have a public affairs department staffed with a number of professional managers tasked with building strong relationships with politicians, media and other key players. In this case the aim is to explain clearly (some might say *sell*) the company vision, its strengths and its value, not only to customers but also to society. They may well comple-ment this with a programme of corporate social responsibility. Similarly, major Telcos will have a regulatory affairs department staffed with economists, financial analysts and, possibly, lawyers. In this case the role is to influence regulatory policy decisions to best reflect the optimum outcome for the company. The dis-cussion between the regulator and the company will be heavily influenced by the quality of analysis and the persuasion of the arguments, but – as with all stake-holder influences – trust and understanding between the parties is vital for a positive outcome (Box 1.6).

Box 1.6   Corporate social responsibility (CSR)

Corporate social responsibility is the idea that a company will monitor and publicly report its compliance with legal, regulatory and, crucially, moral standards. In companies that champion their CSR there is a strong sense that the rules should not only be followed, but that the spirit of the rules should be followed. Often the public reporting of compliance will be accompanied by a programme of socially responsible activities such as environmental protection, support of the arts or educational advancement for the dis-advantaged. Clearly the purpose of such programmes is to advantage custo-mers, citizens, suppliers, etc., but it is also to persuade stakeholders that the company has more, and perhaps higher, aims than just making money for their shareholders.

Examples of CSR programmes are:

- Microsoft YouthSpark, a global initiative to connect young people with opportunities for education, employment and entrepreneurship [10].
- Apple and the Environment, a programme that comprehensively reports the company's carbon footprint and seeks to reduce it through envir-onmentally-friendly design and the use of renewable energy [11].
- Sky Rainforest Rescue, a partnership with World Wide Fund for Nature (WWF) to help save a billion trees in the Amazon rainforest [12].

Note: CSR is variously known as corporate citizen, responsible business, corporate conscience, etc.

Finally, alongside these professional influencers, line managers, directors and even front-line staff can be a useful and authentic voice in explaining the aims of the company and the importance of particular outcomes. Box 1.7 includes a worked example of stakeholder analysis.

Box 1.7   Stakeholder analysis: a worked example

Let's imagine a major national Telco is considering outsourcing their customer service call centres (employing 500 people) to a specialist company operating in Vietnam. Who might they consult and how?

- Their employees and – where appropriate – the trade unions will be highly interested. The company should consult them in detail before the decision is announced; indeed they may have a binding obligation to formally consult on the changes; the company will need good arguments and be ready to clarify how the changes will affect various groups of staff (together with benefits if there are some e.g. keeping the more interesting work in-house).
- Customers will be affected; the company may not choose to inform them in advance but the crucial issue here is to ensure that there are special provisions for particular customer needs e.g. for people with hearing difficulties, frequent enquirers or the wish for a one-to-one customer helpline.
- The government and media may wish to be involved if things go badly. The company may therefore be well advised to inform them at an early stage in the process; they will need good arguments to justify the change.
- Similarly, the big shareholders may be concerned if the change goes badly but, generally, they are likely to be more sympathetic to the cost-saving arguments.

1.6.2.3   Conclusion

I hope it has been shown that in making any major decision, a Telco should consider all the relevant stakeholders and consult them in an appropriate way to ensure the best long-term outcome for the company.

1.6.3   Reducing margins

There many pressures on the prices and margins of telecommunications products are an important element of the business environment for ICT companies (Figure 1.19). In 2007 The Office of Communications (Ofcom) quoted, 'prices continue to fall – a typical basket of residential telecommunications services costs £69.85 a month, £35 less than in 2002' [13]. This trend continued and in 2013 they reported that the total price of the mobile element of the typical household's usage had reduced by 25% and that the combined price for the household's fixed voice, fixed broadband and pay-tv also fell, in this case by 14%, during the year. Indeed, our own experience as

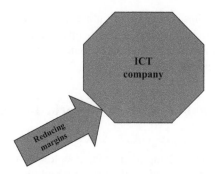

Figure 1.19   The business environment: reducing margins

consumers in the United Kingdom must convince us that overall prices are falling. Mobile companies offer packages including *free* texts; fixed telecommunication providers offer many hundreds of minutes of voice calls within a fixed fee and some providers are now offering free broadband if you sign up to a bundle of services. In addition, we have seen 'The Death of Distance'. No longer do we typically pay additionally for long distance or trunk calls. A minute is a minute. This trend is even beginning to break down national boundaries, albeit slowly. (The problem here is that international mobile roaming fees and international call prices are heavily influenced by settlement agreements between Telcos. These deals are not always fully competitive and not always well regulated. This is a complex issue that we will return to in later chapters.)

Price pressure is a natural business feature that affects all industries. Normally the pressure comes from competition, reduced price of production and growing consumer expectation. All of these pressures have an impact on telecommunications. However, a particularly strong influence for basic services such as calls and lines is a marketing phenomenon called *commoditisation*.

Commoditisation is the phenomenon whereby products (or services) are selected by customers based on price alone, with no differentiation of features, service, brand or quality. In some cases, customers' expectation of what they receive for a given price is constantly rising (e.g. broadband).

A market that is subject to commoditisation would display:

- a highly standardised product such as a standard call minute
- falling prices
- oversupply
- no dominant brand
- low margins
- growing demand

This is shown in diagrammatic form in Figure 1.20.

In the past 10 years, all of these conditions have been applied at the same time to the telecommunications industry. Structural demand rose steeply with the introduction of the Internet and also mobile calling. (How many additional calls do people

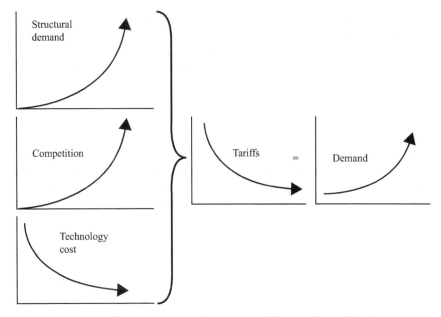

Figure 1.20 Market forces that lead to commoditisation

make that they might never have contemplated if they had not had a mobile phone in their pocket?) Competition has risen in virtually every country in the world through de-regulation and competing technologies; and technology costs have fallen, not only through digitalisation of networks, but, in particular, through the adoption of fibre cables that can carry a vast number of simultaneous calls and data connections.

Prices have fallen as we have seen and this has naturally created more demand. Nonetheless, the growth in supply of infrastructure and services continues to outrun demand, as Telcos invest in fibre in the access section of the network, next generation network (NGN) nodes and fourth generation (4G) mobile data capacity. Therefore, in many areas we are still seeing ongoing commoditisation of basic products such as calls, text messages and broadband services. This trend is clearly good for consumers, although it may well not suit those who want more sophisticated services.

Despite growing demand, this constant pressure on prices can only reduce company margins; make future investment more difficult; and cause less efficient companies to go bust. Nonetheless, there are strategies to address this issue. In a commoditised market a supplier might aim to earn higher margins by:

- bundling products into a single-priced package
- offering creative tariff packages
- seeking innovation constantly in product developments and features
- signing customers to long-term contracts (where regulation allows)
- seeking to become *least cost producer*

We will return to all these strategies in greater detail in Chapters 6 and 7.

1.6.3.1   Conclusion

A commoditised market is one where customers choose products and services solely on price. For a variety of very specific reasons such conditions particularly apply to certain telecommunications products. Prices and margins are therefore, on average, declining. Telcos should address this issue through marketing programmes such as product bundling.

1.6.4   Convergence

1.6.4.1   Product convergence

Convergence is a well understood part of the business environment in ICT (Figure 1.21). It has three aspects. Initially, it was applied by product and marketing managers to identify a customer requirement for converged products and services. That is to say, when customers seek the same functionality from all their tele-communication services and also when they wish to access all their services from whatever hardware they happen to be using. Digital networks are a key here. In fact, convergence is a prime example of technology and markets working together. Whilst we will cover the technology drivers in much greater detail in Chapter 8, we should, at this point, acknowledge that the ability to carry digital signals, irre-spective of the embedded content, is a primary driver of product convergence.

The phenomenon is easily seen if we consider simple voicemail. Many people have a mobile phone which switches to voicemail on busy and engaged; they may have an answer phone at home; and voicemail in the office. Life would be easier and pro-ducts truly useful if these services converged and access to all messages was available from a single *platform* reached by any phone, PC, gaming device or tablet. Similar arguments apply to speed dialling directories, contact lists, Internet favourites' lists, instant messaging, etc. Much of this is possible today, but mainly it is achieved through rather clunky work-around and proves rather expensive. Successful convergence requires a re-think of the whole service to provide intuitive easy management, single log-on access, a common security platform, etc. It also requires agreed standards, open interconnection between services and operators, and common customer services.

To illustrate the final point, if we imagine that we are booking rail tickets using our Apple iPhone on Vodafone's mobile network, logged on to a Wi-Fi hotspot managed by The Cloud, accessing the secure credit card element of trainline.com, at the key moment the screen freezes. Question: Who are we going to contact to resolve the problem and how?

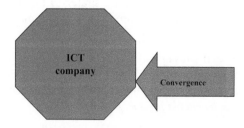

Figure 1.21   The business environment: convergence

Whilst companies are already developing answers to some of these issues (see the earlier discussion on Triple and Quad Play), it is likely that the true answer will only come with the management systems and extra facilities available from the NGN currently being developed and implemented around the world (see section 1.6.4.3).

1.6.4.2   Industry convergence

We have already seen in section 1.2 on value chains that companies acquire, merge and partner with other companies across functions to provide more complete bundles of services. But technology developments are also driving new groupings. For example, IP networks carry telephone calls exactly like any other data; in this world, voice services effectively become an Internet data application. Skype is a key example. Similarly, if we consider the digitalisation of content, publishing a new novel becomes precisely the same as delivering online music or movies. And as we've seen content is increasingly coming to include the information that drives commerce, rather than just entertainment e.g. insurance policies, healthcare records and even road traffic data.

These drivers for convergence across functions are causing some people to change the value chain for ICT into a more layered model where functions are re-drawn to represent:

- content – journalism, film, music publishing
- services – re-purposing, aggregation of content, security
- networks – Telcos, Cable Television (CTV), mobile operators
- applications – Google, YouTube, eBay

This is represented diagrammatically in Figure 1.22.

If we look at a particular example we can see how IBM would appear in the ICT value chain and how its acquisitions, mergers and shift in strategic focus place it centrally in the new model (Figure 1.23).

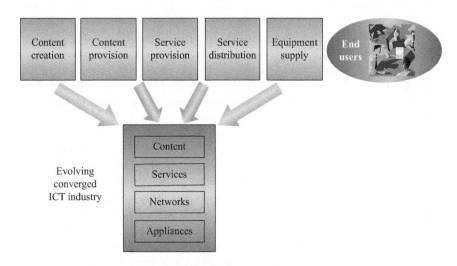

Figure 1.22   Industry convergence

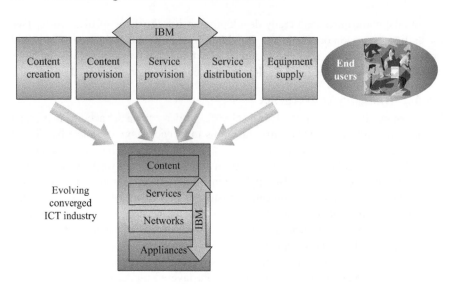

Figure 1.23   IBM in the emerging industry model

Interestingly, the new model somewhat represents a computer that is running applications across its processor, using an operating system and other utilities, storing content and accessing it in real time. This model is still evolving and is only one of a number of directions the ICT industry might take. We will explore these issues in more detail in Chapter 10.

1.6.4.3   Network convergence

Telecommunications networks have developed and evolved over more than a hundred years. So have the services and customer requirements that depend on the network. What started initially as telegram messages sent from a central office, evolved to simple voice calls from the home, to mobile calls and text messages sent from anywhere. They have seen the growth of the Internet, e-commerce, online education and entertainment, etc., all delivered over the evolving telecommunications network. If we add in services aimed at business customers (virtual private networks (VPN), converged data and voice services, etc.) the picture is massively complex. Until recently, all these changes have been accommodated by telecommunications networks constantly accreting more and more networks.

Furthermore, each of these networks has its own set of support systems (for operations and maintenance) – the combination is often referred to as platforms. These new platforms are based on different technologies, different generations and different protocols, each delivering a different service. What's more, to provide usable services each platform needs to interconnect with every other. This world is described as a multi-platform network and is best pictured as a bowl of spaghetti; and expensive spaghetti at that! The cost of investment, maintenance and operation of such a system is huge. Inevitability, many of these technologies are becoming obsolete. Finally, the chances of this myriad of interconnecting platforms delivering the sorts of advanced, intuitive, converged products that customers now demand is low, and certainly not at sufficiently low cost to match the increasing demands of the market.

We will discuss in Chapter 8 the history, technology and business case for the NGN now being deployed throughout the world. The purpose of introducing the concept here is to point out that this is another example of convergence, in this case network convergence. Whereas current services are delivered on a multi-platform system (spaghetti), tomorrow's converged services will be delivered on a multi-service data platforms, based on IP technology.

1.6.4.4   Conclusion

In ICT there are three types of convergence developing simultaneously. Product convergence refers to strong customer demand for ubiquitous services that have complimentary feature sets, are intuitive and have the same look and feel. Industry convergence refers to a changing industry topography that recognises that companies no longer need to be integrated from end to end to deliver services but can specialise in one layer of the delivery system whether it be content, services, networks or equipment. Finally, network convergence refers to the changing technology that will enable a multi-service platform (MSP) to replace many existing bespoke networks and deliver the type of converged services that 21st century customer's demand.

1.6.5   Globalisation

1.6.5.1   Introduction

Another of the key current business environmental influences is globalisation (Figure 1.24). Technology doesn't recognise international boundaries; more and more ICT companies see their market in global terms. There are many reasons for this, but the major drivers are:

- Globalisation of industry and commerce e.g. financial services.
- A squeeze on home telecommunications markets from both increased regulation and greater competition.

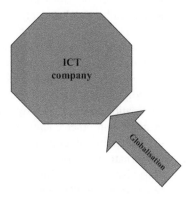

Figure 1.24   The business environment: globalisation

- The demands of multinational companies (MNCs).
- The fact that we are all geographically more mobile than ever before and are travelling widely for business and for pleasure.

These international trends are, to some extent, driven by ICT technology. The Internet for example, enables companies to market worldwide; low-cost, high-capacity connections enables call centres to be situated anywhere in the world; wireless access and laptops enable the knowledge worker to operate from airports and hotels worldwide; and Internet cafes enable young people on world trips ready access to their parent's money!

At the same time as feeding this trend, the telecommunications industry has also evolved and developed in an attempt to serve these developing global needs. They have done this through a range of strategies that we will explore further in a moment.

They have been assisted in their global aspirations by technology developments such as fibre, multiprotocol label switching (MPLS) and (IP). These technologies ensure signals can be sent over vast distances, that huge bandwidth can be carried at low costs, and that different networks interconnect seamlessly. International standard setting has also been vital in this and has proved a great success in ensuring interconnectability. All national fixed, mobile and data networks are interconnected to ensure that (in theory at least) any connection, anywhere in the world, can access any other connection. This is certainly true in terms of the technology, although in some rare cases it may be modified by government policy (censorship) or by commercial propositions. It is worth noting that the interconnected telecommunications networks throughout the world make up the largest single man-made machine on the planet!

These global developments are likely to continue, indeed to increase.

1.6.5.2   Global telecommunications' domains

Global telecommunications business has three distinct domains, discussed in the following sub-sections.

International services
This area of global telecommunications provides public services, ensuring that customers can make international calls, purchase international leased lines and extend their data services overseas. Essentially these services are providing a bridge between the networks (usually the public-switched telecommunications networks (PSTN)) of two countries. The commercial relationship between the relevant Telcos will be either unilateral or bilateral (corresponding). Where a national operator is a state monopoly or faces ineffective competition it is often in a position to charge excessive termination charges, thereby making the cost of calling abroad excessively high for customers in other countries. In recent years this problem has been partly addressed by the introduction of transit calling, refiling, and VoIP services such as Skype. International Services are described in Figure 1.25.

More recently regulation has also assisted in liberalising markets and allowing foreign competition to enter new markets. In the United States, this was heralded by Judge Green's ruling of 1984 that broke up the Bell Corp and enabled cross-territory competition; in Europe, the European Commission has enacted a number of telecom

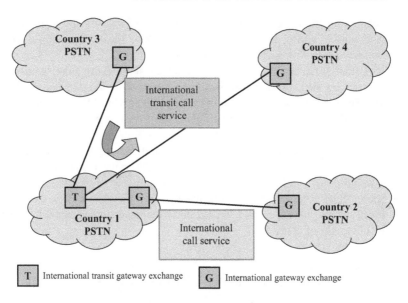

Figure 1.25   International services

directives to open up markets and break down the monopolies enjoyed by national telecommunication companies. Whilst these policies are being copied with a great deal of enthusiasm by some other parts of the world, e.g. various parts of Asia, it would also be true to say that the emerging business models are hampered by the very different forms and degrees of market liberalisation around the globe. Opening telecommunications markets remains a relatively recent phenomenon, but as liberalisation matures we may well see a resurgence of alliance and partnership models. Nonetheless, the success of Vodafone and other global players does suggest that global scale carries significant competitive advantage where the markets are new or fully liberalised.

The infrastructure providing the international links between land masses, such as sub-sea cables and satellites, is provided by consortia of the Telcos served, together with global investment companies. Probably the best known example of a Telco-backed consortia is Intelsat providing the global satellite coverage, but there are several other global (GEO) and regional (GEO and MEO) satellite consortia, as well as many sub-sea cable consortia around the world (e.g. SEACOM serving East Africa).

Global services for multi-national companies (MNC)
The second area of global telecommunications business aims to serve the needs of MNCs. As we have seen, more and more industries are developing on an international scale e.g. financial services, consumer goods, fashion, etc. The 'mega' companies that make up these industries have wide-ranging organisations with operations in several countries. They will typically have voice and data requirements both within country and between countries, and will require comprehensive communications and computing services on an international scale. Despite the breadth of their requirements, these customers may have a single hub, managing all

of their communications needs worldwide. They, therefore, demand a single point of ordering, unified billing and a dedicated service desk. Ideally they would wish to buy these services from a single, global ICT provider.

Telcos, therefore, aim to provide a dedicated service for MNC businesses, normally based on their own network facilities, enhanced by *local ends* provided by indigenous Telcos. However, it is hard to make profit serving MNC's global needs. MNCs are highly demanding and intelligent users, who are cost conscious and globally influential. Not unreasonably, they expect a discount for committing all of their business to a single provider. On the other hand, serving these needs requires the Telco to run complex and costly networks and service management support systems; to build and run infrastructure and service organisations in alien domains; and to successfully contract with local providers when they can't service their customer directly. This final aspect is often difficult to achieve at an acceptable cost and to acceptable levels of QoS.

Box 1.8   Providing a *one-stop-shop* for MNCs

The late 1990s saw a very strong drive towards international business in many industries. Financial services became globally consolidated through the major takeovers that were happening at the time (e.g. HSBC, CitiCorp). Other examples included consumer retail brands which on the one hand became huge global names (e.g. Nike, Apple) and at the same time, moved their production and manufacture to Asia. In an attempt to serve the ICT needs of these huge multi-national companies major telcos set up joint venture (JV) to offer a 'one-stop shop' for all communication needs worldwide. For example, BT and AT&T set up Concert, and France Telecom and Deutsche Telecom set up Global One. These organisations were major ventures subsuming a large part of the major business accounts of both partners and aiming to offer a full range of both basic and advanced services worldwide, across borders and within country. These global organisations made good progress in tackling the market for international telecommunications services but suffered from three major problems. First, there were cultural clashes between the J V partners, exacerbated by the fear of losing existing business to the joint venture without gaining sufficient new business. Second, the target multi-nationals expected a price discount for handing over all their business to a single operator worldwide, and third, costs were often much higher to provide services in far-flung areas than at home. Indeed, it often proved very difficult for Concert or Global One to acquire circuits in territories where de-regulation hadn't yet occurred and where a national Telco was able to protect its monopoly.

For these reasons joint ventures on this scale appear to have gone out of fashion.

Box 1.8 describes some real-life attempts to provide MNCs with a *one-stop-shop* for global services.

Provision of domestic services in foreign countries

The third area of global telecommunications business aims to provide domestic services in foreign countries. Depending on the company strategy this might include corporate customers, consumers or wholesale services. The opportunity to provide such services has arisen through liberalisation and regulatory intervention and continues to occur as more and more countries open up to inward investment. Nonetheless, these ventures are not without risk and different Telcos have approached the challenge in different ways. In general, those Telcos that have decided on a global strategy have approached their expansion in one of the following three ways:

- by strategic partnership
- by acquisition of minority interests in other Telcos
- by full acquisition of companies

1.6.5.3 Strategic partnerships

Partnerships and agreements have always existed since undersea cables first connected the early national telegraph networks. However, with growing competition within country, as well as from outside, Telcos have been working together to share traffic and to ensure that as much business (voice and data) remain on their networks as possible. The move from switched traffic to IP data requires ever more sophisticated peering arrangements that provide the basis of commercial partnerships. Whilst providing some opportunity to defend business, there is little opportunity for new revenue to be achieved through strategic partnership; and the unequal balance of the partners often leads to problems.

1.6.5.4 Acquisition of minority interests

Another approach is to buy minority interests in Telcos in other countries or regions. This was a particularly popular strategy when de-regulation first began because there were often restrictions on foreign ownership of national telecommunication companies (e.g. no more than 25%, etc.). These restrictions were imposed by governments seeking to protect national security interests (defence and security services are renowned users of high capacity and ultra-secure telecommunication services) although there might also have been some covert protectionism involved. Most of the restrictions have been removed (the EU especially have worked hard to open markets to full capital migration), but some restrictions remain, especially in parts of Asia and in the United States.

A further reason for companies to start with minority acquisitions is to gain help in understanding the local market conditions from the target company. Indeed, it is often vital to build relationships with local stakeholders (regulators, shareholders, etc.). This can be best done as a minority shareholder. BT's expansion into Western Europe in the 1990s was almost entirely through minority acquisitions in European operators.

Often, this strategy is implicitly or explicitly linked with a desire to expand partial interest into full control as time and regulation allow. This approach was very successful for BT in Italy and Japan where they initially took minority holdings and later were able to take full control. Telefonica had a similar strategy in

South America which they have now leveraged to include mobile operators throughout Europe and Asia.

However, more often than not, acquiring minority interests has led to problems. There are two main reasons.

First, the target company retains control and will, in the end, exert their preferred strategy, often in contradiction to the global strategy of the newcomer. This was certainly the case when European operators bought minority shares in Indian and Korean mobile operators in the late 1990s.

Second, by purchasing minority interests in companies that are just like you, but that operate in a different country, your whole portfolio is subject to the same global economic drivers. When 2000 saw the *dot.com bust* all Telcos saw major share price reductions and could no longer support their debt levels. The minority holdings were just a further drag on funds.

1.6.5.5 Full acquisition

Finally, perhaps the most straightforward global strategy is to buy companies that extend global reach and compliment market and product strategy. This was highly successful for Vodafone who went from a United Kingdom -only mobile operator to a top world telecommunication company by buying second or third ranking operators in all their target geographies. WorldCom took a similar approach and might have succeeded had it not been for wrong-doing at the top and the dot.com bubble bursting.

In this strategy, a key question for an acquirer is how much rationalisation to apply across territories, and how much autonomy to allow in-country. Clearly, combining product management, marketing, brand, network management, etc., will reduce costs and perhaps offer a global presence to the traveller or multinational, but it may reduce the opportunity to respond to local needs. The two models are shown in Figure 1.26.

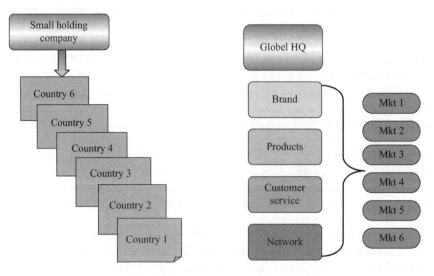

Figure 1.26   Alternative post-acquisition strategies

1.7 Summary

In this chapter we have described the purpose and players of the ICT industry and the role of telecommunications within it. We have explored the industry value chain and how VCA can be used to understand the dynamics of the business. In addition, in section 1.3 we developed a commercial model to describe how Telcos make money and in section 1.4 we explored how this model applies in practice.

Finally, we looked in detail at industry stakeholders, reducing margins, convergence and globalisation, four of the key elements of the ICT business environment. In the next chapter we will look in detail at a fifth, which is regulation.

References

[1] Porter, Michael. *Competitive Advantage*, Free Press, New York, 2004, pp. 33–61.
[2] Valdar, A. *Understanding Telecommunications Networks*. IET Telecommunications, Series 52, Institution of Engineering and Technology, London, 2006, pp. 99–103.
[3] BBC News. 'NTL Renames Itself Virgin Media', 3 August 2015, http://news.bbc.co.uk/1/hi/business/6343341.stm
[4] Naughton, J. 'Our Changing Media Ecosystem'. From 'Communications, the next decade' a collection of essays prepared for the UK Office of Communications, edited by Ed Richards, Robin Foster, and Tom Kiedrowski, London, 2006, pp. 46–47.
[5] Leadbeater, C. 'The Genie is out of the Bottle'. From 'Communications, the next decade' a collection of essays prepared for the UK Office of Communications, edited by Ed Richards, Robin Foster, and Tom Kiedrowski, London, 2006, pp. 263–274.
[6] Communications Market. Interim report, Ofcom, 2006, www.ofcom.org
[7] Marc, B; Liau B: 'Mobile Virtual Network Operators' 15th International Telecommunications Network and Strategy Symposium, 2012.
[8] 'MVNO Classification & Types', www.mvnodynamics.com
[9] 'Ofcom Enquiry into what MVNOs Could Offer Consumers', June 1999, www.ofcom.org
[10] Microsoft '2014 Citizenship Report', 3 August 2015, http://www.microsoft.com/about/corporatecitizenship/en-us/reporting/
[11] Apple. 'Environmental Responsibility', 3 August 2015, http://www.apple.com/uk/environment/
[12] Sky. 'Sky Rainforest Rescue', 3 August 2015, https://corporate.sky.com/bigger-picture/inspiring-action/sky-rainforest-rescue
[13] OFCOM. 'Communications Market Report', 3 August 2015, http://stakeholders. ofcom.org.uk/market-data-research/market-data/communications-market-reports/

Chapter 2

Regulation

2.1 Introduction and history

In general, regulation is one of the key elements of the business environment (Figure 2.1), in most industries, in most parts of the World. However, regulation is a particularly strong influence in parts of the ICT value chain and especially in the telecommunication elements. Historically, this is because telecommunications was seen as a state-owned monopoly. While this is still the case to some extent in some countries, the majority of the world has moved on. The past 30 years has seen a huge amount of liberalisation and privatisation i.e. the floating of Telcos on the stock exchanges and the sale of shares to private shareholders – both big and small. In the United Kingdom this was a key and early aspect of the then Prime Minister Margaret Thatcher's policies of the 1980s. Despite the generally accepted success of liberalisation, competition and privatisation, sectorial regulation, as we see it, e.g. in telecommunications, persists. Indeed, it will persist whenever market power would otherwise put consumer protection or fair competition at risk.

Competition has also been introduced, both by legislation (locally and at European Union (EU) level) and by innovation. By innovation we mean that new technologies are competing with the old ones. Perhaps the best examples are mobile networks which now compete with fixed phones to provide customers with a choice of how they make their calls. Similarly, VOIP (see Box 2.1) is

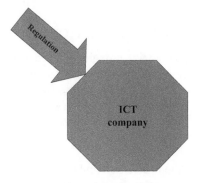

Figure 2.1   The business environment: Regulation

now beginning to displace conventional mobile and fixed calls. There are many other examples.

Box 2.1   Internet, IP and VOIP

The Internet is an open constellation of interconnected data networks across the world, all using a common way of carrying and routing the data in packets – known as the Internet Protocol (IP) and a common addressing scheme. As well as being the basis for email and other services, users can get access to content on remote computer servers (e.g. web sites) through using a special addressing scheme, thus creating the world wide web. The key challenge for regulators is that voice calls, which traditionally were handled by circuit-switched PSTN and mobile networks, and were regulated accordingly, can now be conveyed as a stream of IP data packets – i.e. VOIP. Furthermore, the calls and associated video content can be provided over the Internet without incurring incremental charges for the user (e.g. Skype) [2].

For those wishing to understand the economic and political background to both privatisation and liberalisation would do well to read chapter 1 of John Buckley's excellent book entitled *Telecommunications Regulation* [1].

Nonetheless, it is worth remembering that all industries are subject to some form of regulation, although not all have industry-specific regulators as telecommunications typically does.

If we consider other major industries that have sector-specific regulators we find they have some key characteristics in common.

In the United Kingdom the following industries have sector-specific regulators:

- Energy – gas and electricity – Office of Energy (Ofgem).
- Rail – Office of Rail Regulation.
- Water – Office of Water (Ofwat).
- Financial services – Bank of England and The Financial Conduct Authority (FCA).
- Broadcasters – Office of Communication (Ofcom).

Many similar examples apply throughout the developed world such as the Transport and Technology Agency which regulates the rail industry in Japan and the Essential Services Commission that regulates water provision in Australia.

These are all nationally important large-scale industries where competition is growing but not yet fully developed. Or, in the case of financial services, it may be that a rash of mergers and acquisitions since the mid-1990s has led to a diminution of competition.

But, are industries without specific regulators not regulated? Again it's worth remembering that car manufacturers, supermarkets, the hospitality industry, etc., all have regulations ranging from health and safety to food hygiene, to emission controls and environmental regulations. These regulations are usually based on primary legislation which applies generally. Such industries are regulated, but not through a sector-specific regulator.

So why is the telecommunications industry a special case? It isn't really. Regulation is about market failure. Mostly it focuses on privately owned (i.e. de-nationalised) enduring *natural monopolies* (see Box 2.2) where the economies of scale make it almost impossible for others to enter a market and compete with the incumbent. (An incumbent Telco is usually the old national telecommunication company that originally had a monopoly and was previously owned by the relevant government e.g. BT, France Telecoms, Telecom Italia and AT&T). Such natural monopolies don't only apply in telecoms; they also apply to gas pipelines, local postal delivery, railway lines and water reservoirs.

Box 2.2 Natural Monopoly

The concept of a natural monopoly is generally attributed to John Stuart Mill [3], though he initially applied it to rare skills and the qualification to practice, such as applied to 19th-century lawyers. He later adapted the theory to include network infrastructure such as water and railway services, though he termed them *practical monopolies*. Not all regulators or economists are consistent in their terminology. Indeed the term *natural monopoly* can sometimes be replaced by *essential facilities* or an *enduring economic bottleneck* or even just a *bottleneck*.

The first use of the term *essential facility* was in the 1912 Terminal Railway case. It was not therefore used to describe a piece of telecommunication plant but to open-up access to a railway bridge that was considered essential to competing railway services.

The term *natural monopoly* was first used in telecommunications to justify regulation in the Willis-Graham Act. This piece of legislation was introduced in the USA in 1921 to protect the monopoly rights of the burgeoning AT&T. Today the term is more likely to be used to justify regulation aimed at intervention to protect new entrants as they try to break down monopoly powers of existing incumbent Telcos.

Where sector-specific regulation is judged to be appropriate the form that regulation takes reflects the need for regulation to be objective and professional. Regulators are usually:

- Independent of the vested interests of the industry concerned.
- Independent of government (though constrained to operate within a specific legal framework).
- Empowered to seek the information necessary to understand and deliver their role.
- Able to set the rules by which the incumbent and other industry players operate.
- Required to operate transparently and constrained to impose requirements which are reasonable and proportionate.
- Supported by economic and legal professionals.

Nonetheless the telecommunications industry has a number of inherent characteristics that increase the risk of market failure. For example, it is only since the mid-1980s that there has been a competitive market in telecommunications. In addition, the high cost of investing in national networks, and the economies of scale and scope that favour the big volume players, makes it hard to become a successful new entrant. Couple these issues of market failure with the importance and far-reaching nature of ICT services to consumers and business, and it is clear why governments see regulation of telecommunications as essential. A further reason why regulation is essential is that networks must interconnect successfully. This aspect of regulation is covered in greater detail in section 2.2.4.

Generally, regulators from any industry share the same broad aims. These are called public policy objectives and might typically include the following:

Universal service – The service should be available to everyone in the country at a standard price. This is especially true of postal and telecommunication services. In the United Kingdom, BT has a universal service obligation (USO), which means it cannot refuse to serve any groups of people or areas of the country. A similar arrangement applies in every European country, though in some cases smaller Telcos are required to contribute to the USO costs of the incumbent [4].

Safety – Safety is clearly particularly important in energy provision, railways, water and food.

Public confidence – This is especially true of financial markets.

2.2   Telecommunications

Let us look at telecommunications a little more closely. In the United Kingdom there is a *converged* regulator overseeing the whole of the ICT industry. That is to say, in 2004 the UK Government, recognising the strong power of convergence affecting the industry, pulled together all of the appropriate regulators (e.g. Broadcasting Television Commission, The Office of Telecommunication (Oftel) and the Radio Authority) into a single body to look at the industry right across the value chain, from content to equipment. While this is not a universal model yet it is becoming the norm in most developed countries. The UK Office of Communications (Ofcom) is therefore a good place to start our analysis of regulation.

2.2.1   Principles and objectives of regulation

Ofcom have a very clear vision, as shown in its own charter, which states clearly: "Ofcom is committed to a thriving telecoms sector where companies can compete fairly and businesses and customers benefit from the choice of a broad range of services."

Similarly, its remit, set by Parliament, is clearly stated and embodied in legislation. The principal duties of Ofcom are shown in Figure 2.2.

So a key question must be 'what issues are particularly of interest to citizen and consumer?' For example, can consumers of telecommunications rely on the

The UK's Office of Communications

"3(1) It shall be the principal duty of Ofcom, in carrying our their functions;

- to further the interests of citizens in relation to communications matters; and

- to further the interests of consumers in relevant markets, where appropriate by promoting competition"

Citizens and Consumers

Figure 2.2   Ofcom's regulatory remit [5]

market (i.e. choice through competition) to provide the best service and prices, or do they need some help from a regulator? Similarly, are citizens happy that their personal data is safe, or that they don't face exploitation or misuse by the likes of Google or Facebook?

Ofcom believe that the best outcome for citizens and consumers throughout the United Kingdom will be provided through the following aims and objectives. This seems a pretty good list for any regulator to try to achieve:

- Innovation – whether consumers are getting new, creative services.
- Investment – a readiness to invest in ICT.
- Efficient production – whether the industry is well run, with low costs.
- Low prices.
- Excellent service.
- An efficient world class economy – i.e. does the telecommunications industry support business growth throughout the economy?
- Wide choice of services, companies and price packages.
- No offensive material and no fraud.
- Access for the disadvantaged.
- Elimination of the *digital divide*.

In fact, a survey of a wide range of telecommunications regulators around the world (carried out by the author) identified an extremely similar list of aims.

Variations could be simply explained by the stage of development of competition, liberalisation and levels of investment in the various countries surveyed.

2.2.2   The digital divide

It is perhaps worth pausing to think about the final aim on the list. The digital divide embodies the idea that people who have ready access to the Internet consequently have ready access to low-price shopping, extensive educational services, health information and many other services to make their lives more fulfilled. These advantages are denied to those who don't have access (e.g. through financial limitations, lack of ICT know-how, age or disability). This is of great concern to governments, pressure groups, parts of the media and other stakeholders; it is especially important to regulators as it is a very clear example of market failure.

In addition to the example of digital divide driven by depravation we see a growing issue concerning a different divide, the divide between those with access to high-speed services (normally based of fibre access networks) and those without. This is very much a rural vs. urban divide, but it is equally a sign of market failure because the low density of households in rural areas is not enough to justify far-reaching fibre investment. Unlike the other form of digital divide the rurally deprived people are often influential and well-off financially, with high demand for ICT services. Nonetheless, there simply may not be enough of them in any one area to support a viable business case for fibre investment. We explore this issue in greater detail when we consider regulation of NGN in section 2.2.6.

Let us consider two very different views of contemporary daily life:

Ian, a business consultant is travelling by train. He has just interrupted watching a broadcast of his favourite football team (Fulham FC) to watch a live newsfeed of an unfolding earthquake disaster in Japan. He puts away his paper-thin electronic reader and uses his smart phone to check share prices; he then rings John, a client and chief executive officer (CEO) of Cranes plc, having first checked through instant messaging that John is free and happy to take the call. They discuss flying heavy lifting-gear to the disaster area to support the rescue effort. Ian has just enough time to fire off emails to confirm the offer and to alert the disaster charity organisation before checking that Fulham have indeed beaten Chelsea in their vital cup semi-final game. He gets off the train with a smile on his face. He will watch the goals in Hi-Definition at home at a time convenient to himself and then book tickets online for the cup-final match.

Across town, Victoria, a single parent is struggling to bring up two kids on a sink-estate. She has just got home from her shift in a call centre and picked up the children from their child minder. She is loaded down with the weekly shop. Emma, her youngest child is diabetic and soon they will all have to go out again to take her for her regular health-check. Harry, her son, is hoping to go to university to study languages and is doing an evening class in Mandarin. However, the college is closing next term and Victoria can't afford the fares for him to go to London to continue his studies. Tonight he is playing computer games and Victoria worries that he's not making social contact with anyone. He certainly doesn't talk to her, other than the usual teenage boy grunts. Victoria picks up the post – at last a letter from her sister in Australia; perhaps she's included a photo of the new baby.

OK, a little exaggerated, but not too much. What has it to do with ICT? Ian's example is obvious. The online, anywhere, anytime business warrior is in control of what information he gets, when he gets it and how he views it. He self-configures his services; converges voice, data, and video; and works in real time. He uses ICT to do better business, organise his social life and to get the best deals on holidays etc.

But what about Victoria? How could she use ICT to improve her life?

- She could work from home, logging on to her station in the virtual, distributed call centre that would have all the facilities of the factory call centre, be managed in real time by her online boss and enhanced by video conferencing.
- She could shop online and have her goods delivered. She would get better prices if she used comparison websites.
- Emma could have her blood pressure, pulse, breathing and insulin levels checked remotely through telemedicine. Indeed, her doses could be adjusted in real time to better match changes in her condition.
- Harry could do his language course through distance learning, avoiding the need to travel; he could enhance the experience with teleconferences to the professor in Beijing and chat rooms set up with other students.
- Who knows … online gaming, instant messaging and swapping YouTube clips might offer Harry some social interaction that suits his teenage inclinations.
- And VOIP, enhanced with webcam capability, would certainly offer better and cheaper interaction with the Australian branch of the family than the postage or traditional calling.

2.2.2.1 Causes of digital exclusion

According to Ofcom's latest consumer review there are still more than 20% of households without access to the Internet in the United Kingdom, and this number is significantly higher for the financially disadvantaged and the older members of society. There is clearly a significant digital divide and it's a major concern to governments and policymakers generally. We would also suggest it's a problem – and an opportunity – for companies in the ICT space.

Why then does our imaginary single mother not benefit from the digital society?

There are four reasons cited in the research:

- inclination
- money
- education or training
- the limitation of broadband service in her particular area

In fact Ofcom's research shows lack of interest to be the most cited reason, followed by cost and availability of a PC, tablet or smartphone [6].

It may be that it has never occurred to Victoria that she could improve her life in this way. She may simply not want to have the Internet. Even more relevantly, her company may not have ever considered moving away from a single-location call centre (with its costs of rent and rates, its travel problems and lack of flexibility) to a home/distributed model. The college might not see the advantages of distance learning and the National Health Service (NHS) hasn't the money to invest in telemedicine. Indeed, there may be all sorts of personal and institutional resistance to the idea of moving medicine online, despite significant success in some areas of public service.

Computers cost money, as does broadband. Victoria may not be able to afford £500 for a PC or £10 a month for broadband.

While linked to inclination, there is plenty of evidence to say that many people would like to use Internet services but do not know where to start. These are not complicated services, but the number of people not yet exposed to basic computing remains surprisingly high especially amongst the older generations and the deprived.

2.2.2.2   The limitation of broadband service availability

This is a slightly different issue and is likely to hit the deprived and the prosperous equally. Copper pairs in the access network can deliver broadband service using asymmetric digital subscriber line (ADSL). However, the speed of broadband declines with the line length. This can be overcome by the use of optical fibre to street cabinets with just the final drop to the household being over the existing copper pairs. The technology is improving all the time, but it remains a challenge to offer reasonably fast broadband (at least 2 Mbs) to the remotest 20% of homes and, in some cases, small businesses. This is seen as a further problem for already under-populated and disadvantaged rural areas. At the same time, some of the wealthiest and most influential people live in remote areas. The ultimate solution to the problem is an all-fibre access network, but unfortunately the economics remains challenging – particularly in sparsely populated rural areas. Various studies based around the world show that – depending on population density – fibre access will payback at between 60% and 80% coverage. A further 10% might be covered by subsidy or profit-sharing models, but there remains a significant proposition of every population that won't be covered economically. Most incumbents, certainly in Europe and Asia, have rolled-out fibre to the profitable areas, or have plans to do so. What to do about the balance is proving a political hot potato for governments as they incur the wrath of many – often very vocal – people excluded from fibre access and therefore deprived of superfast broadband. We return to this subject when we discuss regulation of NGN in section 2.2.6 [7].

2.2.2.3   The future challenge of the digital divide

These aspects of the digital divide are an obvious concern for government and policymakers generally. The digital economy will offer growth, modernity and a better quality of life. It will reduce the cost of delivery and improve the quality of many of our public services from health to education to transport. Delivering services online should reduce inequalities, rather than extend them. Governments are therefore likely to develop a series of measures such as:

- Regulations (e.g. rules on roll-out of fibre and ADSL).
- Incentives (e.g. tax breaks for companies subsidising low cost PC for home use).
- Direct investment by governments.
- Resource allocation (e.g. ring-fenced money to develop telemedicine).
- Encouragement (e.g. to charities to support PC swap schemes, training, etc.).

All of these routes are likely to offer either impositions or opportunities to the industry and the players in the ICT value chain. How companies react will depend on their strategies, public affairs position and vision for the future. There are massive business opportunities to Telcos in supporting the public sector to deliver services electronically, as well as developing and supplying other services aimed at the digitally deprived.

We should also consider the wider economic arguments. Metcalf's law states that the more end points to a network (e.g. the number of subscriber lines connected), the more valuable the system becomes. To take a silly example, few phones calls were ever made when only one phone existed – you need two. But a million is better still! So, there is economic value in broadening the market for broadband and online services. Given appropriate incentives, some of these market segments then become profitable.

So in tackling the digital divide we can expect to see:

- regulation and perhaps legislation requiring fair broadband access to all
- some cooperation between government and industry in educating the public about the transformational aspects of ICT
- a greater focus on marketing and selling major programmes to the public sector (although they will have to improve their record on implementation)
- third sector and not-for-profit players focusing on subsidised kit and computer training
- improved technology and investment to tackle some of the limitations of broadband, low-cost PCs, etc.
- Some limited, direct investment in rural fibre by governments.

2.2.2.4 A possible pitfall

Giving regulators responsibility for narrowing the digital divide, or ensuring universal service or services for the disabled are all examples of using regulation to advance social agendas. This sits in contrast with their duties to promote economic objectives, such as competition, innovation, investment and efficiency. Balancing social and economic agendas is a key challenge for regulators, and an area where considerable regulator discretion is required.

2.2.2.5   A final thought on the digital divide

All of the above remain key issues in even the most developed economies. How much more important are the issues in the developing economies where current, traditional services are not yet developed. Even today, China and India have a far greater reliance on mobile phones than fixed phones because there has never been a traditional telecommunications network built in the rural areas. Distance learning, distributed labour models and telemedicine could all transform these economies and the quality of life for their populations.

2.2.3   *Regulation as a proxy for competition*

As we have seen, regulation is – to some extent anyway – a proxy for completion. Most of the issues referred to in the previous section (though not all) would be resolved if the market worked perfectly to hold prices low; provide excellent service; and force companies to be efficient or go bust. However, introducing competition into any market can be tricky. In the early stages regulators have some specific jobs to do. They need to:

– create licensing and investment incentives to attract new entrants
– set the rules and conditions whereby new entrants get access to existing networks
– determine the prices that they pay (referred to as interconnect rates – see section 2.2.4)

As competition starts to grow, the focus of regulation changes (shown in Figure 2.3). Incidentally – as an aside – every regulator says that his/her job is to make themselves redundant; sometimes it is even embedded in their objectives.

As competition grows regulatory focus shifts

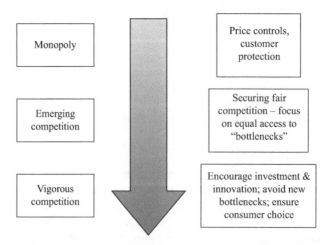

Figure 2.3   As competition grows regulatory focus shifts

As an example Viviane Reding (EU Commissioner) wrote as early as 2006 that

'Effective competition in the markets benefits consumers in terms of increased choice at lower prices. The regulatory framework provides both [a basis] to control the exercise of significant market power and the means to allow national regulators to take steps to allow self-sustaining competition to emerge. As markets become effectively competitive, the framework provides that sector-specific regulation will be removed. Thereafter, competitive markets will be subject only competition law and consumer protection regulation' [9].

Despite this, no regulator has ever made themselves redundant, although it would be fair to point out that some have reduced the activities that they regulate, for example in the abolition of retail price regulation. However, a reliance on the market, combined with competition law and consumer protection legislation, may provide the opportunity for *light touch regulation*, another, rather more realistic, objective of developing regulatory regimes [10].

Typically, when competition is introduced into an industry the new entrants need support, maybe even some form of positive help to break down the incumbents monopoly. This might be in the form of preferential interconnection rates; the mandating of wholesale products; or favourable licence conditions such as a waiver on the Universal Service Obligation. But the regulator will also be aware that consumers need protection from a newly privatised monopoly. This might come in the form of retail price controls (see Box 2.3) and other forms of consumer protection such as guaranteed quality of service.

As competition grows further the market will begin to achieve many of the desired outcomes, such as innovation in services and declining prices. The regulator's focus will first be on securing fair competition and then, as matters progress, to creating effective and vigorous competition that works seamlessly for consumers. For example, in Hong Kong domestic fixed network services were opened up to competition in 1995. By 2005 the Office of the Telecommunications Authority (OFTA) felt able to remove all retail price regulation as they assessed that competition had developed sufficiently to ensure fair prices for consumers. They, therefore, turned their regulatory attention away from management of the retail market to gaining greater traction in the wholesale market [14].

A major concern of regulators has been the role of the incumbent as both a provider of essential wholesale services (such as bitstream, private circuits and unbundled local loops) and the biggest retail competitor of the customers of those wholesale services. New entrants often felt that the wholesale products they received were not provided in a timely way, to an acceptable level of quality, or at reasonable prices. Regulators became concerned that incumbents might be tempted to protect their retail businesses by giving sub-standard services and products to wholesale customers. This problem is often described as the incumbent having both the incentive and the ability to unfairly discriminate in the provision of wholesale services.

Box 2.3   A word about price controls

The first stage of liberalising a market is to privatise the incumbent operator and, at the same time, to introduce competition. The intention is that competition (i.e. the market) will control the dominant power of the incumbent and bring consumer benefits, such as reducing prices, better service and greater innovation. However, these effects don't happen immediately and therefore a form of price control may be introduced in the form of a price cap.

Price controls prevent an incumbent with market power from abusing that power by charging more than is fair or by failing to realise or pass on to customers achievable improvements in efficiency.

Price cap regulation was invented by Stephen Littlechild whilst working on utility privatisation under Mrs Thatcher in the 1980s. It is now widely applied, not only in telecommunications but also to many types of utilities (water, energy, transport) throughout the world.

The principle is that a company with significant market power should have to reduce the price of a defined *basket* of products by CPI $- X\%$ (where CPI is the relevant annual consumer price index. In the UK this is often the retail price index (RPI)). If $X >$ CPI then prices must fall in real terms. Normally price caps are set for 3 or 5 years, and the aim of the regulator is to set X at a level where the incumbent is earning no more than their long-term weighted average cost of capital (see Box 2.5 for an explanation of the calculation of WACC) by the final year, i.e. they are earning no more than a fair rate that would encourage future investment, but would squeeze out any super profit. Price caps can be applied to retail or wholesale prices, or both.

The benefits of a price cap are that new entrants have greater certainty about likely prices in the market; consumers get a better deal; and incumbents are unable to earn super-profits and thereby consolidate their position. Price caps should therefore encourage efficiency, competition and growth in the market.

Once market forces have grown sufficiently to squeeze out any abuse of market power the price cap can be removed [8].

Not all wholesale services suffer from this problem. Interconnection and call termination services, for example, are both bought and sold by incumbents and new entrants alike. The scope for discrimination is absent or small. The problem is limited to what are termed 'enduring economic bottlenecks'. These are the products and services which realistically can only be supplied by the incumbent but which are essential to the businesses of competitors. They include unbundled local loops, essential infrastructure services such as access to ducts and poles, (in some regimes) wholesale line rental, next generation access products and bitstream services.

To take our example of Hong Kong; the regulator withdrew from retail price regulation in 2005 but at the same time introduced Type 2 Interconnection, a measure that enables operators to interconnect to the customer access networks (e.g. the local loops) of their rivals. This further enabled the successful establishment of competition, while also encouraging self-built. Indeed, Type 2 Interconnection has subsequently been withdrawn as OFTA now believes faculties-based competition to be sufficiently embedded to enable it to withdraw from the regulation of the wholesale market [14].

To ensure fair and equitable access to these enduring economic bottlenecks regulators introduced the concept of *equivalence*. Equivalence is defined as a requirement for wholesale customers to be able to use exactly the same set of regulated wholesale products, at the same prices and using the same systems and transactional processes, as the incumbent's own retail activities.

This ensures that here is no discrimination, since identical products, systems and processes are used. This can be confirmed by requiring the publication of key performance indicators showing the service standards delivered to the incumbent and to its wholesale customers.

While equivalence clearly resolves the issue of discrimination, it can be a very costly change for the incumbent to make. As a long standing vertically integrated business the incumbent will have developed its own systems and processes for retail service delivery before the requirement to introduce wholesale products arose, and those systems and processes will not generally use the wholesale products and wholesale transaction and service mechanisms. Transforming all retail service delivery to full equivalence would be very costly, and, as we have seen earlier, regulatory decisions must be reasonable and proportionate. For this reason regulators tend to consider imposing equivalence on new services, such as next generation access products, where the use of wholesale services and mechanisms can be built in from the outset. For long existing services regulators prohibit undue discrimination and use case-by-case tests as to what is reasonable and proportionate.

As the industry moves to full and vigorous competition the regulators will wish to reduce the frictional costs of moving between providers, and particularly in moving away from the incumbent. Examples of measures favoured by regulators in this area are number portability (to ensure you can take your number with you to a new service provider) (see Box 2.4); a limit on the length of contract i.e. service providers shouldn't be able to tie people in for excessive periods; and controls on unfair bundling. Such regulations make it easier for consumers to switch suppliers to enable them to truly benefit from competition. Many of these regulatory initiatives arise both from the Strategic Telecommunications Review (STR) and regulation of next generation networks (see sections 2.2.5 and 2.2.6).

While every country is at a different stage of evolution, the good news is that since the mid-1990s regulation of incumbent Telcos have generally been judged to be a success, with effective competition developing in most markets. Consumers around the world have seen significantly lower prices in real terms; a high degree of new services; and improved customer service.

Box 2.4   Number portability (NP)

Number portability is a facility provided by one Telco to another which enables customers to retain their number when switching between those operators. It is normally reciprocal, and may be mobile to mobile or fixed to fixed (but not between mobile and fixed).

In 1991 Oftel (the predecessor of Ofcom) believed that competition was slow to take-off in the UK because customers were reluctant to leave BT and join, for example, a cable TV company. They believed the need to change numbers was one of the biggest inhibitors of switching. They, therefore, modified BT's licence to require them to provide number portability [11]. This is believed to be the first time that NP had been mandated anywhere in the world.

Despite the fact that this new measure clearly posed a commercial threat, BT accepted the condition. However, they disputed that they should incur the extra costs associate with onward routing of calls (sometimes called *tromboning*) i.e. the inefficient call-routing associated with NP [12]. The case was therefore referred to The Monopolies and Mergers Commission (MMC) who, after extensive enquires and analysis, ruled that NP was in the public interest; that it would generate additional competition which, in turn, would generate benefits for customers. They believed that the additional costs should be shared between the operators and that BT should be penalised if they did not adopt more efficient call-routing as time went on [13].

This landmark ruling paved the way for similar pro-competitive measures around the World and has greatly increased both competition and customer benefits.

[As an aside, the financial services equivalent of NP is the mandated ability to move accounts, standing orders, etc., quickly and easily. This was introduced by Don Cruickshank, architect of NP when he was the director general of Oftel and later invited to review financial services by the Government!]

2.2.4   Interconnection

A competitive market for telecommunications implies two or more operators competing for customers in the same area. Historically since the 1980s this has meant an incumbent Telco (i.e. a large existing operator, previously owned by government but now probably privatised) competing with mobile operators, possibly a cable TV company (CTV) and some new entrants. All of these operators will need to be licensed or qualified to set up in business through some other process. Normally, the terms and conditions required to operate are embodied in national legislation and are enacted by the regulator [15]. As a part of this process, and mindful of the principle of 'any to any', regulators will require networks to link together. Put simply, every phone must be able to ring every other phone, no matter

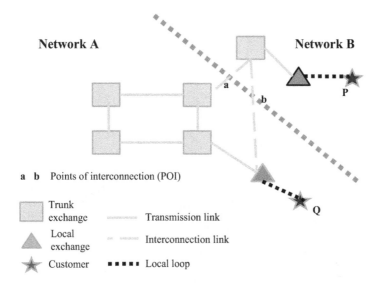

Figure 2.4   Interconnection

which network originates the call and which terminates it. This linking together of networks is technically called interconnection and is achieved by physically connecting the networks at pre-defined points of interconnect (POIs). There may be a single POI for each pair of networks (say at a high-level trunk exchange) or alternatively, there may be many hundreds of POIs (at a variety of trunk and local exchanges). A simplified version of network interconnection is shown in Figure 2.4, with POIs at **a** and **b**.

The technical aspects of the interconnection of all the different types of networks are beyond the scope of this chapter but are covered in detail in Understanding of Telecommunications Networks (Valdar) [16].

It is for the new operator to determine where and how they wish to interconnect and they make that decision based on financial grounds. We will return to this subject in a moment.

2.2.4.1   Setting interconnection rates

A key role for the regulator in this sphere is to set the interconnection rates, or prices. These prices will be based on the calculated average product costs for every type of interconnect call, together with a reasonable rate of return (profit). Product costing is exceptionally difficult in any industry but especially so in telecommunications where there are many products; they are subject to rapidly changing technology; and common costs are an extremely high proportion of overall costs. We will return to product costing in far more detail in section 8.2.2, but it is important here to understand the challenge that faces a regulator when setting interconnection prices.

The analysis requires a detailed knowledge of the existing costs of the network, the amount of telecommunication traffic that it carries (connections, calls, data, etc.) and the average routing of every type of service. Given that there may be

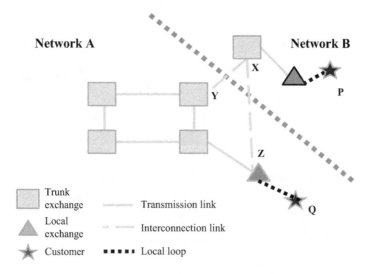

Figure 2.5   Interconnection – routing calls

many interconnected networks connecting at a wide range of POIs, this is a huge amount of data to collect and to analyse. Normally the network operators are required to collect the information and to provide it to the regulator so that both the incumbent operator and the regulator can determine their views on a fair price for interconnection. The Telco will do much of the work and collect much of the data as part of their normal financial analysis, but, nonetheless, there can be a heavy burden of extra data extraction and manipulation.

A simple example serves to illustrate how the calculation works. In Figure 2.5 a new entrant (network B) has requested interconnection with the incumbent operator (network A). They have requested a single point of interconnect (**X–Y**) and have supplied the link themselves. There is a shared responsibility to provide and pay for this link, but in our example network B has done the physical work themselves. We now need to consider the revenue flows. If customer P calls customer Q the retail revenue is collected and retained by the origination network (that is network B). However, network B will then need to pay to have that call delivered to customer Q. This is to say they will need to pay network A for the part of the call (from **Y** to customer Q). In our simple example they need to pay network A for the use of:

- a trunk exchange
- transmission link (connection)
- transmission link (by length)
- a second trunk exchange
- a local exchange

These segments are called network or engineering elements.

Clearly it would be impossible to do the sums for every single call, so the regulator starts by setting a price, on average, for each engineering element. Certain elements are priced per minute of use e.g. local exchange use, other elements by a single price per connection and some by length e.g. transmission. Not every interconnect call is the same and not every call uses one of each engineering element. Depending on the number of POIs and the network topography some calls will use more or less links, and of various types. It is therefore necessary to calculate the average route of each type of call and the average number of each engineering element used. These are known as route factors. Once the route factors are calculated it is a relatively easy job to calculate the cost of each type of call minute by combining the relative route factors with the average cost per engineering element. (Route factors and other aspects of interconnection pricing will be covered in detail in Chapter 8.)

Having calculated the cost of each type of interconnect call minute it only remains to add an element of profit or, more accurately, the cost of capital, for the recipient network. One might ask why a network operator should make profit from just terminating calls. To answer this we should return to the aims and objectives of regulators. One aim is to ensure investment in existing and future networks. If operators receive only their minimum current cost of operation they will not be incentivised for future investment or rewarded for the risk of past investment. To ensure an incentive to invest the regulator will try to calculate a fair weighted average cost of capital (see Box 2.5 for an explanation of the calculation of WACC) that will reward risk, incentivise future investment but not allow for any super-normal profit. They may also wish to incentivise greater efficiency from the terminating network operator and so may include either a built-in deflator (see Box 2.3) or they may reduce the actually incurred costs to something that they calculate might represent 'most efficient operator' status.

Clearly calculating interconnect rates is complicated and time consuming. It requires judgement as well as analysis and is highly political. Nonetheless it's a vital role for regulators if markets are to be opened up, new entrants encouraged and incumbents fairly rewarded for the use of their assets by other operators.

While we have concentrated here on fixed to fixed calls exactly the same principles apply to mobile, international calls and data interconnection.

2.2.4.2 A final word on interconnection

It was mentioned earlier that new operators have to decide where and how many POI's they require. Having decided initially, they will continue to evolve new connections and links as their network grows. Given what we have now learnt about the pricing of interconnect calls it is obvious that interconnect costs will, overall, be reduced if the call is carried further on the originating network. This is called far-end-handover. If we look again at our simple example (Figure 2.5), network B might seek a POI at the relevant local exchange (Z). They would need to provide the link XY but it would then reduce the original interconnect charge from YQ to ZQ (just a local exchange and local loop segment). Operators, therefore, try to find the optimum balance between building more of their own network and paying to route their originating traffic over other operator's network. (This is often referred to as the operator's 'make or buy' decision.)

Box 2.5   Weighted average cost of capital (WACC)

WACC is a calculation of a company's long-term cost of capital in which each type of capital (shares or equity; bonds and other long-term debt) is weighted by its relative share of total capital. The formula is normally expressed as:

$$\text{WACC} = \frac{E}{V} * \text{Re} + \frac{D}{V} * \text{Rd} * (1 - \text{Tc})$$

where

Re = cost of equity
Rd = cost of debt
E = market value of the firm's equity
D = market value of the firm's debt (i.e. the interest they pay)
$V = E + D$
E/V = percentage of financing that is equity
D/V = percentage of financing that is debt
Tc = corporate tax rate

NB: Interest is paid before tax and so we reduce the cost of debt by the tax avoided.

However, knowing a firm's historic WACC is only a part of the story. A regulator would wish to restrict a firm's future earnings to no more than WACC, i.e. to squeeze out any current super-profit, normally over a number of years (often called a *flight-path*). To do this, regulators try to forecast an acceptable WACC that should apply in, say, 3 or 5 years' time.

Forecasting the future cost of borrowing, i.e. likely interest rates, is relatively simple; for example central banks and independent economic analysts all publish such forecasts.

Predicting the cost of equities is, however, much harder. Here regulators often turn to the capital asset pricing model (CAPM). This aims to predict future earning by looking at a risk-free return and adjusting it by both the general level of risk in the market and the relative additional risk associated with the company or market in question (i.e. the telecommunications market). In short:

Cost of equity $\text{Ke} = \text{Kf} + \beta(\text{Km} - \text{Kf})$

where

Kf is return on risk free asset e.g. treasury bonds.
Km is return on the stock market, often measured over 30 or even 50 years of history.
β is the volatility of a particular company or share.

Whilst the model brings order to the calculation it still requires considerable judgement. In particular, the level of beta (β) is always contentious, with incumbent Telcos arguing that any investment they make brings considerable risks ('what if we are left with unused network assets', etc.) whilst regulators and new entrants view the risks as limited. In fact fixed line telephony is usually judged to have a β of just less than 1, i.e. to be slightly less risky than the general stock market whilst, say, internet services would be rather more risky at $\beta = 1.15$.

Regulators will consult, do economic studies and consider the overall impact of their decisions before finally fixing an appropriate WACC to apply. Despite the contention and uncertainties, WACC remains a vital tool in the regulators toolkit.

In fact, in the early days of interconnecting it became clear that some operators, e.g. cable TV companies, could reduce their interconnect charges by delivering local traffic locally, rather than routing all interconnection through just one POI at a distant trunk exchange. Initially, incumbent Telcos tried to resist POIs at local exchanges, arguing that it increased their costs and the complexity of interconnection. Regulators, therefore, had to examine the arguments, balancing reduced efficiency for the incumbent with lower overall costs for new operators and challengers. On the whole, regulators ruled in favour of POIs wherever they could be economically justified. Nonetheless, this proliferation of POIs – and the complex mesh of interconnection of many networks connected at many points – has severe implications when the industry wants to move to a new network architecture. We will examine this issue when we look at developing next generation networks (see section 2.2.6). Once again the regulator has a job to do [17].

2.2.5   The Telecommunication Strategic Review (TSR)

As we have seen, in 2003, the UK Government combined the responsibilities of a large number of specific regulators, each looking at a limited aspect of the ICT industry, into a single regulator to be called Ofcom. This was a bold move, one of the first in the world to recognise convergence of the ICT industry and the extensive links that run throughout the value chain (both issues that were covered in Chapter 1). On taking up office the team at Ofcom commenced a strategic review of telecommunications that has had far-reaching consequences. The outcome has been closely watched by regulators around the world, many of whom are looking to emulate the outcomes.

In essence, after a long and detailed analysis Ofcom concluded that there remained market failure in parts of the telecommunications value chain in the United Kingdom and that BT still had some residual competitive advantage from its previous incumbent position. They published their findings in 2005 [18] and found specifically that:

- Competition was restricted in wholesale markets for access and for backhaul services.
- This limited competition in certain directly related markets.
- BT retained substantial wholesale market power.
- BT remained a vertically integrated provider serving both wholesale and retail markets.
- This combination gave BT the ability and the incentive to discriminate against its competitors.

Given Ofcom's concerns, BT offered some significant voluntary undertakings to overcome the perceived market failure. For example, one such undertaking was to create functional separation whereby BT would operate two completely separate businesses: one serving their retail customers (BT Retail) and the other serving other telecommunications operators (openreach). Figure 2.6 shows the technical boundaries of functional separation.

A summary of the undertaking can be found in Box 2.6. Although these undertakings were made voluntarily, they are, nonetheless, binding and backed by significant powers if BT fails to comply. Breaches can lead to:

- directions from Ofcom and/or court enforcement
- reference to the Competition Commission
- third-party actions for damages

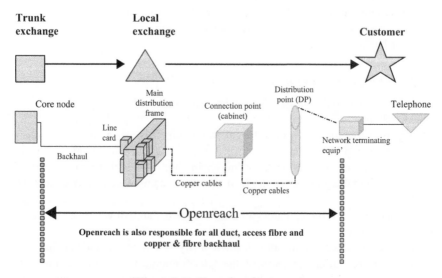

Figure 2.6   Functional separation

Box 2.6 Summary of BT's voluntary undertakings

- *Functional separation*
- *No foreclosure* of network access
- Providing wholesale access on *equivalent inputs* basis
- Timely provision of network access
- Setting charges on basis of *efficient design*
- Compensation
- Industry collaboration

2.2.6 Regulation of next generation networks (NGN)

One of the main drivers of the UK Government creating a converged regulator in late 2003 (see section 2.2.5) was to tackle the issues that arise from the development of next generation networks (NGN). We discuss the detailed design aspects of NGN in section 5.2.2, together with the case for its introduction. For the purposes of this chapter a high-level summary will suffice: An NGN provides data and voice services on a new IP data platform (not the Internet) – and it will eventually replace the PSTN. (See Box 2.1 for an explanation of the differences between the Internet, VOIP and IP service.) A similar transition will also occur when full-scope 4th generation (4G) mobile networks are introduced. It is clear that there are fundamental differences between NGN and PSTN; these differences change both the business and regulatory landscape. If we mention just some of those differences it quickly becomes clear that regulation will be required to encourage investment in the new risky technology; to safeguard national and personal security; and to ensure the best deal for customers. While regulatory principles and practice already apply to current services (e.g. price controls; unbundled local loops) they are not all transferable to an NGN world. Similarly, it might be argued that investing in new networks is open to all on an equal footing, so competition will regulate the market without intervention. However, most regulators around the world view the successful development of national NGNs as far too important to leave to chance, or even to market forces. They set clear aims to bring the benefits of NGN to the national economy that they serve.

Key differences between NGN and the networks that went before are therefore:

That there should be incentives to invest in the new technology. Most previous investment in new generations of technology has taken place when monopoly providers were guaranteed a captive market and an assured revenue stream. What is more, these network providers were normally government-owned and so the investment was in the form of public assets, rather like roads. In the new world, multi-operators are investing in interconnecting networks and competing in an uncertain market. Regulators, therefore, have to ensure that operators are incentivised to invest, or, as a minimum, that they are not disincentivised.

That next generation networks should be multi-operator. That is to say they should be a system of interconnected networks, owned and operated by a number of service providers. What is more, they should accommodate value-added service providers who haven't necessarily invested in infrastructure but who wish to use NGN to deliver their services (perhaps through application program interfaces, known as APIs). This means regulators have the task of ensuring industry-agreed standards, clear and shared network design and a published and forward-looking implementation plan.

That there should be no anti-competitive advantage. In particular NGN technology and the new architecture should not be seen as an opportunity to re-introduce new economic bottlenecks or anti-competitive traffic management. Regulators will therefore consider measures such as unbundling (see Box 2.7), mandated wholesale products and net neutrality (see Box 2.8).

Box 2.7   The case for unbundling

We have seen, earlier in this chapter, the importance of recognising and dealing with enduring economic bottlenecks. There is probably not a clearer case for action than that of the local access network or local loop. For example, as the internet took off and home broadband became a reality for many people (around the turn of the millennium) it became clear to regulators, both in the United states and in the United Kingdom, that the measures that they had taken to introduce competition would not be enough. In particular they feared that the incumbent Telcos (the ex-Bell companies and BT) would use their near-monopoly of the copper wires that delivered ADSL broadband to control the market for broadband. Following an extensive period of technical and commercial investigation, the Federal Communications Commission (FCC) in the United States and Oftel in the UK, passed a series of measures that required unbundling of the local loop [1]. In essence this required the incumbent Telcos to open up their local telephone exchanges to other operators who could then install their own Digital Subscriber Line Access Multiplexers (DSLAMs) and take control of their customers' physical local loop [2]. In this way the new entrant service provider could sell broadband and internet access services to any customer in a location where they had access to the local exchange (in reality, the densely populated high-value urban market). What is more, the price that the new entrant pays for the unbundled loop is regulated, normally at a long-run incremental cost plus forward looking WACC (see section 2.2.4 and Box 2.5 for the principles of setting regulatory prices).

The policy of mandating local loop unbundling was a radical one, but proved highly effective. Broadband penetration grew very rapidly wherever it was introduced, retail prices for broadband fell dramatically and new entrants, e.g. Covad, Bulldog and Netplus, all prospered and helped change the competitive landscape. For example, by 2010 around a third of all broadband lines in the UK were provided by *unbundlers* and prices were 50% lower than before.

With the undoubted success of copper-loop unbundling we might have expected a similar regulatory approach to be applied to next generation access – e.g. the replacement of copper with optical fibre. However, this is unlikely to be the case. First, it is far more difficult technically to unbundle fibre loops and where it can be done, the economics is far less advantageous. Second, whereas copper access lines had been installed and often written-off many years ago, the new fibre access requires significant investment. Therefore, Telcos have argued that the threat of unbundling fibre increases risk, to the level where they might not be able to afford to make the necessary capital injection. This argument has been notably successful in the United States where the FCC have stepped back from their previous determination to tackle the enduring economic bottlenecks; and in Germany where Deutsche Telecom (DT) has been granted a *holiday from regulation* to fibre five major cities. In the UK and in a number of other European countries, regulators have held tight to the desire to defeat economic bottlenecks but, in recognition of the technical problems, have favoured mandating wholesale products such as *high-speed bitstream*, as well as giving greater access to the physical poles and ducts that enable greater opportunity to build competing infrastructures.

Finally, we should acknowledge the success of pure completion in Japan, where a combination of intelligent regulation, highly concentrated markets and low cost of capital has led to competing physical fibre networks. In Japanese cities most consumers now have a choice of low-cost, super-fast broadband from a range of providers (e.g. The city council, a power company, Telco, etc.) who provide the service on their wholly-owned fibre network.

These are all issues of key importance to regulators around the world as they seek to control developments in NGN – whether it is encouraging infrastructure investment by incumbents; incentivising existing IP network operators, or granting fair access for service providers who are seeking to build services on other operators' NGNs. Indeed, an important survey of key industry leaders around the world identified NG access most important regulatory issue in next 10 years [21].

These aims are key, not only to regulators, but to governments. This is because governments often see modern ICT services as essential to economic growth and the business success for their country. We only have to look at some published national targets to see how important they are. Let us consider just four examples summarised from press reports in 2014:

- The US Government publically pledges that by 2020 that there will be 100% coverage at >4 Mbps and 100 m homes will have >100 Mbps, with 1 Gbps for schools.
- In China the president has said that by 2015 4 Mbps will be available *generally* and that city-dwellers can expect at least 20 Mbps. China has also a published plan to provide superfast broadband (up to 100 Mbps) in five key cities.

Box 2.8   Net neutrality

One of the founding principles of the internet (if something that has grown so organically can be described as having founding principles) is that it is open i.e. anyone can develop a new application and use the internet to access the worldwide market. This openness has been a mainspring of innovation, encouraging a very wide range of services and uses that would never have been thought of in a closed or walled-garden environment. Consequently, the internet has become essential to most people's daily lives, providing entertainment, shopping, information, public services, communication and social interaction.

However, openness brings its own problem. The rampant success of the internet has led to a massive growth in the use of data and the growing demand for more bandwidth, which in turn requires more and more investment in capacity and better utilisation. Many networks already experience congestion and blocking and consumers are becoming increasingly used to a poor quality of service, slow-running and frozen screens. This might not be seen as a problem if Telcos and service providers were well rewarded for carrying more and more bandwidth but, regrettably, the business model doesn't work that way. On the contrary, as broadband speeds increase and data usage goes through the roof, service providers are locked into fixed monthly fees. What is more, some of the growth in bandwidth is actually VoIP, a low-price (or free) telephone service replacing the service providers' existing lucrative voice calls business.

For traditional fixed and mobile Telcos this is a *perfect storm* of disadvantages, potentially leading to a future of low margins, providing just *dumb pipes*. So, what to do? Increasingly service providers are deploying traffic management techniques to control the ever increasing bandwidth. Rather than relying on *best efforts* where all internet traffic competes on equal terms (and services fail at random), they are increasingly applying *capacity management services*, where services are prioritised depending on the value or importance of that service e.g. guaranteed voice services or video services where stutter and flicker are unacceptable.

Clearly, this is an appropriate business and technological response to a problem, but the use of traffic capacity management brings regulatory concerns. In particular, regulators around the world are beginning to fear that traffic management may lead incumbent Telcos and other service providers to act anti-competitively; that they might favour their own high-value services over their competitors. This is clearly demonstrated where mobile companies are already barring VoIP applications from their handsets.

In principle regulators generally believe there should be *net neutrality*, that traffic management should only be used in an open and transparent way that sets priorities based on value and the needs of the service, not on the commercial value to one operator or another. To date, regulators have not come up with a uniform answer to this problem but some limited action has been taken. The European Union has passed a revision to its Framework for Electronic Communications that prevent anti-competitive blocking or slowing of services; in the United Kingdom, Ofcom require service providers to provide clear consumer information about the effects of traffic management and are monitoring other potential concerns; the Federal Communications Commission (FCC) in the United States have agreed a set of measures ensuring smooth interconnection of services and they are considering the need for more consumer information; whilst a number of middle-eastern states e.g. United Arab Emirates (UAE) have taken a contrary view by protecting the commercial advantage of existing Telcos by severely restricted use of VoIP services.

There is some way to go in this debate. What is clear is that the open nature of the internet has brought great benefits; that service providers need to be properly rewarded for investing in infrastructure and carrying digital traffic of all types; and that a new business model needs to develop. Regulators will aim to ensure that *net neutrality* is at the heart of any new business model.

- The *E U Digital Agenda* includes 13 separate targets e.g. that by 2020 all homes should have access to >30 Mbps, with 50% having >100 Mbps.
- Finally, the UK Government intends to provide its citizens with the *best* broadband in Europe by 2015, with *virtually* 100% having more than 2 Mbps and 90% of the population having access to superfast broadband (>40 Mbps).

In looking at how regulators are tackling some of these challenges and ensuring that government aspirations are met, it is helpful to look at NGN core and NGN access separately.

2.2.6.1 NGN core network

As operators replace the existing hotchpotch of existing interwoven networks, built-up over a century, with the far-more coherent and layered network of NGN, regulators will wish to ensure that services and functionality aren't lost. Customers will seek continuity of service, both from a quality of service/reliability point of view and from the point of view of functionality. Initially regulators will probably leave this to be resolved through competition, but in extremis they may have to step in.

A second concern for regulators will be the changing topography of the networks. In section 2.2.4 we explored the relationship between interconnection of networks and costs, especially the costs incurred by challenger service providers. It was clear that the number and topography of POIs were crucial. As major Telcos move to an NGN, the topography of the network will change dramatically. The number of local exchanges or nodes will decline dramatically and the large number of trunk exchanges will be replaced with far fewer multi-service nodes. The number of possible POIs will decline, far-end handover will be less cost advantageous and smaller operators could suffer stranded assets and financial write-off. The concern for regulators is that these changes will significantly increase the financial risks for challenger service providers and that incumbent Telcos might use network design as an anti-competitive weapon.

The regulatory solution is to try to create an industry consensus: with plans published in advanced; compensation paid for unjustified increases in interconnect charges and penalties for non-delivery. These solutions put great emphasis on network planning and implementation as an industry-wide activity, with appropriate consultation rather than the domain of just one or two operators.

2.2.6.2   NGN access

We have seen elsewhere the great success of copper local loop unbundling (Box 2.7) in opening up and stimulating the market for broadband. One might therefore expect this approach to be followed as Telcos progress towards an all-fibre local access network and deliver superfast broadband (up to 100 Mgbs). Certainly, fibre in the local access network is a clear candidate for economic bottleneck status. However, the nature of optical fibre systems means that no direct equivalent of copper unbundling is currently possible with fibre access. Ofcom in the United Kingdom have instead pursued a 'virtual unbundling solution' where by the incumbent's (BT) optical fibre services are, in effect, resold by other operators. Nonetheless, some regimes, such as the Netherlands, have pursued the idea of physically unbundled fibre, but it has proved uneconomic as a basis for alternative providers to challenge the price of superfast broadband in the retail market.

The regulatory solution may also include making available the physical infrastructure that would help the civil engineering of new-build (e.g. planning-rules, poles and ducts) and to mandate wholesale products that could be purchased by challenger service providers. Such wholesale products would be enhanced with brand, service, etc., to produce a retail product that might compete with the incumbent.

Given the importance placed on superfast broadband as a stimulant of any national economy a study by Olswang in 2008 identified next generation access as one of the most important regulatory issues during the next 10 Years [21].

2.2.6.3   Many regulators and many solutions

While the regulatory challenges on NGN are reasonable clear, the detailed measures to overcome them are not. Regulators around the world are beginning to tackle these issues, in a variety of ways. We can only touch on a representative sample here:

- The EU Commission completed a major consultation in 2010 and concluded that despite the problems, the regulation of fibre access networks should be *consistent with copper regulations*. They encouraged member states to look at both physical unbundling (where economic) and wholesale broadband access. They also believed that, where market power existed, Telcos should be made to provide access to poles, ducts and in-building termination points. They specified that prices for wholesale access should be cost-orientated, with allowance for risk and cost of capital. Finally, they recognised that in certain circumstances competing fibre networks might deliver a viable market without the need for regulatory intervention. In these cases geographic sub-regional analysis could be employed [22].

- In the United Kingdom the Telecommunications Strategic Review (see section 2.2.5) prompted a number of voluntary undertakings that are relevant here. For example, the agreed principles of *No foreclosure of network access* and provision of wholesale products on *equivalent inputs* basis should ensure that the benefits of fibre are available for all service providers. Beyond this, however, are a number of measures introduced by Ofcom specifically aimed at NGN and in line with the EU Directive. These include measures to make available BT's poles and ducts to other operators at regulated prices and a requirement for BT to provide virtual unbundled local access (VULA). VULA is a wholesale bitstream product that is delivered over fibre and provides faster speeds that available from physically unbundled ADSL copper lines. The price of VULA will be regulated at a cost-oriented level but with an allowance for risk and cost of capital [23].

- It is perhaps surprising that, following the amazing success of copper unbundling in the United States, the FCC have not sought to emulate that success in the new fibre world. Nonetheless, on this occasion they took the view that requiring any form of shared access (either virtual or physical unbundling) would inhibit investment in the expensive city fibre rings required to quickly deliver superfast broadband. They have therefore not mandated a high-speed bitstream product. This failure to act is reputed to have reduced competition and consumer choice, and increased average prices [24]. Despite the EU's Directive, Germany has taken a similar approach, granting Deutsche Telekom a holiday from regulation in return for a guarantee to fibre five major cities. This *forbearance* of regulation has brought the German government into conflict with the European Commission [25].

- A very different approach is being pursued in Australia. Frustrated by what they saw as Telstra's intransigence the government threatened to break up the incumbent in a, as-yet unspecified, form of structural separation. They have also announced the National Broadband Network investment programme which aims to deliver 100% broadband by 2018, at the cost of 23 billion Australian dollars [26].

- Finally, we should acknowledge the success of pure completion in Japan, where a combination of intelligent regulation; highly concentrated urban populations mainly living in high-rise apartments; and low cost of capital has led to competing physical fibre networks. In Japanese cities most consumers now have a choice of low-cost, super-fast broadband from a range of providers (e.g. the city council, a power company, Telco, etc.) who provide the service on their wholly owned network. Although there is provision for mandated wholesale products they are seldom requested, as infrastructure-based competition ensures ubiquitous low-cost retail products and a highly saturated market.

2.2.6.4 A question remains

Despite the perceived importance of superfast broadband; despite all of the proposed and actual regulatory intervention aimed at producing a vibrant market; and despite some governments being willing to underwrite investment in next generation access (NGA); a problem remains. Often described at the final 10%, the problem highlights the fact that there isn't a viable business case to deliver NGA to the final portion of the country in question. The exact portion of the population varies, depending on the topography of the country, its urban and rural density and the willingness of customers to pay a premium to gain access to lowly populated regions. Nonetheless most countries, however well-developed their plans for NGN, have a problem with vociferous rural communities that are unlikely to be properly served with broadband until a new, as yet undetermined, business model is found. This digital exclusion is a new form of digital divide (see section 2.2.2).

2.2.6.5   Summary of NGN regulatory challenges

It is clear that the design, implementation and the use of NGNs bring significant regulatory challenges. Solutions will depend on the particular market, demographics or topography of the country in question. It is also likely that previous answers will not necessarily be relevant, e.g. unbundling may not be possible in a fibre world. We have also seen that the core and access networks present different problems and require different approaches to solve. Risk assessment is important if investors are to be encouraged to invest, but it's exceptionally difficult to do accurately; its assessment will always be controversial. Finally cooperation across the industry is vital if NGNs are to be well planned and implemented to the satisfaction of all the players.

2.2.7   The politics of regulation

In Chapter 1 we explored the pressure that various stakeholders exert on Telcos. One such stakeholder is the regulator. However, we should also acknowledge that regulators are subject to similar pressures, not least from politicians and governments. Balancing all the various conflicting interests of all the interested parties ensures that regulators do not have an easy time.

In an ideal world regulators would be independent of politics; that they are able to make decisions purely based on their published objects e.g. to promote the interests of consumers and citizens. However, in countries where Telcos are wholly or partly owned by the government, it is hard to achieve full independence.

Putting on one side the ownership issue, it remains hard for regulators to avoid undue influence. Let us take some examples:

- Competitive markets normally drive lower prices and better services, but potentially isolate non-profitable parts of the market e.g. rural areas, low-calling customers, etc.
- Regulators will aim to encourage investment (e.g. in fibre access) but will not wish to create monopoly suppliers of retail services.
- Incumbent Telcos will have strong opposition to the challenge that innovating new entrants will bring and may threaten to curtail investment if their profits and market share are challenged.
- Sections of society will want freedom of expression while others will want children protected from inappropriate material and the suppression of terrorism.

- State-funded or subsidised broadcasting is challenged by multi-channel commercial broadcasters, while free access to iconic sports events reduces commercial opportunities to fund those same events.
- And there are many other areas of the regulatory universe that lead to similar conflict and the need for careful judgement.

So what do we conclude? Well, that regulating is difficult; that major players in ICT, especially the media companies, exert significant power; that independence from politicians and government is important, but hard to achieve; and that decisions are likely to be compromises at best (and fudges at worst).

2.3   The future of regulation

Although regimes around the world differ, there is no doubt that generally things have progressed over recent years, as consumers and citizens have benefited from more open markets, greater competition, wider choice of products and services and lower prices. Nonetheless, there remain some challenges that are common to all. A few of them have been described in the following sub-sections.

2.3.1   To ensure investment and innovation

This is essential to encourage the building of NGNs and fibre access networks. As we have seen in the previous section Telcos around the world can expect some encouragement, perhaps through regulation holidays, decent returns that recognise risks or first-mover advantage, subsidises, or direct investment by governments. But they can also expect to suffer greater regulatory intervention in the form of access rules, inter-working standards, penalties for late delivery, etc. Many domains around the world have started to tackle these issues and many companies have already made multi-billion pound investments in both next generation access and next generation core networks, but there remains a long way to go. We can expect to see far more regulatory developments before NGN becomes universal.

2.3.2   To avoid new bottlenecks and monopolies

While regulators are keen to encouragement convergence of services and greater integration across the ICT value chain, they will wish to avoid the development of new bottlenecks or monopolies. This concern should apply equally in the provision of content; or at the service level (e.g. Sky, Apple or Google); or at the network level e.g. fibre to the home. In regard to content it's interesting to note that only three companies (EMI/Universal, Sony and Warner) control the distribution of 90% of all music – a far greater concentration of market power than in, say, network provision. So content players can expect a range of new regulatory burdens, such as the need to police piracy; measures to stop peer-to-peer use of networks; or functional separation as now applies to BT.

Again, we are beginning to see action in these areas, such as the net neutrality provisions (see Box 2.8) and the challenges to Sky's dominance in provision of live football (see Box 2.9).

Box 2.9 Is premiership football an enduring economic bottleneck?

Economic bottlenecks normally refer to scarce infrastructure or capacity, such as local access lines or backhaul. However it's an intriguing question to ask if scarce content can be viewed as an economic bottleneck and if so, should regulatory action be the consequence. Let's consider Sky's contract to show premiership football around the world. What are the market conditions? Is the advantage enduring? Well, the contract is renegotiated at regular intervals (3 or 5 years). It's contestable, in the sense that other companies have the opportunity to bid. There are several different packages of games to bid for (although in 2009 Sky won two out of three packages and the other company (Setanta) went bust).

So is it a bottleneck? This depends on market definitions. Is watching premiership football part of a bigger market? People can go to the match or watch lower division football. Is that a reasonable substitute for watching on TV? Most armchair football fans (i.e. the majority) would say no. Is watching televised rugby a reasonable substitute? Again, probably not for most football fans. So the contract probably would be judged an economic bottleneck, but maybe not an enduring one.

If regulators were concerned, what remedies do they have available? Well, they could require the contract to be renegotiated more often (annually); split the packages differently; disallow any one company winning more than one package; or they could require Sky to provide access to the matches to other companies at a regulated price. However, a regulator would have to be aware that such regulation would be likely to reduce the value of the contract and therefore the money going to the rights owners (the Football Association). For example, when it was suggested that the international cricket matches between the MCC and Australia (the iconic Ashes series) should be available free to air the England and Wales Cricket Board were horrified, arguing that the loss of revenue from selling the rights would set the sport back 50 years.

In fact, this case is not hypothetical. In 2010 BT challenged Sky's dominance of premiership football. Ofcom ruled that Sky's contract with the FA did constitute an enduring bottleneck and required Sky to provide access to other service providers (notably BT Vision) at a regulated price. However, in 2012 Sky appealed the decision and Ofcom were forced to modify their finding so that they continued to require Sky to offer a wholesale product but they did not specify a price, leaving that to commercial negotiations. At the same time BT won a package of matches in the 2012 contact auction and so have less reliance on wholesale services from Sky.

2.3.3   Managing convergence

Linked to the development of possible new monopolies are some existing elements of the ICT value chain that become far more important to regulation as convergence occurs. For example, music companies (see above) or broadcasters, such as TV stations (BBC, CNN, etc.), are now in direct competition with Internet service providers (ISPs), Internet protocol television (IPTV) services, etc. Again, we can expect to see regulators setting conditions to ensure that the dominant position these players currently enjoy is not carried across into a far-wider range of ICT services.

2.3.4   Consumer protection

As communications become far more ubiquitous and even more important to consumers and businesses, we can expect to see regulations to ensure consumers do not become tied to a single supplier; that they can switch easily between providers; that they aren't overcharged; or they are required to buy additional services that they don't require. We will see the extension of the principle of 'any to any' to more services. For example, in limited areas, we will see required standards such as an open architecture to ensure inter-working of services across all domains and suppliers.

Finally, not all ICT services will be focused on leisure and enjoyment. Many public services will be delivered virtually, from healthcare, to the benefit system, to education. In future, people without ready access to services (either through cost, location or training) will become even more highly disadvantaged. As we have already seen (section 2.2.2) the divide between those that are online and those that are not (the digital divide) is of great concern to governments. We can expect to see significant regulation focused in this area.

2.4   Summary

In this chapter we have learnt that regulation is mainly focused on market failure and aims to deliver benefits to citizens and consumers. However, the emphasis and priorities of regulation change over time, as competition develops. Going forward, regulation will focus on encouraging investment in new networks; ensuring that consumers benefit from innovation and convergence; and that new economic bottlenecks do not develop. Finally, it is clear that regulation will continue to be a very important part of the commercial environment for telecommunications companies and the ICT industry generally.

References

[1]   Buckley, J. *Telecommunications Regulation*, IET Telecommunications Series 50, Stevenage, UK, 2003, Chapter 1.
[2]   Valdar, A. *Understanding Telecommunications Networks*. IET Telecommunications Series 52, Stevenage, UK, 2006, Chapter 2, pp. 24–27.

[3]   Mill, J.S. and Riley, J. *'Principles of Political Economy and Chapters on Socialism'*, Oxford World's Classics, July 2008, Introduction, p. xi.

[4]   Buckley, John. 'Telecommunications Regulation', IET Telecommunications Series 50, Stevenage, UK 2003, p. 39.

[5]   Ofcom. 'What Is Ofcom', http://www.ofcom.org.uk/about/what-is-ofcom/

[6]   Ofcom. 'The Communications Market Report', August 2014, p. 278, http://stakeholders.ofcom.org.uk/binaries/research/cmr/cmr14/2014_UK_CMR.pdf

[7]   Valdar, A. *Understanding Telecommunications Networks.* IET Telecommunications Series 52, Stevenage, UK, 2006, Chapter 5, pp. 107–111

[8]   Buckley, J. *Telecommunications Regulation*, IET Telecommunications Series 50, Stevenage, UK, 2003, pp. 75–81.

[9]   Reding, V. 'Foreword' to 'Communications, the next decade' a collection of essays prepared for the UK Office of Communications, edited by Ed Richards, Robin Foster, and Tom Kiedrowski, London, 2006, p. 3.

[10] Buckley, J. *Telecommunications Regulation*, IET Telecommunications Series 50, Stevenage, UK, 2003, pp. 201–202.

[11] Buckley, J. *Telecommunications Regulation*, IET Telecommunications Series 50, Stevenage, UK, 2003, pp. 134–155.

[12] Valdar, A. *Understanding Telecommunications Networks.* IET Telecommunications Series 52, Stevenage, UK, 2006, Chapter 5, pp. 260–261.

[13] Monopolies and Mergers Commission. 'Telephone number portability; a report by the on a reference under section 13 of the Telecommunications Act 1984', HMSO, London, November 1995.

[14] Au, M.H. 'Possibilities for Deregulation: A Case Study of Hong Kong' from 'Communications, the next decade' a collection of essays prepared for the UK Office of Communications, edited by Ed Richards, Robin Foster, and Tom Kiedrowski, London, 2006, pp. 310–321.

[15] Buckley, J. *Telecommunications Regulation*, IET Telecommunications Series 50, Stevenage, UK, 2003, p. 42.

[16] Valdar, A. *Understanding Telecommunications Networks.* IET Telecommunications Series 52, Stevenage, UK, 2006, Chapter 5, pp. 24–27.

[17] Buckley, J. *Telecommunications Regulation*, IET Telecommunications Series 50, Stevenage, UK, 2003, pp. 85–109.

[18] Ofcom. 'Final statements on the Strategic Review of Telecommunications, and undertakings in lieu of a reference under the Enterprise Act 2002', http://stakeholders.ofcom.org.uk/consultations/statement_tsr/

[19] Buckley, J. *Telecommunications Regulation*, IET Telecommunications Series 50, Stevenage, UK, 2003, pp. 157–174.

[20] Valdar, A. *Understanding Telecommunications Networks.* IET Telecommunications Series 52, Stevenage, UK, 2006, Chapter 5, pp. 107–112.

[21] Olswang. 'Forward to the Past? Olswang Telecommunications Industry Survey', Olswang, 2008, p. 3.

[22] The Commission to the European Parliament, the Council, the European Economic and Social Committee and the Committee Of The Regions. 'A Digital Agenda for Europe', European Commission, Brussels, 19 May 2010 (245 final)

[23]   Ofcom. 'Review of wholesale local access market-Statement', July 2010, http://stakeholders.ofcom.org.uk/consultations/wla/statement

[24]   Wikipedia. 'Unbundled Access', http://en.wikipedia.org/wiki/Unbundled_access.

[25]   Computing. 'European Commission says no to regulatory holidays', 9 March 2012, http://www.computing.co.uk/ctg/news/2158268/european-commission-regulatory-holidays

[26]   Government of Victoria, Australia. 'The Australian National Broadband Network', 2010, http://www.invest.vic.gov.au/Assets/1375/1/NBN_UK_web.pdf

Chapter 3

Business strategy

3.1 Introduction – the philosophy of strategy

The global ICT industry represents a massive, highly complex and fast moving market. Conservative estimates put revenues at around $3.5 trillion in 2013, with growth rates in excess of 5%, doubling every 15 years; and there has been an unprecedented degree of dynamism throughout the industry's life [1].

Technology companies have come from nowhere to be amongst the biggest global brands ever know e.g. Google and EE, while national incumbent Telcos have been privatised, subjected to high levels of competition and heavily regulated. Many Telcos have shrunk dramatically, or ceased to exist altogether.

In such a fast-moving environment it is essential that a company has a firm grasp on the key strategic issues that they face and a clear vision of their future. Having a firm strategy provides managers with a shared vision of the direction, intent and purpose of the company. It provides confidence to make difficult decisions – to launch new initiates, to reallocate resources and to withdraw from activities.

This chapter aims to explain the basis of strategic analysis and to provide a practical approach to preparing a business strategy, whether it is at the business level; the infrastructure level; or the market, product or customer level. The practical models, tools and techniques covered here are only a small selection of the many available, but they serve to illustrate the overall approach to developing a winning strategy and to ensuring its successful implementation. In developing strategy we should always remember what Professor George Box said:

> Essentially, all models present an approximation of the system or situation. They all have limitations and shortcomings, but some can be useful. Like all tools, particular models should be applied only if and when they are appropriate and helpful [2].

We hope that the models described here are, at a minimum, useful.

Too often strategy is seen as a high-level activity undertaken by company boffins (or even worse – management consultants!) in the rarefied atmosphere of head office. Nothing could be further from the truth; if strategies are to be effective, practical and relevant they need to be developed by those who are close to customers and the market.

Nonetheless, before we start looking at the practicalities, it is worth spending a moment to reflect on the philosophy behind strategy development. One of the best books ever written about strategy is *The Art of War* by the Chinese philosopher and warrior Sun Tzu Wu (Box 3.1). Indeed Sun Tzu defined strategy as 'The Art of War'. Although Sun Tzu Wu was writing in the 6th century BC, his ideas and analysis are just as relevant to today's businesses as they were to the warring armies and nations of ancient China. Indeed, while his focus was on military strategy, and his aim was to win battles, its remains a key principle of business strategy that it should always aim to achieve competitive advantage.

Box 3.1   Sun Tzu: Military strategist turned business consultant

The historical background to one of the world's greatest strategists is, in fact, quite uncertain. He is generally believed to have been born in Wu, in what is now eastern China, and to have lived in the 6th century BC. This places him in the Spring and Autumn period of ancient Chinese history, a period when the power of the Zhou dynasty waned, as more and more strongholds challenged royal authority and set up city and provincial fiefdoms. King Helu of Wu was one such challenger; Sun Tzu served him as a military general and strategist.

In 506 BC the Wu armies, led by King Helu and his brother Fugai, fought a decisive battle against the Chu dynasty, another of the kingdoms of eastern China. This was the Battle of Boju and it is said to be Sun Tzu's greatest victory. He acted in the battle not only as a key commander but also as the chief strategist. Despite being heavily outnumbered, the Wu armies won the day and the Wu kingdom ruled the Chu lands for years to come. Many believe that Sun Tzu wrote *The Art of War* as a result of the successful lessons that he learnt in the conduct of that battle, as well and many others. The book became a vital guide to war in the period that followed.

Today the book remains widely read throughout the world and its lessons are still applied to business, politics and sport.

Sun Tzu set many rules that are totally relevant today. In particular, he said that before going into battle it is essential to understand the battleground or circumstances in which the battle will be fought. In business, this means analysing the macro and micro environment; the industry; the market; the resources available; and your own capabilities. Further, if strategy is 'The Art of War' then companies need to understand the enemy – the competition. As we have already seen, a key development in the industry has been the growth of competition, driven by technology advances, regulation and globalisation. Most telecommunication and ICT markets are now massively competitive, so a systematic approach to competitor analysis is vital for success.

The remainder of this chapter, therefore, suggests ways that the *battlefield* can be analysed and how the resultant data can be marshalled into a practical strategy; finally we will look at some of the key aspects of implementing the strategy to ensure competitive advantage is actually achieved i.e. that we win the battle.

3.2   The battleground – the macro business environment

3.2.1   PEST analysis

There are many techniques used for analysing the macro business environment but one of the most effective and easy to use is PEST. Although it first came to prominence in the 1960s, it is unclear who used it first; certainly it has developed significantly since then.

A PEST analysis is a good way to view the major features of the macro-economic environment in which a company operates. PEST stands for political, economic, social and technology (see Figure 3.1). (There is an extension – PESTLE – that adds legal and environmental (in the sense of *green*) issues). The analysis may be applied to a national economy or to a geographic region; it may be an environment in which a company already operates, or a new country or area that they may be considering entering. Initially, there is no need to consider the

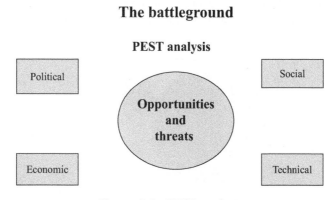

Figure 3.1   PEST analysis

specific product offering or any particular customer segment; the analysis should be far wider than that. It should include only a few major points against each heading (the really key issues) and should categorise them as positive or negative. In some cases, analysts will weigh the positive and negative scores to provide a sense of scale of importance.

Obviously one is seeking a greater balance of positive items over negative, but identifying the negatives may still be of value in helping to approach the business opportunity in a way that overcomes or reduces negative impact.

3.2.1.1   Political

Where and how governments choose to intervene in business is important to any new commercial venture. Clearly legislation, such as employment law, health and safety rules, green initiatives and tax policy all need to be understood. Perhaps most relevantly, the government in question may have initiative to encourage technology or more general ICT programmes. Finally, investment in public services such as education and health may be relevant.

3.2.1.2   Economic

In Chapter 1 we identified economic conditions as one of the most important external pressures on an ICT company (Figure 3.2).

Professor Joe Nellis, Professor of International Management Economics at Cranfield School of Management, identifies four key requirements for a successful economy and therefore for a successful business environment:

- A high and sustained level of economic growth.
- Full employment of economic resources, including labour.
- Low and stable inflation.
- A sound balance of payments, coupled with stable currency values in the foreign exchange markets.

Taken together these factors provide the business and consumer wealth to create a successful market; to enable businesses to grow; and to ensure that they can operate efficiently. A stable currency and sound balance of payments will ensure healthy exports and the opportunity to import raw materials at fair prices.

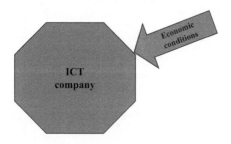

Figure 3.2   The business environment: economic conditions

The economic element of a PEST is perhaps, more than any other, quantifiable. Measures, such as gross domestic product (GDP) and the consumer price index (CPI), enable undisputable evidence of growth and inflation, respectively. However, in completing a PEST it's important to remember that the future, rather than the past, indicates how successful we might be. It is therefore more important to look at forecasts, rather than history. In addition, economists and strategy analysts categorise certain measures as lead indicators. That is to say, certain measures relate to an activity that predicts future growth. For example, manufacturing levels, retail sales and growth in share prices all tend to indicate growing confidence. Therefore, a pick-up in these indicators will herald greater economic activity and therefore economic growth. Lag indicators, such as GDP or company profitability, measure after-the-event activity and are more useful in measuring the level of success than in predicting it.

Equally, it's important to consider the relevant measure. For example, forecast growth in consumer spending may be more relevant than whole-economy GDP.

3.2.1.3 Social

Social factors include three broad headings – demographics, social trends and employment. If we look at demographics, clearly issues such as an ageing population or declining birth rate will be important to the size of markets and customers propensity to buy. Social trends, such as a growing interest in home entertainment, increased fragmentation of society and increased tourism, will all contribute to a company's view of a potential business opportunity. Finally, employment will have a direct impact on a company's view of its ability to produce, sell and maintain a new product or service.

3.2.1.4 Technical

Finally, technological change may be the most important aspect for an ICT company. Innovation is essential for growth in business. Clearly, levels of research and development (R&D) and emerging technology are relevant and will source future products, customer demand and reduced costs. In some cases there will be proprietary breakthroughs enabling a single company to steal a march on the completion. On other occasions, industry-wide standards will provide incremental improvement for the whole industry and success will depend on good implementation.

3.2.2 An example: home working

It is best to explore an example. Perhaps a company is keen to consider expanding their business to serve people who wish to work from home. That is to say, many people travel to work each day; sit at a computer to fulfil tasks ranging from answering sales enquires; to designing cars; to managing building projects; to offering legal advice; etc.. When they are not operating a computer they may be speaking to other people, presenting to teams, or thinking! Much of this can be done remotely (and is being done by a minority of people). Our example company is considering providing a package of solutions (secure virtual private networks (VPNs); online conferencing; remote mobile access; etc.) to help companies disperse their workforce. Before considering the market and potential customer needs in this area the company might complete a PEST analysis. They are considering launching the service initially in the United States and so would apply their thinking specifically to the political, economic and social aspects of that domain. (Technology tends to be universal, though there may be exceptions to this.) Figure 3.3 shows a possible outcome to their analysis. Determining the key aspects in each section requires judgement, but that judgement should be backed by data and analytics wherever possible.

Major political issues that might impact – positively or negatively – the market for home working would certainly include the 9/11 attacks on the World Trade Centre in New York. A growing fear of terrorism certainly causes people to be less inclined to commute into major cities. On a wider scale, as new major markets open up (e.g. China), it will be easier to serve them at a distance, rather than travel to the far location. Data, such as the growth of international trade, could be produced to support this positive aspect of the PEST.

A similar analysis should be applied to the other elements of PEST. If we complete an abbreviated PEST for home working it might look something like Figure 3.3.

Clearly, the balance is favourable towards more home working. Security of company data might prove a problem, so in addressing this market the company should pay particular attention to the security features of any products that they might develop.

In a real-world PEST you would need data, quotes and references to support your analysis. As mentioned earlier, you may choose to develop a scoring system to help rank the relative strengths of the positives and negatives.

PEST: an example – Home Working

❑ **Political**:   +   International unrest; terrorism; opening up
                         developing countries e.g. China, Brazil, etc.

❑ **Economical**:   +   Globalisation; growth in service based
                         industries; oil prices

❑ **Social**:   –   Corporates don't keep pace with possibilities;
                         individuals need to adapt their routine

                      +   Work/life balance; greater acceptance of home
                         working

❑ **Technological**: +   Developments e.g. 4 G, Ubiquitous WiFi and
                         Broadband

                      –   Security

Figure 3.3   An example PEST – home working

3.3   The battleground – the industry analysis

3.3.1   Introduction

Stepping one level down from the macro business environment, ICT managers will want to know something about the condition of the specific industry in which they operate. In particular, they will want to understand the industry's attractiveness (i.e. its profitability). Michael Porter (a professor at Harvard Business School) developed a way of assessing the competitiveness of any particular industry and hence its potential profitability. This is his 'Five Forces' analysis [3]. His theory posits that the more competitive a market, the less profitable the whole industry will be, although he believes that even in a relatively unprofitable industry, well-managed firms will earn more than the average – indeed in some cases well above the average. He believes that five specific forces contribute to the overall attractiveness or unattractiveness of an industry.

3.3.2   Five Forces analysis

The five forces are:

Competitive intensity or rivalry: This is a feature of not only a number of companies competing in the industry, but also the aggression with which they compete. For example, are price wars a regular feature of the industry; do companies spend a lot on promotion, etc.?

Substitutes: It is necessary to consider if there are easy and relevant substitutes for the products. For example, there are a number of substitutes for a voice call e.g. email, post, instant messaging (IM), visiting; but there are very few substitutes for, say, electricity.

New entrants: How easy is it for firms to join the industry? Does it require massive investment, or a licence? Are there regulatory hurdles to climb? Is a reliable or known brand essential for success? Are there other barriers to entry?

Suppliers: The supply of inputs (raw materials, components, expertise) to the industry can have a significant impact on the attractiveness of an industry. Companies should ask themselves: Are there unique components required to offer relevant services? Are raw materials widely available or rare? Are there substitutes for components? Are there many firms offering relevant expertise?

Buyers: It is important to assess the power of the customers of the relevant industry. Can they easily change supplier? Do they have bargaining power and how strong is their power?

Once again, like PEST, completing a '5 Forces Analysis' requires both judgement and relevant data. A practical approach is to draw a radar chart with each leg measuring the strength of one of the forces (perhaps as a score out of 5). The wider the plot, the greater the competitiveness; the greater the competitiveness, the lower the attractiveness of the industry. A narrow plot implies low competitive intensity and therefore, an attractive industry in which to compete, or to enter (see Figure 3.4).

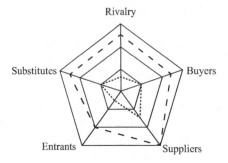

Figure 3.4   Five forces radar chart

3.3.3   An example of applying 5 forces analysis: voice over Internet protocol (VOIP)

Consider a situation where the marketing director of a *New Entrant* Telco might be considering entering the VOIP market. If he or she were to analyse the five forces on a scale of 1–5 and draw a radar chart the results might be something like this:

- *Rivalry*: Although Skype dominates the market there is a lot of rivalry in the market with many ISPs offering a service; innovation is used to promote new services (4/5).
- *Buyers*: The customers (buyers) of VOIP have a great deal of power; switching costs are low – just a little time to download software; users may have several services available on their desktop or smartphone (4/5).
- *Substitutes*. There are many substitutes for a VOIP call e.g. fixed calls, mobile calls, instant messaging, social networking, post, email, etc. (5/5).
- *Entrants*. It is cheap and relatively easy to develop a service but much harder to market it. This is because success requires a strong brand and a lot of traffic to your site to persuade customers to use you (3/5).
- *Supplies*. There is weak power for suppliers in the market because there isn't any special software or capability behind the product; VOIP services can be easily replicated (1/5).

In summary: This isn't a very attractive/profitable market to enter. However, the weak power of suppliers offers a possible opportunity to a market entrant, as input costs will be relatively low, compared to the customer value of the end product. Nonetheless, in the absence of any other particular strategic approach, it seems unlikely that a worthwhile business can be built in VOIP (shown diagrammatically in Figure 3.5).

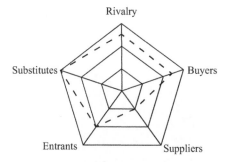

Figure 3.5 Five forces analysis: an example – VOIP

3.3.4 Value chain analysis

Before leaving the subject of industry analysis it is worth remembering the benefit of value chain analysis. We covered this is detail in section 1.2.1.

In essence the value chain starts with the customer and works back through the various activities required to provide the product or service to the customer, drawing each activity or component as a block on the diagram.

In gaining a strategic view of the industry, or developing a strategy, use of the value chain will enable a company to analyse where they have particular strengths and capabilities; where they might create new opportunities; which elements of the chain they might add to their capabilities or whether they should work with others. It is of equal use when considering the current and future strategies of competitors.

So VCA enables the companies to analyse the current situation, but it also enables them to develop new strategic moves for the future.

3.4 The battleground – competitive analysis

3.4.1 Introduction

In any battle it is important to understand the opposition. Indeed, competition figured highly in section 1.4.1 when we discussed the current major influences of the ICT business environment (Figure 3.6). Information is not difficult to obtain; there are a huge number of sources of important information – they include company reports, analysts' briefings (often available from corporate websites), brokers' reports, regulators and news feeds. In terms of ICT, Ofcom, in the United Kingdom, produce quarterly competition reviews and very useful annual market reports. Similar data is available from regulators around the world.

This information will identify relative market shares, industry leaders, best-selling products, and the financial health of competitors, their R&D plans and much more.

In addition, it is important to think laterally about who our real competitors are. Does Ferrari compete with other car companies or with speedboat and jewellery companies? Do restaurants compete with other restaurants or with the cinema and other leisure activities? In ICT it is especially important to think about where future

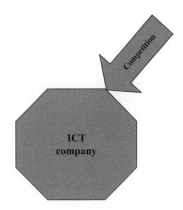

Figure 3.6   The business environment: competition

competition is likely to occur. A company's biggest threat over the next 5 years may be from an ICT student working in her parents' garage!

3.4.2   SWOT analysis

Simply identifying the key players isn't enough. It is important to understand their current competencies, the areas where they are not so strong, and where they may have strengths that they will exploit in future. A good approach to this analysis is to complete a SWOT analysis – strengths, weaknesses, opportunities and threats. Again, it is important to stick to the few, vital issues and to support your analysis with data.

Competitor analysis should be dynamic. That is to say, you should use the output of your SWOT to assess the likely future actions of your competitors. For example, every strength represents an opportunity to develop new products, attach new markets, etc. Similarly, a weakness may represent an opportunity for you to enter the market and to compete head to head against that weakness.

3.4.3   An example: SWOT

Let us look at an example. 'Flexible computing' is a computing service that enables a company or public service organisation (the customer) to outsource all of its ICT system needs (day-to-day running, future development, etc.) to a third-party pro-vider (the supplier). IBM has been a leader in this market since it was first devel-oped in the early 2000s. Let us imagine that we now wish to enter this market and that we wish to assess IBM as a competitor.

Clearly IBM has a successful track record and recognised brand in this market; we might also discover from reading analysts' reports that they have deep technical capacity to implement systems. On the other hand, an in-depth review of the IBM product offering might identify a weakness in the range of business agility metrics provided to the customer. A more complete example SWOT is shown in Figure 3.7.

IBM's Position in Flexible Computing

Strengths	Weaknesses
• Market positioning & aggression	• Lacking in business areas (see opportunities)
• Experience & investment	• Poor working relationships within partner
• Technical execution	• Lack of business agility metrics
• Industry focused	
• Global presence	
• Volume of partnerships	
Opportunities	**Threats**
• Business consultancy, applications development & maintenance, business integration and business process outsourcing	• From the first vendor to provide real business agility metrics (the only attribute that IBM significantly lacked)
• Improve working relations with partners	

Figure 3.7   An example SWOT: flexible computing

3.4.3.1   In summary

Clearly, IBM is a formidable competitor in this space, but there may still be opportunities to compete against them. For example, based on the imaginary SWOT (Figure 3.7), a new entrant would either need a well-known and highly regarded brand, or promote themselves as an energetic challenger brand – much as Avis used to represented themselves against Hertz in the car-hire market with their *We try harder* campaign. Similarly, in developing a product range to compete against IBM, product managers would be sure to provide an extensive range of business agility metrics.

3.4.4   *Product-specific competitor analysis*

Much of the previous analysis has focused on company capability – finance, R&D, brand, etc. Nonetheless, the importance of the competing offerings of each company should not be underestimated. That is what consumers focus on and what ultimately ensures success in the competitive marketplace. (As an aside: it is interesting to note that the age of online shopping has made product comparison far easier for the consumer, especially through the growth of comparison sites and consumer reviews.) It is therefore vital that analysts developing company strategy should always compare competing product and service features. There are many techniques developed to do this, such as snake charts; perceived use value; and benchmarking. All these techniques for comparison start with formal market research such as customer surveys, focus groups, questionnaires, and retail and dealer feedback.

We will return to this subject in far greater detail when we consider market and products in Chapters 7 and 8.

3.5   The battleground – own company capabilities

3.5.1   Introduction

As we have seen, in a competitive market it is important to understand the competition, their strengths, weaknesses, capabilities, etc. But it is even more important to understand your own capabilities. There are a number of ways to analyse internal capabilities. Some examples would be:

- SWOT analysis
- a skills audit
- strategic asset analysis
- an internal value chain review

3.5.2   Strategic skills audit

As well as a *SWOT analysis* (see section 3.4.2) a company might carry out a strategic skills audit. This would consist of reviewing the workforce, its general skills, specialism, training, etc. The key is to match the current capabilities against the needs of the business and especially any new requirements that occur with a revised strategy. A classic issue for Telcos is to find themselves to be very strong in the strategic skills of network management (design, planning, building and maintenance) at a time when their strategy requires very different skills, such as greater customer intimacy; sales techniques based on relationship-building; and rapid product development. The problem might be as simple as being strong in hardware management at a time when software development is required. Clearly any gaps call for an action plan to develop, re-train or recruit the missing capabilities. Here, too, lies a problem, because companies often find it hard to recruit large numbers of new people with desirable new skills at a time when they are making existing, long-serving employees redundant. In this situation retraining and re-assignment often appear the fair approach and may prove successful to some extent. Nonetheless, it must be remembered that a new or revised strategy often calls for staff with different skills and experience, and time can be short to achieve successful implementation.

Also, the company exists in a competitive world, so here is another dimension that must be considered. In completing a strategic skills audit we should ask the question 'how do we measure up against the competition?' In effect, how do we *benchmark*? This is not a general or fluffy question. We should take a detailed and rigorous look at our key competitors (and maybe some coming up on the rails) and compare ourselves to the best.

3.5.2.1   Outsourcing

The value, or otherwise, of outsourcing (or off-shoring, or both) is one specific strategy that can be informed by an extensive skills audit. All too often a company sees, what looks like, significant cost benefits by outsourcing parts of their operations and hands over control without looking at the wider, strategic impact. ICT companies have often been guilty of this, though they are certainly not alone – banks, hospitals, local government organisations, etc., have all been seduced by apparent short-term financial gains, without full assessment of the longer term impacts. The problem is that it is not always clear exactly what parts of the value chain should be considered strategically important for a company, and hence not outsourced (see Box 3.2).

A more strategic way of looking at the problem is to start with a full analysis of the strategically important elements of the business and the company's ability to deliver them i.e. their competence. An important aspect of this analysis is that it should relate to the future vision and mission i.e. the company's strategic aims, and not to the past.

The first part of this analysis can be deceptively difficult. For example, a traditional Telco might believe that network engineering, traffic planning and field-force management are all strategic to their success. More forward-thinking ICT companies might see these functions as easily replicated by other people and that market analysis, customer intimacy, and innovation are the strategically important capabilities. The second element i.e. the company current competence can be achieved through a full skills audit, coupled with some assessment of efficiency and effectiveness in applying the employees skills, training and experience. Having completed the review the results can be plotted onto a 2 × 2 matrix, showing the strategic importance of the function on one axis and company's current competence on the other (see Figure 3.8).

It then becomes clearer what can be outsourced:

A strategically important function shouldn't be outsourced, though if a company is not very effective in delivering it they might wish to partner, buy-in competence or up-skill their employees. But remaining in control is clearly vital. In 2013 Apple decided to move into high street retailing. They recognised it as a new, strategically important move for them. Rather than outsource running their shops they bought-in high-end retail competence in the form of Angela Ahrendts, the chief executive of Burberry (Figure 3.8 top-left sector).

A function that isn't strategically important, but where the company has considerable expertise, might be spun out or, if it remains in house, offered to others at commercial rates. Many companies have developed standalone training or field delivery businesses on this basis. For example, Amazon now delivers groceries for other retailers (Figure 3.8 bottom-right sector).

A function that is strategically important and executed very well should be left alone, apart from the usual business of continuous improvement. Why not offer to provide a service for others? To share strong operational competence in a strategically important function is to invite competition in a key

Box 3.2   To outsource or not to outsource: a cautionary tale

The case of Dell and its component supplier ASUSTek is an interesting one, and serves to illustrate the opportunities and pitfalls of outsourcing. In the early 1990s ASUSTek was a small, Taiwanese company manufacturing electronic components. They were therefore delighted to win a significant piece of business from Dell to produce circuit boards. They won the contract largely on price, but also by guaranteeing an agreed and contractually binding, level of quality. On the basis of doing a good job with the circuit boards, ASUSTek went on to win a further contract to provide all of Dell's motherboards. Clearly this was good commercially for ASUSTek, enabling them to build a significant revenue base, earn what were fair margins for a manufacturing business and enhance their worldwide reputation. But it was also good business for Dell; their revenues continued to grow, their operating costs were reduced, their profit increased and their profitability (return on investment) also went up, because they were able to divest themselves of a lot of manufacturing investment.

Now that ASUSTek were manufacturing most of the components of the computer they offered to assemble it as well. Not all of the components were made by ASUSTek; some (e.g. screens, power leads, etc.) were manufactured by other companies, but ASUSTek also offered to manage the supply chain on Dell's behalf. Dell confirmed that component assembly wasn't a strategic function for them, far from it. It was low-margin work, required employment of a lot of low-skilled people and called for investment in plant and machinery. Nor was managing the supply chain considered strategic; it was just a management headache and it tied up money in the form of stock and work-in-progress. Dell decided that the strategic thing to do was to outsource everything to do with component manufacturing, supply chain management and assembly to ASUSTek.

The final stage was when ASUSTek proposed actually designing the computers for Dell. After all, the computer is an assembly of components, selected from the supply chain. ASUSTek already dealt with all of those aspects of Dell's business. They could easily select the components and carry out the design work. Again, Dell saw the point and decided that design wasn't strategically important to them, either. The important aspect of their business was the Dell brand, their customer relationships and the sales processes. It, therefore, outsourced the design work to ASUSTek as well.

Throughout this process Dell continued to enjoy commercial benefits by reducing costs and increasing profit, and to do so with far fewer people, overheads and assets. Their profitability increased and their standing with shareholders increased commensurately.

But, what was happening at ASUSTek. Throughout this period they had been moving up the value chain. They had moved from a low-volume manufacturer of basic components to a world renowned company that could design and build some of the most competitive PCs in the market; and to do it at volume and for low cost. They were therefore able to offer these competencies to others. In particular, they approached retailers and offered to build own-brand models for them. This enabled the high street and online retailers to offer their own models to compete with the established brands. Working for Apple, Sony and HP, as well as Dell, ASUSTek provided everything from components to fully assembled computers. They manufactured the Microsoft Origami and the Nexus 7 range for Google, and in 2005 they launched their own range of machines under the ASUS brand.

All of these brands enjoyed low-cost, high-quality inputs from ASUSTek and increased competitive pressure on the leading brands. And one of those brands of course was Dell! Since 2005 Dell has lost market share, top-line revenue and profit and has, consequently, suffered significant share price decline. Clearly, in outsourcing work to ASUSTek they took sensible and strategic decisions, but were they the right decisions? At what point did a low-value, basic operation (component manufacture, assembly, supply chain, design?) actually represent a strategically vital element of Dell's business. It's an impossible question to answer, but it does point out some of the dilemmas involved in outsourcing [4].

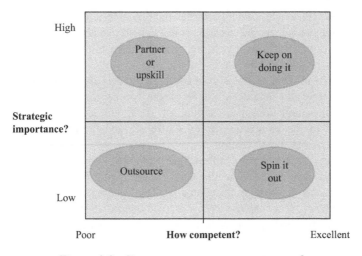

Figure 3.8   To outsource or not to outsource?

area of your business. An example might be Google's search capabilities, which they guard fiercely and use to continuously build their own value proposition (Figure 3.8 top-right sector).

Finally, a function that is not strategically important, and isn't well fulfilled, can be outsourced, so long as standards of delivery are clear and contractually enforceable. An example might be a Telco outsourcing the design and installation of exchange equipment or the day-to-day operation and maintenance of server-farms (Figure 3.8 bottom-left).

3.5.2.2   Conclusion

It is important that a company takes a hard and realistic view about their internal skills and capability, but also that they do so in the light of any new strategic direction that they intend to take. In this way they can assess what will be important to their future and learn how to best deliver it. It is particularly noticeable that many organisations, in all sorts of industries and public service roles, espouse the strategic importance of customer intimacy and then outsource customer service, sales and many other forms of customer contact. If companies where to carry out the sort of analysis recommended here they might avoid that trap.

3.5.3   *Strategic asset analysis*

Similarly, a company might complete a strategic asset analysis, covering both the current and the required assets. In this context, assets might include property, networks, systems, licences, R&D capability and even partnership and subsidiaries, as well as skills and know-how. Cliff Bowman, Professor of Strategic Management at Cranfield School of Management, recommends a framework that identifies the following categories of strategic assets:

- *Tangible assets*: This would include buildings, machinery, location, equipment, patents and even information if it was held in a physical form and was of value to doing business e.g. a well-populated customer relationship system (CRM) or health records.
- *System assets*: This includes systems in the widest sense, not just computer systems. It would include, e.g., the production processes; systems to assure quality; the induction and training of staff; and any systems especially important to the company's product offering and target customer segment e.g. billing or proposal writing.
- *Knowledge assets*: We would include technical know-how, entrepreneurial insight and knowledge architecture (the way experience is formally captured and made available to others in the company). This heading also includes the sort of *tacit knowledge* that enables employees to do things in a superior way, but which can't be written down, such as the impact of one teacher compared with another.
- *Relational assets*: These assets are far harder to quantify, but are exceptionally important. They include trust of the community, respect within the company, reputation, brand, significant contracts with certain high-value or long-term customers, and social capital with, e.g., the government, the media or the stock market.

- *Cultural assets*: Finally, every company or organisation has cultural values and behavioural norms. This is sometimes abbreviated to 'the way we do things around here'. These assets might include creativity, cooperation, responsiveness, professionalism and learning.

In assessing strategic assets there are three considerations:

- To what extent is this particular asset important in delivering our preferred product offering to our target market segment?
- How do we benchmark i.e. How good are our assets, relative to our competitors?
- And how easily can these assets be replicated by others in the industry?

This final question is vital when it comes to achieving and sustaining competitive advantage. Broadly, the ability of others to replicate assets depends on the physical aspect of the asset in question (anyone can buy a building or piece of machinery); how much is written down (a well-documented process is easy to follow and to replicate); and how much is embedded in the tacit routines and know-how of the employees (an experienced head-brewer is vital to a successful brewery but he or she will be total unable to write down or describe what makes them successful). Better still is to have the competences actually embedded throughout the company, in a strong and unique culture i.e. *How we (all) do things around here*. The reason that Apple is so successful is not that they have good outsourcing relationships with producers or well-located shops (though they do); their success is built on a brilliant brand and a culture of creativity and innovation.

If we take the strategic assets discussed earlier, they become more difficult to replicate as you move further down the list, from tangible to cultural assets. We will revisit the question of the ease of replication when we look at achieving sustainable competitive advantage (see section 6.2) [5].

3.5.4   The internal value chain

Finally, a company can adapt the value chain analysis described in Chapter 1 to a company viewpoint (in fact, this is where Michael Porter actually started). A hypothetical example of major Telco is shown in Figure 3.9. By drawing up an internal value chain a company has the opportunity to review the whole operation and to tune it to match the new strategy. In this approach it is essential to identify:

- Core competencies – things you are good at and that are essential to the operation of the business.
- Non-core competencies – things you are good at but maybe are not essential to your business.
- Capabilities – the component activities that are required to offer your services.

The value and strength of competencies and capabilities are only relevant to the industry, market and segment that you hope to conquer. Excellent telesales will be vital to sell a low-value product to a mass market but of little use to sell a bespoke, high-end product to multinational companies. In that case you would

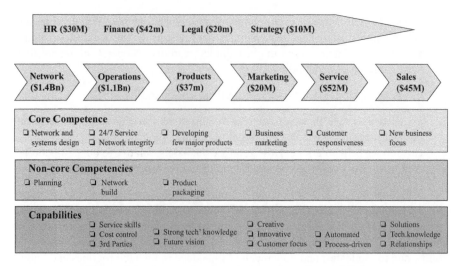

Figure 3.9   Internal value chain: a hypothetical Telco

need face-to-face relationship sales force. Both might be described as an excellent sales competency.

Competencies and capabilities embody many of the strategic assets discussed earlier and they are subject to the same questions. Can they be easily duplicated? To what extent are they written down? Are they supported by a strong and unique culture? As we have seen, the answers to these questions are vital when we come to consider achieving and sustaining competitive advantage (see section 6.2).

As well as assessing every part of the internal value chain for competence, resources should also be identified in the form of cost, people and capital invested. In this way today's operation can be compared to a preferred or *to-be* position that reflects tomorrow's needs. For example, a company that has identified the need for greater innovation and agility can assess if they are allocating enough resources to R&D to achieve these aims. Similarly a company that understands the need for far greater customer intimacy can consider the level of resources applied to market research, customer service and their online presence.

3.5.4.1   A final thought

The internal value chain, at its highest level, describes the organisational structure of the company or business unit. This is relevant to the delivery of the strategy, as well as the efficiency of operations. As strategies change, so should the organisation. This may go well beyond reallocating resources; it may extend to new lines of business; changes to where profit is measured; new regional headquarters; combining certain functions into a single division; or new accountable business units. Often making these changes is the hardest, but most important, part of implementing a new strategy. We will discuss this in more detail in section 3.7.

3.6 Creating sustainable competitive advantage

3.6.1 Introduction

We have spent a considerable amount of time investigating and predicting the conditions that will apply as we go into battle. We have looked at our competitors and have mustered the resources available in our own company. We must now consider how to use this information to create a strategy. This requires rigorous analysis, using some well-developed tools. There are many such techniques available, but we will describe just three here – perceived use value (PUV); strategic thinking; and scenario planning. (Reader may wish to explore some of the many others techniques available such as *as is/to be analysis*; *war gaming*; *benchmarking* and *strategic business mapping*.)

Let us take stock for a moment. We have considered the business environment and created a PEST; we have looked at the market dynamics using Porter's five forces and considered what strategic assesses are available to us and to others. We have looked at our strengths and weaknesses and those of our competition. We have assessed both opportunities and threats. But what is it that we are seeking to achieve with our new strategy? We would suggest we need to create sustainable competitive advantage, and that each word in this phrase is essential in judging a successful strategy. What is more, essentially there are only two ways to achieve sustainable competitive advantage. That is to be either:

- the industry's least-cost producer or
- to display a culture of innovation

A useful technique, again pioneered by Professor Cliff Bowman at Cranfield School of Management, clearly demonstrates this truth.

3.6.2 Perceived use value (PUV) and the customer matrix [6]

As we have already said, all strategic plans should be focused on winning the competitive battle. In effect, this means creating (through products and market focus) a sustainable competitive advantage; the key word here is *sustainable* as markets work in dynamic ways and any short-term gain quickly gets eroded (a bit like an army that makes a successful surge on the western front, only to be left exposed elsewhere). Two key levers available to businesses are price and functionality (the features, technical capability and styling of the product). Price is easily quantified, but functionality is far harder to measure.

Nonetheless, Professor Cliff Bowman has shown how functionality can be specifically quantified in regards to its value to customers (its perceived use value) and how PUV can be plotted against price in a customer matrix to provide an extremely valuable strategic tool. In an ideal world you would assess your product's attractiveness, not by the features that you believe to be valuable, but by the value attributed to them by the customers themselves – perhaps by surveying people. You would start by asking customers to identify the key value features, e.g. for a mobile phone they might identify style, memory, ease of use and weight as the

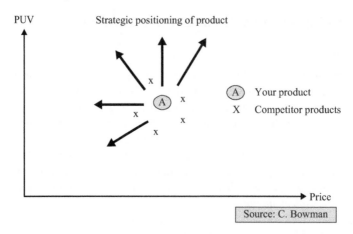

Figure 3.10   Customer matrix: the strategic options

key features that they value. You would then weight the key value features to reflect their relative importance and then ask users to rank your product compared with equivalent products offered by your competitors. It is then possible to plot all the products in the market by PUV and price, e.g. as we have done in Figure 3.10. This is the customer matrix. (See Box 3.3 for a more detailed explanation of how to calculate PUV.)

Normally, you might expect the products with the greatest PUV to command the highest price, although this is not always the case. But we can now try some strategic moves to attempt to capture a greater portion of the market: in doing so we will be mindful of the dynamic nature of markets. What will happen if we reduce the price of our product without changing the functionality? This is represented the arrow pointing due west in Figure 3.10.

All other things being equal, we can expect to win more business, but unless we can also reduce our unit costs, our *profit per unit* will be reduced. Not only that, in a dynamic market other companies may also reduce their prices, thus wiping out our advantage. If we reduce our price again, our competitors might follow again, and again, ad infinitum. So unless our costs are *always* lower than the other companies, we are back where we started relative to our rivals but at lower profit levels for everyone. Not a good idea!

What if we increase the functionality of our product (perhaps by adding features, introducing a new operating system or reducing weight) without changing the price? That is we follow the arrow pointing due north.

Again, we would expect to win increased market share in the short term as people choose to buy a better phone for the same money. However, again, unless we have exclusive access to the new feature, perhaps because we have registered the intellectual property, our competitors will also improve their product. So now we risk being back to stage 1 but with the industry providing phones that are more expensive to produce but at the same price i.e. companies have lower profits.

Box 3.3   Calculating PUV: A worked example

Let us assume we intend launching a new laptop into a crowded market. We might be a start-up company; an established company thinking about diversifying; or a laptop provider who wishes to add to our portfolio.

First, we need to start with the customer. We should use professional market research to survey them (always ensuring we get statistically valid results) and carry out focus groups or otherwise understand their preferences. We are seeking their views, not only on the key features but also the weight they attach to each. While these answers are, by their nature, averages, it is important to get to the true feelings of the consumers and not what we might expect their views to be. For example, the reason that someone might prefer one car over another may have more to do with image and ego, than it does to performance, safety or comfort. It's important to understand this. Similarly, results will differ between different demographic factors or market segments. We will cover this in greater detail in Chapter 7.

Having clearly identified the key features and their weights we need to ask consumers to rank each product (or a representative sample of products) against these features. We do this by taking our proposed product as the benchmark, and asking our respondents to rank the others as better or worse than the benchmark, perhaps on a scale from +3 to −3 (where +3 is far better and −3 is much worse).

By simply listening to customers we will learn that laptop 1 is far more stylish than our product, that laptop 3 is exceptionally fast and that our product is especially slow. We will glean that laptop 2 is particularly heavy to lug around and that it has a poor screen. We can see all of this far more clearly if we plot a graph such as this.

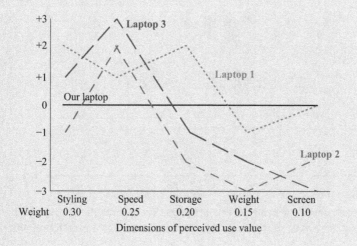

The table below than shows the example answers with the scores multiplied by the proposed weights. By accumulating the scores for each feature we are able to arrive at a single value for PUV for each laptop. Because our proposed new laptop is the benchmark it always has a PUV of zero. It is important to note that price is considered separately from the product features the customer values; we introduce the price of the product only after identifying every other feature of the product's usability or value to the consumer.

Features (weighting)	Laptop 1		Laptop 2		Laptop 3		Our laptop benchmark
	Score	W'd	Score	W'd	Score	W'd	
Styling (0.30)	+2	+0.60	−1	−0.30	+1	+0.30	0
Speed (0.25)	+1	+0.25	+2	+0.50	+3	+0.75	0
Storage (0.20)	+2	+0.40	−2	−0.40	−1	−0.20	0
Weight (0.15)	−1	−0.15	−3	−0.45	−2	−0.30	0
Screen (0.10)	0	0.00	−2	−0.20	−3	−0.30	0
PUV		+1.10		−0.85		+0.25	0

There is then a final stage. We are now in a position to show each laptop by its total PUV compared to its price. Again, this is best done on a graph where the horizontal axis represents price and the vertical axis represents total PUV. This is called the customer matrix. In our example it might look like this:

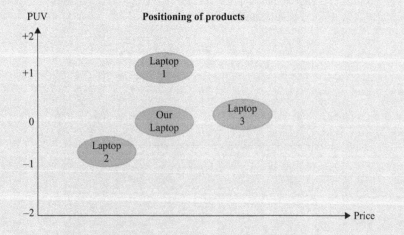

We are now in a position to start considering strategic moves, such as increasing our price, adding features, lowering the cost of production, etc.

Clearly we can try any number of alternative strategies – such as reducing the price somewhat and at the same time increasing functionality (northwest arrow); or trying to increase functionality and increasing price as well (northeast arrow). We can even try reducing functionality i.e. lowering the quality of our product. While this is a rare and normally dangerous strategy, it can be successful. Indeed, reduced product features, coupled with very significant cost reductions (southwest arrow), succeed in both winning market share and expanding the overall size of the market, for the, so-called, no-frills airlines in the early part of the 21st century.

However, as we have seen, in a dynamic market all of these approaches risk prompting competitor response that lowers the overall profitability of the industry, unless we can create sustainable competitive advantage i.e. make these moves in ways which rival firms would find difficult to imitate. If we wish to follow a strategy fundamentally based on price, we must become the *least-cost producer* in the industry. That is to say, we can only sustain a competitive advantage by cutting prices if we have the lowest cost of production in the long term. A short-term advantage will quickly be eroded, unless we have a relentless focus on always being the cheapest producer.

Typically a company can become the least-cost producer by investing in the latest production technology, having the highest labour productivity or the lowest labour wage rates, or by having high volumes of product sales (economies of scale). However, a number of academic studies have also identified a further quality – longevity. The longer a company has been in production, the greater the company's ability to contain costs and achieve productivity. This is the accumulated benefit of collective experience.

Alternatively, if we intend to create a strategy based on increasing product functionality and attractiveness we must demonstrate a culture of continuous innovation, exactly as Apple has famously done. Again, in a dynamic competitive environment, sustainable advantage can only be achieved if other companies cannot easily replicate your improvements (or worse still, better them). There are a number of ways to achieve this type of sustainable advantage e.g.:

- The best R&D.
- A culture of initiating and rewarding innovation, wherever it comes from.
- A rigorous policy of seeking and protecting intellectual property.
- Market research that regularly identifies the key elements of customer value before others.

Together these approaches create a culture of *Systematic Innovation.*

3.6.2.1 Conclusions

We can draw some very clear conclusions from our review of PUV and the customer matrix; they are essential in developing any business strategy.

- To be successful it is vital for a company to understand the competitive positioning of its products, compared with others in the market.
- There are numerous techniques to help, but they all depend on a clear understanding of customer preferences and therefore require research.

- The customer matrix enables optimisation of the price/feature trade-off and shows that there are three routes to sustainable competitive advantage:
  - o   to become and sustain the position of least-cost producer
  - o   to innovate to achieve increased PUV in ways that cannot be easily copied
  - o   to constantly increase PUV through a culture of continues innovation

3.6.3   Strategic thinking

So far in this chapter we have looked very hard at data collection, analysis and the insight that strategic models can provide. However, if you have the impression that this is all you need to do to build a strong strategy, then you have been misled. If that was all that was required then a computer could provide the perfect strategy for every company. Even a management consultant could do it!

Producing a strategy requires thought, vision and judgement. It also requires the experience, knowledge and buy-in from the people who are going to be most affected, most involved in the implementation and whose commitment is going to be vital to success. Data and models can help inform strategic thinking, but they cannot substitute for it.

In Box 3.4 we explore the role of the charismatic leader in setting strategy and describe an example where Microsoft relied on more than just Bill Gate's innate genius to set a radical new course for the company.

Implementation of a new strategy requires difficult decisions, disruption to company organisation and attacks on individual fiefdoms and powerbases. It will not be successful unless people are committed to the change; and the best way to ensure their commitment is to involve them in the formulation of the strategy. This is why strategies developed by external management consultants or internal strategy departments seldom work – though they certainly can be a valuable input to the process.

A company should therefore ideally involve as many of its key leaders as possible. They should come together, with the data and the analysis that we have described, and debate the issues and the possible options. They will inevitably bring their prejudices and preconceived ideas, based on a successful business career, but this will be helpful. Nonetheless, they must be prepared to discuss difficult, even unthinkable, versions of the future and to be open-minded. Such discussions will not be successful as part of a normal management meeting and often the most successful events happen *off-site*, perhaps with a facilitator. As well as the data and analysis, the group might also be challenged by an external expert who might provide some uncomfortable and unconventional views of the future. The process will not be complete in a single meeting; indeed, the best results, and the greatest level of ownership, will arise overtime, perhaps with a series of discussions scheduled over several months. Such a schedule will also allow for the collection and refinement of information, views and data, as well as an opportunity to revisit radical ideas and decisions.

There are occasional exceptions to this collegiate approach, e.g. when a successful new strategy is associated with the arrival of a new CEO. A new leader

Box 3.4 Microsoft does a U turn

The role of charismatic leaders is often quoted as critical in identifying radical new strategies. This is what might be called *strategy by flashes of blinding inspiration*. Larry Ellison, Richard Brandon and Steve Jobs all fit this media image. Whilst there are cases where single-minded determination to change has been enough to set a company on a new direction, the reality is far from that easy. Nonetheless, the perception of the omnificent boss persists.

Microsoft was famously founded by Bill Gates and Paul Allen when they realised that microcomputers were *the next big thing*. Therefore, excellent software to operate the fast growing personal computers – initially in the office and later at home – would be highly successful. This proved the case and by the 1990s Microsoft Windows had 90% of the global market. Throughout this period Bill Gates paid no attention to the emerging web or to online services. His expressed view was that they would never challenge software running on the desktop. So it appeared to be a flash of inspiration when, on 26 May 1995, Bill suddenly issued an all-company memo entitled the 'Internet Tidal Wave' [7]. In the memo he described the capabilities of the internet, the way it would transform people's lives and how it would provide revenue and profit for ICT companies. What is more, he said that "the internet has the highest strategic importance" for Microsoft and he included a large number of objectives and action for his managers to ensure rapid progress in pursuing the new strategy.

Despite implying that the memo represented a *Road to Damascus* moment for the charismatic leader the reality is a little different. In fact, Microsoft has secretly been collecting data, analysing strategic models and debating (among a very select group of trusted leaders) the pros and cons of embracing the internet. Only when the answer was clear, and the change inevitable, did Bill issue his personal call to change direction. Despite this backstory Bill Gates inspired insight continues to be quoted as the saviour of Microsoft and the source of its success since 1995.

is appointed and immediately shows the company the way to a radical new strategy. The likelihood is that the new CEO has arrived specifically because the old strategy wasn't working. Importantly, the new CEO arrives from a different place and with different experience. He or she doesn't share the current group-think. They might already have experienced the successful outcome of the new strategy that they are proposing. This was certainly the case when Ben Verwaayen joined BT in 2003. Faced with lacklustre take-up of broadband in the United Kingdom he realised that BT's strategy was to achieve a premium price for the new service and to limit roll-out. He immediately enacted a market development strategy, pricing broadband as low as possible and accepting a very long payback period on all new broadband investment. Indeed, he was ready to override the existing company network investment rules in pursuit of mass take up of broadband. He didn't do this because he was a genius. He had seen that this strategy was already having a great impact in the USA, where he was previously been CEO of Lucent Technologies.

Strategic issues are typically complex and difficult. Discussion of them is often best stimulated by addressing a series of, apparently simple, questions. Again, Cliff Bowman has provided an excellent framework [8] with his five key strategic questions:

1. Where should we try to compete – which markets or segments of markets?
2. How should we try to gain competitive advantage in these markets and segments?
3. What key competencies and capabilities are required to compete, and how should we organise ourselves?
4. What key competencies and capabilities do we have?
5. How should we change to be successful?

These questions, and no doubt others, should be addressed in the light of the data and analysis already available. Without a doubt, judgement is required (especially for question 5) and its worth remembering that there is no single, perfect answer. Many routes can be followed to achieve competitive advantage. However, we are aiming to identify a strong, viable strategy that can be implemented and that has the broad support – ideally the commitment – of the senior team.

The aim of the strategy process is to produce a clear, written vision and summary of the strategy – ideally in a few, clear pages and written in language that will inform and inspire the whole organisation. It will be supported by a series of actions and, in time, it will be broken down into objectives, measures, targets and plans. We will discuss this further in implementing strategic change in section 3.7.

3.6.4   Scenario analysis

Once a company has a strategy, or perhaps is feeling its way towards a strategy, it is useful to test assumptions against some possible future outcomes, or scenarios. Much strategic thinking is based on projections of today's reality. For example, demographics can be easily forecast based on the current birth rate, etc; economic growth tends to follow well-researched models; and even social trends, such as the greater use and acceptability of technology, follow a pattern. But not all trends are linear. And not all aspects of the future are predicable based on today's observations. Indeed, it was Niels Bohr, Danish Physicist and Nobel Prize winner, who said 'prediction is very difficult, especially about the future'.

Scenario analysis helps overcome these problems by creating a number (3 or 4 is optimum) of different possible futures based on a mix of today's observable facts projected forward, together with some binary or unexpected events, such as the outbreak of war, a major political shift, discovery of new technology or a severe economic recession, perhaps prompted by a loss of confidence in the world banking system, such as occurred in 2008. In some cases the analysis will include an optimistic scenario, a pessimistic scenario and the most likely case. In other situations it might include a scenario where world leaders embrace globalisation, free trade and inward investment, contrasted to a scenario where political and economic tensions lead to retrenchment.

Indeed, this was the case in the mid-1990s when liberalisation of telecommunications markets around the world led major Telcos to develop global strategies. They reasoned that, as they lost business at home to fledgling new entrants, they would seek to compete in overseas markets, previously closed to them. For example, BT launched Concert, a company that aimed to serve global customers anywhere in the world. They also bought interests in alternative operators in France, Holland, Japan and South Korea. In a similar scenario, France Telecom and Deutsch Telekom formed a joint venture called Global One, to compete with BT Concert and other cross-border global players. At the same time, AT&T built a global network by buying small companies around the world and stitching them together.

What none of these companies did was to test their strategy against a scenario of future global retrenchment and trade barriers. So when, in 2001, the dot.com bubble burst, the world economy went into recession and governments were no longer willing to open up their crucial ICT industries to foreign entrants. Share prices slumped and senior industry leaders lost their jobs.

The purpose of scenario analysis is not to design one perfect outcome that we think will definitely happen. The purpose is to pick a wide range of plausible outcomes (often with quite uncomfortable or even *unthinkable* variants included) and then to test the new strategy against each. The aim is not to create the perfect fit to a single most likely outcome, but to future-proof the strategy against a range of possible scenarios. So, for example, if a company needs to create top-line growth i.e. increase revenue, it might marry a strategy based on expanding into overseas markets with a strategy based on market development at home. In this way, if the scenario that predicated global retrenchment were to occur, growth could still be achieved from the increased market penetration at home.

Similarly, a company designing and manufacturing mobile phones might construct different scenarios based on their views of different potential technological developments (e.g. the future capability of 4G and Wi fi). They would do well to adopt a strategy that was flexible between these possible outcomes. An important part of scenario analysis is the identification on a timeline of future events that need to occur for the described scenario to happen. A company should then apply the agreed strategy, while keeping watch for the events to happen. Ideally, a suitable response strategy can be developed and kept available for use when required. Box 3.5 describes a real-life example where scenario planning was used to explore possible outcomes of eCommerce.

Scenario Planning is explored in greater depth in section 8.2.5.

Box 3.5   Scenario planning – a real-life example

Back in the early 1990s the use of the Internet was still confined largely to academic researchers and data professionals in the IT departments of companies. One of a set of scenario planning workshops in Boston, USA, was asked to consider all the possible futures for the Internet, and what would cause them to come about. The workshop comprised 10 representatives from a range of ICT companies in the United States and Europe (the *experts*), together with a facilitation team who prepared descriptions of possible future events written out on index cards. Each event was considered in turn by the experts – with the aim of evaluating the likelihood of it happening, and the consequences and impact it would have on the Internet. Of the 100 or so events considered, only 20 were deemed significant. Four scenarios for the Internet were then constructed during group discussions by the experts – these ranged from mass adoption of the Internet by all parts of the community at one extreme to relegation to use by IT enthusiasts (often called *nerds*) at the other end. The group then spent considerable time allocating the events to each of these scenarios – identifying which events would cause, or hinder, the futures portrayed in the scenarios occurring.

The one scenario which was closest to today's mass adoption of the Internet, with the purchase of goods and services by businesses and consumers of all socio-economic groups and ages, depended on one big event, namely: the availability of secure payments systems. Modern Internet savvy people may find it hard to believe, but before such systems (e.g. enabling secure credit card payments) existed there were almost no purchases made over the web. It took a scenario planning exercise to help characterise today's radically different world and what would make it happen.

3.7   Implementing strategic change

3.7.1   Introduction

Strategy is nothing if not implemented. The business world is littered with well-written, well-argued strategic reports, all gathering dust on shelves. Worse still are the companies that have set out on strategic change, only to find it all too difficult and abandoned the attempt, normally with disastrous consequences. While disappointing, the story of failure is not surprising to those who have been charged previously with major cultural and strategic change.

Implementing strategic change is difficult and complex. It calls for difficult decisions; it challenges fiefdoms and power bases; and it can appear to be a leap into the unknown. It requires strong nerves, tenacity and inspirational leadership. It certainly won't happen successfully if it's left to evolve or if it is squeezed into the gaps in *business as usual*. It, therefore, requires a framework that ensures specific steps are taken to introduce and underpin the change. Incidentally, implementing a strategic change is not the same as project management (which we cover in the appendix), though there are occasions when project management will prove a useful tool to compliment strategic change. There are numerous approaches to strategic change suggested in the management literature, such as:

- the change Kaleidoscope, by Balogum, Hope and Hailey
- the change curve, by Kubler-Ross
- the ADKAR model, by Prosci
- the eight-stage process of leading change, by John Kotter [9]

They are all legitimate, useful approaches and they all have a lot in common. We don't have a room here to explore them all (the reader can do that at their leisure), but we will focus on one. John Kotter's eight-stage process has many advantages. It is clear, intuitive and provides a means of monitoring progress. It is based on deep academic research but can easily be followed by a busy and harassed CEO.

3.7.2   The eight-stage process

John Kotter was a professor of Strategic Change at Harvard Business School when he developed his approach based on the study of a wide range of change initiatives, applied across many industries. Initially focused on why strategic changes fail, his work later broadened into a roadmap for successful change. The eight stages need not be strictly applied sequentially, though they should be applied broadly in the order suggested. They will certainly overlap and some will require more time and effort than others. Circumstances will differ from initiative to initiative, so the emphasis for each stage should vary. In particular, change initiatives vary in scale and timescale. Nonetheless, the eight-stage process is relevant across the board, so long as the end-result that you are seeking is significant change to both the business and culture of the organisation.

3.7.2.1   Establish a sense of urgency

There is no stronger argument for the status quo than 'we are doing OK as we are'. There is therefore a need to explain to the whole company the need for urgent change and the more viscerally the call to arms, the more the effect will be felt. For example, in the 1980s the sight of newly licensed operators digging up the streets and laying cables galvanised incumbent Telcos far faster than any number of dry emails about the coming competitive treat.

The need for urgent change may already be clear. The strategic review of the market, customer attitudes, competition or technology advances may already have identified a major problem. Potentially, the need to issue a profit warning, or a significant fall in the share price, may put the company in crisis. Alternatively, and rather more positively, the urgency may arise from a major opportunity that will be missed if change isn't rapid. For many ICT companies, the growth of the Internet or the invention of 4G represented an opportunity and a call to arms.

Whatever the reason, senior leaders should widely communicate the need for urgent change and provide a sense of optimism about the new strategy.

3.7.2.2   Create the guiding coalition

It is important that the change should be led by people who are committed to the new strategy and have the energy and authority within the business to lead the change. In this regard, authority may be formal or informal. Every business has people that others follow, whether it's through force of personality, their good sense or their track record. These people aren't always in very senior roles, but nonetheless they have influence and should be key members of the guiding coalition. There is value in having members of the guiding coalition who represent a variety of levels, functions, business units, etc. The guiding coalition will certainly include the authors of the new strategy and will grow in numbers as change reaches out to every part of the organisation. Similarly, very senior members of the management team who are not committed, or even obstructive, need to be dealt with. This might take the form of isolation, demotion or even exit from the business.

Finally, the guiding coalition needs to act as team, albeit a dispersed one. It is therefore essential to give some thought to team-building activity, shared values, etc.

3.7.2.3   Develop a vision and strategy

We touched on this earlier when we discussed the output from the strategic development process (section 3.6.3). Ideally a vision should be inspiring. It should describe a new future for the company and the direction of travel to get there. By describing the benefits it should motivate people to take on-board the new strategy and to change the way they work. We must remember that, in the short term, the change may well be painful for many, so the vision should make clear what the long-term advantages are. Finally, by providing clarity, the vision should help to align action, right across the company. There are many companies with excellent visions and many that miss the point, or that demotivate. Kotter believes that a good vision will embody six characteristics. They should be imaginable, desirable, feasible, focused, flexible and communicable.

In the next section we will see how some existing Visions measure up.

The vision should be supported by a clear and rather fuller description of the strategy, covering the reasons for the change and how it can be achieved. It will include high-level objectives, measures, targets and timescales. Beyond that, the company will develop no end of plans, budgets and personal objectives; this is the management role. But the leadership role is to develop a compelling vision and strategy.

3.7.2.4   Communicate the change vision

The power of a vision isn't for senior managers to tell one another where the company is headed or to inspire them to action. Hopefully the senior team helped to develop the vision and are committed to it – certainly if they are in the Guiding

Coalition. No, the power of the vision is to align, inform and motivate everyone in the company. This is easy to say, far harder to do, but there are some clear guidelines. For example, it should be possible to explain the vision and the supporting strategy to an interested, but non-expert, member of the team and to do so in just a few minutes. The vision should be written in simple language and to paint a picture of the future business. Let us look at some examples.

The first example is from BT:

> *Our vision is to be dedicated to helping customers thrive in a changing world. The world we live in and the way we communicate are changing, and we believe in progress, growth and possibility.*
>
> *We want to help all our customers make their lives and businesses better with products and services that are tailored to their needs and easy to use.*
>
> *This means getting even closer to customers, understanding their lifestyles and their businesses, and establishing long-term relationships with them.*
>
> *We're passionate about customers and are working to meet the needs they have today and innovating to meet the needs they will have tomorrow [10].*

This appears to be a collection of very worthy, but fluffy, sentiments with little clarity about how things will be in the future or the benefits for the individual or the company. It is written in simple English but is rather long and hardly inspiring. In fact it rather neatly fits one cynic's definition of a vision statement as: *A long, convoluted paragraph that clearly demonstrates that a company has no clear vision!*

The second example is from the early days of Microsoft: 'a PC on every desk' ... which later became ..... 'a PC on every desk and in every home' is short, clear, aspirational and ambitious and it implies benefit, certainly to Microsoft. Clearly it would be supported by a more complete strategy, but as a vision to inspire and to gain action it is hard to beat [11].

Communicating with a whole company can be a challenging task. It can often be facilitated by having one version that concentrates all the key elements into a diagram or graphical representation – 'the plan on a page' or 'strategy at a glance'. With careful thought and clever presentation the vision, top-level strategic objectives and key measures can all be formatted into a single memorable page. Finally, having crafted the words, it is important to communicate them broadly, frequently, in many different settings and in many different ways. Consistency in the message will be vital to the vision's credibility and leaders must be seen to *walk the talk*. One wrong promotion, one inconsistent action, one failure to listen, will detract more from the message than any number of words.

3.7.2.5   Empower broad-based action

It is important to communicate the vision widely, not to show off a shiny new strategy, or to point out just how clever the top team are; the reason to communicate it is to enlist the whole workforce in making the necessary changes. Major strategic change does not take root unless it engages and influences many people to do many things differently. The irony is that, quite often, companies make it hard for this to happen. It may be that the structure of the organisation is no longer fit for the new strategy and that, for example, rapid time-to-market is hampered by a stove-pipe organisation, which was great for cost control but ineffective for the new strategy.

Alternatively, it may be because the old strategy, and the culture that supported it, didn't empower people to act independently. For example, perhaps the formal systems of authority didn't allow front-line staff to show discretion e.g. to write-off a charge or to vary a contract. If the new strategy involves providing the industry's best, most responsive customer service, it cannot include systems that require the smallest of decisions to be signed-off by three levels of management. Take a simple example. If you are having a bad day the barista in Pret-a-Manger is empowered to give you your coffee for free. OK, he or she probably can't do it more than once a day, or maybe a week. It's just a small thing. But think what it means for the company's strategy to develop customer intimacy. *The barista talked to me, asked me how my day was, showed real interest and empathised.* How many people will I tell about this apparent act of random kindness and how much more likely am I to stick with Pret for my morning fix of coffee. And, what about the barista – how nice to be empowered to make the gesture, rather than just froth milk all day. All this traction to the company's vision for the price of a coffee!

Of course, if a company is going to empower people to act differently they will need to think carefully about the skills required; to offer appropriate training; and also to recruit people with a rather different outlook and capabilities. In our simple example above, they had better recruit and train baristas who are outgoing, chatty and empathetic (as well as able to froth milk).

Finally, there needs to be consideration of the supervisors and middle managers. In the past these people were empowered to run the systems, to make decisions and to authorise and question other peoples' decisions. If power has been devolved lower in the organisation their role needs to change (perhaps to coach and supporter) or to go altogether. When a certain Telco spun-off its mobile arm, the new company found that they had inherited a thick layer of supervisors and middle managers which was inhibiting fast communications between leaders and the front-line staff. Even more of a concern was the fact that the supervisors resented and obstructed the delegation of decision-making to the customer-facing staff. The mobile company's answer was to dismantle the hierarchy and to flatten the organisation, thereby reducing costs, speeding up decision-making and revolutionising customer service.

3.7.2.6   Generate short-term wins

The 2012 London Olympics was a 7-year project, costing £9 billion and, in the early days, it certainly had its doubters and distractors. To build momentum and to gain credibility with the public (whose taxes were paying for the games) Seb Coe and the leadership team needed to show some tangible short-term gains. They, therefore, designed a series of pre-game events called 'London Prepares'. Most eye-catching of these early win was to complete the iconic Velodrome a year early and to inaugurate it with a world series cycling event. This had the effect of making the change real; convince the doubters that they were going to get something brilliant for their money; allowing everything to be tested; and gave an opportunity to congratulate the planners, builders and volunteers who made the event a success – galvanising them for more hard work to come. (Interestingly, 4 years earlier, Beijing had done the same thing with their new transport systems built for the 2008 games.)

It is essential in any change programme that there should be early wins. First, to build momentum; second to prove that the change is real and worth the sacrifices; third to provide benefits and quell any doubts; and finally to enable the strategy to be fine-tuned.

To catch the eye, a short-term win needs to be big, well communicated and clearly related to the new strategy. It needs to succeed and, ideally, implemented as part of a series of wins that builds into the strategy itself. For all these reasons they need to be well planned and have a dedicated team to ensure that they are delivered. Finally, everyone involved in the early successes should be celebrated and thanked; in this way the company builds a positive atmosphere as they go into the next – potentially harder – stage of implementation. It certainly worked for the London Olympics!

3.7.2.7   Consolidate gains and producing more change

Short-term wins are essential to build momentum, but even with a number of successes on the scoreboard, strategic change can peter out all too easily. For example, the forces of resistance may reappear claiming that the short-term wins represent success and now the company can get back to business as usual. It may be that the original advocates of change move on or lose heart. It may be that 'it's all too hard'.

This last point is very relevant. As we have already seen, ICT companies are highly complex with many inter-dependent functions and departments. They also face significant competition and rely heavily on technology to do business. This means that any significant change will quickly snowball and will require change, not just in one department or line of business, but right across the firm. This interconnection adds a high degree of complexity to any programme of strategic change. One initiative, big in its own right, quickly spawns five or ten or more new initiatives. This is a good thing, as it embeds change more completely in the organisation and produces more change, but it requires a new form of change management.

So Kotter's seventh step is to push the management and ownership further down the organisation, engaging more people in its implementation and moving it even closer to the customer. To do this, the original band of change agents, the guiding coalition, have to stop managing change and start coaching others. The strategy team will become an advice centre and a monitoring point, supporting possibly thousands of change agents and holding together the overall programme, perhaps made up of hundreds of threads of change.

If this sounds like organised chaos, well it maybe. The success of such an approach is dependent on two things:

- A shared vision and strategy that is clear and well communicated. This will be the guiding light for all those individual threads and ensure they remain consistent with one another.
- A culture of openness, trust and respect throughout the firm. We will explore that in more detail in the next section.

3.7.2.8   Anchor new approaches in the culture

Finally, transformation will be so widespread that it becomes embedded in the culture which is shorthand for the shared values and behaviour norms of the organisation. Some people might feel that cultural change should come first. Indeed, it might be argued that people can't do business differently if they still believe the same things and behave the same way. Kotter would argue that changing culture is such a long-term and difficult project that it can never precede strategic change or comply with a pressing business need. After all, the first stage of the change process is to engender a sense of urgency. Furthermore, people don't change their beliefs and behaviours easily or quickly. They don't do so just because the boss has asked or told them to. Broadly, people only change their personal characteristic after a long and sustained period of exposure to the benefits of change. In other words, cultural anchoring can only occur once the change has taken root and provided significant and proven advantages, both to the company and to the individuals employed in it.

3.7.3   Competing forces

In implementing significant change it is vital to assess not only the forces of support (e.g. customer benefit, service improvement, financial advantage, stakeholder support, etc.) but also the forces of resistance (e.g. an economic recession, customer resistance, technological delay, competitive response, political interference, etc.). These forces may be either internal or external; they may be within the company's direct control, or they may not. Either way, they need to be assessed and dealt with.

One good way to model these forces and to help achieve change effectively is to complete a force field analysis (see Figure 3.11). The change team would first list the enabling forces, those in support of the change, and then they would list the

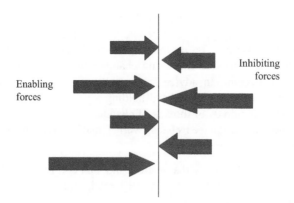

Figure 3.11   Force field analysis

inhibiting forces, those who might oppose the change. They would assess the power of each force and how it might be modified. The forces would then be shown in a chart, notionally applying all the forces to the current position – however, that might be measured. To aid the visual impact of the chart, the size of the arrow should represent the power of the force. You do not need an exactly matching number of forces on either side, nor do they need to be of equal weight.

The aim of the programme is to move the current position to a better place, i.e. to the right of the chart. This is achieved by increasing the power of the enabling forces (e.g. by improving the customer offer) or reducing the power of the inhibiting forces (e.g. by negotiating with the government). In analysing force fields it is as important to increase the impact of the good things as it is to reduce the power of the blockers. Force field analysis is mainly a presentational tool, not an exact science. However, often, by simply identifying the competing forces the company can tackle some of the obstacles to progress. Beyond that, the chart offers the opportunity to rigorously analyse each force and to have a development plan.

3.7.4 Monitoring progress

There is one final aspect of the strategy process, that is monitoring whether the desired outcomes in deploying the new strategy are being achieved as expected. That is, monitoring benefits to see that they have been achieved. A number of key principles apply. It is essential that the monitoring process is set up early in the development process; that the performance indicators that are chosen are relevant to the new strategy; and that targets are set against each, with timescales associated. Clearly it's important that each aspect of the new strategy can be measured and assessed for progress. While this sounds easy, and is for certain aspects (e.g. revenue, volumes of sales, product launch dates, market share numbers, etc.) other aspects are far less easy (e.g. customer loyalty, staff empathy, brand awareness, etc.). Nonetheless, it's important to develop what Robert Kaplan and Paul Norton call a balanced scorecard of measures [12]. This might include, e.g., sections that cover financial measures; efficiency and productivity; customers and the product portfolio; and people. The final choice of measures will depend on the organisation, the strategy and what the Board feel best represent their strategic aims. Some of the more esoteric measures, such as staff satisfaction, may require regular surveys, focus groups or exit interviews, etc.

3.8 Summary

Strategy is the *Art of War*. All successful strategies are based on finding *sustainable competitive advantage* and this can only be achieved in two ways:

- to be the industry's least-cost producer or
- to display a culture of continues innovation

Global economy
Regions
Country/domain

Economic
Indicators
PEST

The industry

Value chain

The market

Porters 5 forces

Competition

SWOT
PUV/Snake chart

Company

Strategic assets
SWOT

Bowman's 5 questions

Agree
strategy

Figure 3.12   The strategy development funnel

In this chapter we have seen how successful strategies can be practically developed. Strategy should always start with a clear understanding of the market and competitive environment (the battlefield) and this requires research, data and analysis. There are many strategic tools/models available including PEST; Porters five forces; SWOT; skills and asset audits as well as divisional value chains.

Strategic analysis should always be data-driven, but successful strategies also require discussion by engaged managers and judgement by the company's leaders. They can be prompted to discuss the right issues in a logical way by addressing Bowman's five questions. Inevitably choices must be made – we will enter this market, target this segment, make this investment and develop this product range; or, alternatively we will exit this market; pass-up this opportunity; sell-off this part of the company. Failure to address these strategic questions successfully can lead to the spectacular demise of even the most, apparently successful company (see Box 3.6). This part of the process can be likened to passing down a funnel, starting with broad-brush analysis of the widest possible view on the global economy and market, down to a narrower view of the particular market involved, then down to the narrowest point of the funnel, which produces the specific company or divisional strategy (Figure 3.12).

Having arrived at a proposed strategy it should be stress-tested against changes in assumptions and a variety of possible future outcomes. Scenario analysis is useful here. Finally, strategy is nothing if not implemented. Directors of strategy have too many dusty (and expensive) reports sitting on their shelves as it is. A systematic change management process, with rigorous progress monitoring, is required to ensure full realisation of the wonderful new world, represented by your strategy.

Box 3.6 Nortel: failure to change

Throughout the 1980s and 1990s Nortel was a darling of the technology industry, not just at home in Canada, but around the world. It had a reputation for excellent R&D, technological know-how and close customer relationships. Their engineers led on the transition from analogue to digital; they manufactured the equipment that customers wanted and they were recognised as providing market-leading customer support. This reputation enabled them to charge *top-dollar* and to become a favourite of the stock exchange.

But there was a cataclysmic shift coming in the ICT industry and Nortel failed to recognise it or to react strategically.

By 2000 the global equipment industry was in turmoil. The internet had created new value for consumers, but Nortel failed to acquire and integrate the fledgling dot.com players in an effective way. Nor did they have the new competences to develop products for these new markets. When the recession bit Nortel incurred significant financial losses.

Similarly, technology was progressing fast and Nortel, confident in their previous good choices, failed to react. Their stronghold of selling fixed-line exchanges to carriers was being eroded by mobile services; a growing sector of enterprise customers; and saturation of the optical fibre supply business. What is more, Nortel backed the wrong mobile standard. Their concentration in North America led them to back code division multiple access (CDMA), rather than the global system for mobile (GSM), which was the eventual global winner.

Finally, all this turmoil in the equipment supply industry, as well as the financial downturn suffered by Nortel's customers, produced a period of intense competitive pressure, exacerbated by the arrival of very serious players from China e.g. Huawei and ZTE.

Nortel failed to react strategically to this severe challenge. Instead, they focused internally and made incremental changes based on an arrogance that emanated from their previous success. On the whole, these changes eroded Nortel's competitive advantage. They looked for cost savings by slashing sales capability. They dismantled the centralised R&D and downgraded the importance of innovation. They favoured legacy products over new developments and became risk-adverse. Perhaps most tellingly, they dismantled their intelligence and strategic planning team!

Nortel was no longer the customer-responsive, agile and technology-leading company that it had been. Not surprisingly their major customers began to notice and to complain that they no longer received visits from well-informed sales executives who were empowered to get things done. They noticed that Nortel wasn't making new software releases and they noticed that, whilst prices remained high, customer support was poor. Naturally, in these circumstances, customers were not quick to renew long-term contracts or to sign new ones. Revenues fell and Nortel's traditionally high cost base (a legacy of the good days) was hard to tackle. In these circumstances it was not possible for Nortel to remain solvent and in 2009 they sold their separate business units and ceased trading.

What does the sad case of Nortel's demise teach us? Well, a company that had been highly successful stopped talking to their customers and learning their needs. They misunderstood who their competitors were or what they could offer; they took as a given their technological leadership without refreshing their capabilities and they failed to innovate. Most importantly, they failed to recognise that the changes in the industry called for a strategic response and so they responded by doing *more of the same* and ultimately paid the price of their arrogance [13].

References

1. Organisation for Economic Co-operation and Development (OECD). 'Size of the ICT Sector', OECD iLibrary, 2012, http://www.oecd-ilibrary.org/sites/factbook-2011-en/08/02/01/index.html?itemId=/content/chapter/factbook-2011-72-en
2. Box, G.E.P. and Draper, N.R. 'Empirical Model-Building and Response Surfaces', Wiley, 1987, p. 424.
3. Porter, M.E. 'Competitive Strategy', Free Press, New York, 2004, Chapter 1.
4. Christensen, C.M. 'The Innovator's Prescription', McGraw-Hill, New York, 2009, pp. 263–266.
5. Bowman, C. 'Strategy in Practice', Financial Times/Prentice Hall, London, 1998, Chapter 3.
6. Bowman, C. 'Strategy in Practice', Financial Times/Prentice Hall, London, 1998, Chapter 2.
7. Gates, Bill: 'The Internet Tidal Wave', Microsoft internal Memo, 26 May 1993, http://www.justice.gov/atr/cases/exhibits/20.pdf
8. Bowman, C. 'Strategy in Practice', Financial Times/Prentice Hall, London, 1998, p. 5.
9. Kotter, J.P. 'Leading Change' Harvard Business School Press, 1996, pp. 20–23.
10. BT. 'About BT. Purpose and strategy', https://www.btplc.com/Thegroup/Ourcompany/Ourvalues/index1.htm
11. The Economist. News article 'After Bill', 26 June 2008, http://www.economist.com/node/11614315

12. Kaplan, R.S. and Norton, D.P. 'The Balanced Scorecard: Translating Strategy into Action', Harvard Business School Press, Boston, 1996, Chapter 1.
13. Calof, J., *et al*: 'An Overview of the Demise of Nortel Networks and Key Lessons Learned: Systemic effects in environment, resilience and black-cloud formation', Telfer School of Management, School of Electrical Engineering and Computer Science and Faculty of Law, 2013.

Bibliography

Tzu, Sun: 'Sun Tzu: The Art of War', translated by Lionel Giles, Prohyptikon Publishing Inc, Ontario, 2009 (first published in 1910).
Adams, Scott: 'The Dilbert Principle', Boxtree, London 2000 (first published 1996).

Chapter 4

Corporate finance and governance

4.1 Introduction

In Chapter 1 we looked at a very high-level model of how Telcos make money (see section 1.3). It is now time to look at the financial picture in more detail, starting with the financial accounts that every company has to publish. The financial accounts of a company consist of three main statements:

- balance sheet
- profit and loss statement (sometimes called the Income or revenue statement)
- cash flow statement

These statements are accompanied by a wide range of other information, which is required, either by company law, or by the various accounting standards set throughout the world.

The financial accounts of a business have a number of purposes, including ensuring probity, enabling financial planning and providing information on which to base an analysis of the financial management and prospects of the company. For this reason they are of great interest to many stakeholders but especially to shareholders, potential shareholders and the organisations that advise shareholders e.g. fund managers. We will look at a company's relationship with its financial stakeholders, some approaches to analysing financial accounts and their importance in determining the share price.

The second part of the chapter then takes a more internal view of financial analysis and focuses, initially, on the high-level financial and business planning required to run a successful company. This includes capital planning and authorisation, budgeting and financial control.

Finally, we focus down at the level of programme or project authorisation and look at writing business cases, the techniques of investment appraisal and risk management.

4.2 Financial accounts

Companies are required to produce three financial statements – a balance sheet; a profit and loss statement and a cash flow statement – on an annual basis; the three

Table 4.1   Generally accepted accounting concepts and conventions

Going concern	It is a requirement that accounts are prepared in the belief that the business is continuing. This has significant implications on the value of company assets and liabilities.
Matching	Income should always be accounted for in the same period as the relevant expenses were incurred. This is achieved though the accruals process.[a]
Consistency	Consistent application of accounting policies, both for similar items and through time.
Historic cost accounting	Financial accounts are prepared on the basis of the value of money at the time.
Prudence	The accounts should reflect a *prudent* view. For example, income should only be recognised when the sale has been concluded. Also, possible future costs e.g. possible bad debts, should be provided for.
Materiality	The application of accounting policies requires a certain level of judgement. The *materiality* convention requires that accountants should take into account the importance or *materiality* to the user of the accounts when exercising judgement.

[a]An accrual is an accounting entry that represents any income earned in the period in question for which the cash has not been received, or an expense incurred for which the cash hasn't yet been paid out. A practical example of applying accruals is described in section 4.2.2.

statements, together with a considerable amount of supporting information, make up the statutory financial accounts of the business. Typically they are published annually and must be prepared in line with generally accepted accounting concepts and conventions; these include things such as 'matching' – the idea that the accounts should include only revenue earned in the period of the accounts and should be 'matched' with the actual costs incurred to earn and service that revenue (rather than the costs paid for in the period). A fuller explanation of the major accounting concepts and conventions is included in Table 4.1.

Before publication, the accounts of the business are audited by a firm of accredited auditors who confirm that they represent a 'fair and reasonable view' of the year's trading and financial value of the business; failure to achieve a 'clean' set of accounts has far-reaching consequences, both for the company's share price and, potentially, the officers of the company (see Box 4.1).

Each statement has been described in the following subsections.

4.2.1   The balance sheet

The balance sheet summarises the company's financial worth at a single point in time, typically on the last day of the company's financial year e.g. 31 December or 31 March. It seeks to exactly match the total value of the assets of the company, less the liabilities, with the sources of the funds that paid for them (normally shareholder funds, bank loans or profit retained from previous trading). As we have discussed previously, an asset of a company e.g. a factory or a telephone exchange

Box 4.1   The rise and fall of Worldcom

Worldcom was once the second largest Telco in North America; it started life in 1983 as Long Distant Discount Services (LDDS), challenging AT&T's monopoly profits by offering America's corporations low-cost, long-distant calling. It grew rapidly throughout the 1990s by acquiring many of the other, smaller challengers in the telecommunications market. It did this by offering to buy out shareholders, not with cash, but with shares in Worldcom. A virtuous circle developed. Because it was growing fast and appeared to have a winning strategy, its shares were attractive, so it was able to continue to acquire smaller companies, and so it continued to grow; the Worldcom share price continued to rise and so it could continue to acquire even more companies. In 1998 Worldcom acquired MCI through a further issue of shares and by the following year it was planning to acquire Sprint to make it the largest Telco in the world.

However, in 2000 the so-called dot.com bubble burst and share prices that had been sustained by top-line growth alone dropped dramatically. Worldcom was unable to sustain its growth, or provide returns for shareholders. Bernie Ebbers, the maverick CEO, and some of his close colleagues, decided that they had to sustain an appearance of top-line growth and turned to fraud to achieve it. In June 2001, following disclosure by a group of internal auditors, the company was charged with inappropriate accounting to the tune of $3.8 billion. Naturally Worldcom lost customer confidence and saw a massive turndown in business. It appeared that Worldcom had been:

- Accounting for interconnect expenses as though they were capital costs. That is to say, costs that were incurred to achieve today's revenues were being spread over future years, thereby breaking the *matching* convention and misleading the market about future earning potential.
- Inflating revenue by what was, at best, overly optimistic future revenue and, at worst, completely fictional revenue. Clearly this contravenes *prudence* and *matching*.

Unsurprisingly, Worldcom was forced to file for bankruptcy in 2002, creating the largest company failure in the US history up to that time. (Coincidentally, the company that it knocked off top spot was Enron, an energy company that had been found guilty of very similar malpractice!) However, there was more to come. In 2005, Bernie Ebbers was convicted of fraud, conspiracy, and filing false statements; he was sentenced to 25 years in jail. Many other Worldcom officials were found guilty of similar offences.

It was a sad end to a major telecommunications challenger that turned to fraud to sustain the illusion of growth and success.

must have future value and to be reasonably expected to provide revenue earning capability. So, to give a very simple example:

A company founder asked his 10 friends to give him £1,000 each to start a coffee business. He promises them a share of any future profits, so they are shareholders. He uses the £10,000 to buy coffee beans, a stock of paper cups and a coffee machine. Even with his purchases he has money left over in the company's bank account. Our coffee entrepreneur hasn't started trading (he hasn't sold a single cup of coffee yet!), but his day one balance sheet looks like this:

Balance sheet: Coffee World 1st April 2000 (pre-trading)

Fixed assets	
Coffee machine	£6,000
Current assets	
Cups	£500
Coffee	£2,000
Cash at bank	£1,500
Total net assets	£10,000
Capital and reserves	
Total share capital	£10,000[a]

[a]Ten ordinary shares at £1,000 each.

Clearly, balance sheets get far more complicated than our simple example. This is particularly true once a company starts trading and, hopefully, making profit, so let us look now at the profit and loss statement and return to our coffee company example a little later.

4.2.2   The profit and loss statement (sometimes known as income or revenue statement)

Unlike the balance sheet, the profit and loss statement (P&L) is not at a point in time, but rather it aims to identify the profit (or loss) made in a complete period (perhaps a month, quarter or year). Profit is, of course, income or revenue earned less costs incurred in earning and servicing that revenue. The P&L, therefore, includes all the income or revenue earned in that year, whenever that income might have been billed or paid for. Any income earned, but not yet billed at the end of the period, is included in the P&L account as an *accrual*. Any income billed but not yet paid for is included in the balance sheet as a trade debtor (also known as accounts payable).

Similarly, applying the *matching* convention, the P&L includes all the costs incurred to achieve that income. Any costs incurred for which the company hasn't received a bill are accrued; any bills received but unpaid are shown on the balance sheet as a trade creditor (also known a as accounts receivable).

Let us return to our simple example.

After one year's trading Coffee World has sold 10,000 cups of coffee, at an average price of £2.50 each. Some customers have accounts and pay for their coffee after a month. At 31 March 1,000 cups of coffee were unpaid. The company has used its initial stock of coffee and bought three more bags costing £3,000 in total. However, not all of the coffee was needed to make 10,000 cups, so half a bag remains in stock. The company bought milk (£2,000) and more cups (£500). There are no cups in stock at the end of the year. The coffee machine has been serviced (£500) but the bill hasn't been paid yet. The owner does all the work himself and earns a salary of £10,000. Finally, the company must account for the fact that it has used its fixed asset (the coffee machine) to produce the cups of coffee and to earn the revenue. It must therefore include some depreciation in the P&L. This is classed as a non-cash cost. See Box 1.3 for a fuller explanation of depreciation. In this case our coffee entrepreneur believes his coffee machine will have a useful, revenue-earning life of 3 years and that he will use a simple straight-line convention to calculate depreciation i.e. the asset value divided by the life.

So, Coffee World's first years P&L looks like this:

Profit and loss statement: Coffee World, 31st March 2001

Turnover			
	10,000 @ £2.50		£25,000
Cost of sales			
	Coffee	£4,500	
	Milk	£2,000	
	Cups	£1,000	
		£7,500	
Gross profit		£17,500	
Operational expenses (pay)		£10,000	
Coffee machine service		£500	
EBITDA		£7,000	
Depreciation		£2,000	
Operating profit (before interest and tax)			£5,000

EBITDA stands for earnings before interest, tax, depreciation and amortisation. Broadly it measures the current operating profit (or trading profit), essentially the profit being made by employing today's assets. Although it is not a formally recognised financial term, it is widely used by ICT companies because the high levels of depreciation associated with network investment can otherwise mask successful trading. Although not exactly equivalent to cash flow, it is often used as a close proxy for cash generation.

More conventionally though, our entrepreneur has earned £5,000 before interest and tax, and for the sake of simplicity we will assume a 0% tax rate.

We will also assume that he distributes half the profit (£2,500) in the form of dividends to his shareholders. Finally, he has plans to expand next year and so has very recently bought a second coffee machine (£5,000) and, despite strong cash flow, he has taken a bank loan to pay for it.

We can now consider the company balance sheet at the end of the first year.

Balance sheet: Coffee World 31st March 2001 (after one year of trading)

Fixed assets	
Coffee machine (after depreciation)	£4,000
Second coffee machine (un-depreciated)	£5,000
Current assets	
Cups in stock	£0
Coffee in stock[a]	£500
Trade debtors[b]	£2,500
Cash at bank	£6,000
Current liabilities	
Trade creditors[c]	(£500)
Total net assets	£17,500
Capital and reserves	
Total share capital	£10,000
Retained profit (after dividends)	£2,500
Bank loan	£5,000
Total capital and reserves	£17,500

[a]Half a bag @ £1,000 per bag.
[b]1,000 unpaid coffees at £2.50 average price.
[c]The unpaid bill for the coffee machine service.

4.2.3   Cash flow statement

The cash flow statement is probably the easiest of the three statutory statements for the non-accountant to understand. Effectively it represents the ins and outs of the company's bank account through the period. In that sense it is no different to our own personal bank statement (though the sums may be higher!). It spans a period of time between two balance sheets as it starts with the 'cash at bank' at the start of the period; includes all the transactions that occur during the accounting period in question (a month, quarter or year); and finishes with the closing balance of 'cash at bank' which appears as a current asset in the new balance sheet. In completing the cash flow statement we need to ask: where did the money in our bank account come from, what did we spend it on, how much is left in the bank? We will need to include the net cash from day-to-day operations; the cash we spend on buying capital assets e.g. coffee machines, telecoms networks or servers; tax payments; loans received or share capital raised. Indeed, we need to include everything that involved a cash transaction throughout the year. If we look at our simple example we can construct the cash flow statement as follows:

Cash flow statement: Coffee World 31st March 2001 (after one year of trading)

Opening cash at bank on 1 April 2000		£1,500
Revenue	£22,500	
Salary	(£10,000)	
Coffee	(£3,000)	
Milk	(£2,000)	
Cups	(£500)	
Net operating cash		£7,000
Loan received		£5,000
Fixed asset purchased (second coffee machine)		(£5,000)
Dividend Paid		(£2,500)
Closing cash balance on 31 March 2001		£6,000

Two vital truths of accounting have become clear.

First, when we review even this simple example set of accounts, it is clear that profit does not equal net cash flow. Indeed, it's quite possible that a company with positive cash flow may still make a loss, while a company that is making a profit may still be starved of cash. (Indeed, as an aside, a common reason for new start-ups in ICT to fail is not that they are making a loss, but they don't have the cash flow to sustain their growth).

How can this be? Well, we have encountered a number of reasons already:

- In buying a major fixed asset a company will pay out considerable cash but will only include a fraction of that cost in the P&L as depreciation.
- Not all cash revenue will be received in the year that it was earned (the difference makes up trade debtors (accounts payable) on the balance sheet).
- Not all purchases will be paid for in the year that they were incurred (the difference makes up trade creditors [accounts payable]).
- Loans received and calls on shareholders do not count as profit (but appear on the balance sheet); similarly dividends – although paid in cash – do not reduce profit (they are a form of distribution of profit).
- And there are many more technical other reasons why Profit $\neq$ Cash.

The second truth is that there is discretion in determining profit, but this discretion must be in line with accounting standards, applied with prudence and to the satisfaction of the different levels of auditors. Again, we have encountered some of the relevant questions already:

- What expenditure can be legitimately capitalised and what should be written off in the year that it was incurred?
- What depreciation policy should we apply and what is a prudent life for this equipment?
- How should we value the year-end stock?
- Should we write-off certain debts that may never actually get paid?
- And there are many more points where an experienced finance director needs to apply his judgement to strike a prudent view of profit and satisfy the auditors.

4.2.4   *Statutory disclosures and other information provided in the financial accounts*

We have now discussed the three main accounting statements of any set of financial accounts. Although there are small differences throughout the world, so many companies are now global, or at least have a presence on more than one stock market, that the generally accepted accounting standard (GAAT) as required in the United States; the statement of standard accounting practice (SSAP) as required in United Kingdom and Europe; and the international financial reporting standard (IFRS) are all converging and becoming the *de facto* worldwide standard.

These standards call for disclosure of company information well beyond the three key statements. For example, companies are required to disclose director salaries; depreciation policy; income from investments; pension costs; and much more.

Similarly the company will wish to provide a commentary on the financial results, the future prospects of the company and its governance. Normally this will be in the form of statements by the chairman, CEO and audits. They are also obliged to disclose anything that has materially affected the company since the accounts were prepared.

4.2.5   *City relationships and the share price*

In most developed countries it is a requirement of company law and the stock market rules to publish a set of audited financial accounts on an annual basis. This is to safeguard the financial probity of the company, to ensure that it is paying appropriate levels of tax and that potential investors are well informed, etc. However, there are many external stakeholders interested in the success of the company and its likely future prospects. In particular this will include current and prospective shareholders, and the wider investment and financial community who are collectively known as *The City* in the United Kingdom or *Wall Street* in America. Investors will encompass large investment banks, hedge funds, pension fund holders and private individuals – big and small.

There is great advantage for a company in having an attractive share price. It determines the overall value of the business; it protects the company from becoming a take-over target; and it is vital if they wish to raise further investment. A strong share price performance reflects *The City's* confidence in the future profit potential of the company and so will attract new investors. This might be key if the company made a call for new funds to, for example, fund a new venture. Similarly, a strong share price will help if it wants to become a predator and take over other companies by offering shares as part of the offer.

The published accounts are a key determinate of the share price and the overall attractiveness of the company. For that reason, the publication of the financial results will normally be accompanied by a series of briefings for members of *The City* i.e. key current and potential investors, business analysts, the business press, etc. As well as summarising the current results, these meetings will also cover any fundamental changes to the market, exceptional influences on the company and the key future plans. These briefings are seen as a vital aspect of stakeholder relationship building.

And like so much in relationship building, trust is vital. If companies take these briefings seriously, e.g. by sending senior people, being open and consistent and by explaining the problems as much as selling the positive story, then they will be rewarded with loyalty and enthusiasm from investors. Such companies become *darlings of The City* and their share price consistently outperforms the market averages. However, the converse is also true. A management team that hides away or fails to convince the city that it has a winning strategy will suffer poor share price performance.

4.2.6 Conclusion

We do not have space, nor is it our purpose, to provide a detailed guide to accounting standards or to compiling books of account. We must leave that to the qualified accountants. However, we hope that these simple explanations and examples help to provide non-finance managers with the ability to read and understand company accounts and to draw some conclusions. Indeed, a key purpose of published accounts is to offer interested stakeholders information on which to analyse the company's past success and future prospects. Such stakeholders would include current employees and shareholders; prospective employees and future shareholders; competitors; etc. It might even include companies considering a take-over, merger or acquisition.

With that in mind we will now look at some standard approaches to analysing financial accounts [1].

4.3 Financial analysis

In considering how Telcos make money (section 1.3) we discussed the difference between profit and profitability. Profit is expressed as a single figure, unrelated to the size of the company, market opportunity or the assets or capital employed. Profit-ability, on the other hand, is expressed as the amount of profit in relation to the assets and capital required to earn that profit. It, therefore, tells us something about the efficiency, or otherwise, of the use of the company's resources. In addition, we have now seen the importance of cash flow in the development, growth and sustainability of a company. The availability of cash to enable a company to operate efficiently is described as its *liquidity*. Finally, we saw in the previous section the importance of the financial results in determining a company's attractiveness to investors.

Consequently, in analysing financial accounts four aspects are important:

- profitability
- liquidity and gearing
- asset utilisation
- attractiveness to investors

4.3.1 Profitability

A stakeholder will be far more interested in the profit of a company, in relation to the size of the company – perhaps measured by the total sales or net assets employed. They are therefore likely to be interested in a series of ratios relating these factors:

- gross profit ratio = gross profit ÷ total sales
- operating profit ratio = operating profit ÷ total sales
- return on net assets = profit before interest and tax ÷ net assets

In our Coffee World example above the results are as follows:

- gross profit ratio = gross profit ÷ total sales = £17,500/£25,000 = 70%
- operating profit ratio = operating profit ÷ total sales = £5,000/£25,000 = 20%
- return on net assets (RONA) = profit before interest and tax ÷ net assets = £5,000/£17,500 = 29%

There are many other ratios that help to illuminate the profitability of a company, but these are the key ones. We should therefore ask the question: how profitable is *Coffee World*; are they making enough money for their size and for the assets (and therefore investment) employed? Gross profit of 70% and operating profit of 20% certainly looks healthy, but should be put into the perspective of the industry in question. Before drawing any conclusions we should ask: what percentages do other coffee makers achieve and how are the ratios changing over time? Similarly, 29% RONA looks good (and probably is) but what is the cost of capital for *Coffee World* and what is the level of risk associated with their business. Again, what is the average RONA for the industry as a whole?

Once again, our simple example has demonstrated a key lesson. Ratio analysis of a set of accounts is helpful, but it isn't enough on its own to provide all the answers. Other comparative data is required, together with some experience and judgement in concluding the success and likely future prospects of a particular company.

4.3.2   Liquidity and gearing

When we look at a set of accounts we need to consider more than profitability. Is the company well run; does it marshal its resources well; does it have enough cash to pay the bills; does it have the right financial structure to develop and grow? Once again there are many different ratios that we might use to judge the financial sustainably of a company but the four most important are probably as follows:

- gearing = long-term debt ÷ equity

Gearing is quite a technical financial issue, but it is vital to ensure the company is sustainable. Key to gearing is the fact that loan capital earns interest, at a predetermined rate; and that interest is a cost to the company and therefore reduces profit. People who loan money to companies (banks, etc.), therefore, bear limited risk. Shareholders, on the other hand, are paid a share of the profits and therefore carry a lot of risk. If the company does well and chooses to distribute some or all of the profit, then shareholders do well. If the company losses money or chooses to retain the profit to re-invest, they might earn nothing, though the value of their shares might go up notionally. A company with a lot of debt and very little shareholder equity is said to be highly geared (or leveraged). That is to say, the company bears a lot of risk. In a downturn, when profits are squeezed, the interest on the debt still has to be paid, and a small operating profit may become a loss after tax and interest. A less highly geared company will have less interest to pay and can choose how much profit to distribute to its shareholders in the form of dividends. In such a company, shareholder equity, therefore, provides a cushion against difficult trading conditions.

Gearing is therefore a measure of indebtedness. One reason companies take on debt is to improve the returns to shareholders. How this works can be illustrated by the example of using a mortgage to buy a flat:

Say you had $100,000 to invest. You buy a $1,000,000 flat using a $900,000 mortgage, so your initial $100,000 is the deposit (ignoring fees and costs). Assume the flat was let out and the rental income just covered the interest-only mortgage payments, so the ownership cost was zero.

You sell the flat for $1,050,000 a year later, which represented a growth rate of 5% in the underlying asset.

After repaying the $900,000 interest-only mortgage, you are left with $150,000.

You, therefore, made a profit of $50,000 on an original investment of $100,000.

Your profit is therefore 50%, even though the underlying asset only appreciated by 5%.

This is the basic concept of gearing.

In our example:

- Gearing = Long-term debt ÷ equity = £5,000/£10,000 = a ratio of 1:2

Once again, acceptable levels of gearing vary from industry to industry, but this is a comfortable ratio in any circumstances.

Incidentally, it is worth considering the question of risk from the other side for a moment. People who provide loan capital bear little risk because their interest payments are guaranteed (except in the extreme circumstances of the company going into liquidation). Shareholders, on the other hand, bear far more risk, especially when the company is highly geared, because there are few shareholders to provide the cushion. In good years all the profit is theirs, but in bad years they bear all the losses.

Nonetheless, interest has to be paid and should it become an unsustainable burden (perhaps because rates are forced up by the economic circumstances) the company will go bust. It is therefore important to check interest coverage, which is the ability to pay the year's interest liability from earnings.

- Interest cover = earnings before interest and tax (EBIT)/interest

In our example, no interest is due on the recent loan, so it is unnecessary to calculate the ratio.

A vital element of a company's liquidity is how well they manage their working capital.

- Current ratio = current assets ÷ current liabilities

This ratio will show how well debtors, creditors, stock and other items of working capital are managed. In our example:

- Current ratio = current assets ÷ current liabilities = £9,000/£500 = 18

This is a high ratio and might suggest that stock levels are too high or that the company is stockpiling cash in the bank. Equally, they might be slow to collect debts that are owed or paying their bills too early. While none of these things will kill the business, the high level of money tied-up in working capital might otherwise be used to invest in future growth or reduce debt

- Acid test ratio = [current assets − stock] ÷ current liabilities

  Using this formula for our example:

- Acid test ratio = [current assets − stock] ÷ current liabilities
  $$= £8,500/£500 = 17$$

This ratio is also high and suggests the problem isn't with stock management but the way that the debtors and creditors are managed. We should therefore consider how well our working capital is being managed. We can do this by looking at the third area of investigation which is asset utilisation.

(Incidentally the acid test ratio is so called in reference to acid's historical use to test metals for gold. If the metal stood up to the acid it was pure gold. If the metal corroded it was of no value.)

4.3.3   Asset utilisation

We might now want to drill down into each of the key elements of working capital. The following ratios show the *burn-rate* of stock and also what we owe and are owed, in relation to the company's daily sales level.

- Stock Days = Stock ÷ Cost of Sales × 365 = £500/£25,000 × 365 = 7.3 days
  Such a level would not indicate excessive working capital tied-up in stock, but might imply an unacceptably small buffer against problems in the supply chain.
- Debtor Days = Trade debtors ÷ Sales × 365 = £2,500/£25,000 × 365 = 36.5 days
  Normal trading terms tend to be around 30 days, though one might expect rather less in a retail business like coffee. 36.5 days suggests that debts are not being chased aggressively and might eventually become *bad* i.e. never get paid and have to be written-off, reducing profit.
- Creditor Days = Trade creditors ÷ Cost of Sales × 365 = £500/25,000 × 365 = 7.3

This is a low figure and might suggest we are paying our bills faster than we need. This burns working capital that might otherwise be put to better use in the business.

4.3.4   Investors

Potential investors (see section 4.2.5) will take the opportunity to analyse the accounts in a great deal of detail before investing their funds. Apart from the ratios described earlier there are many more, specific to investment opportunities. The main two are:

- Earnings per Share (EPS) = Net Income/Average Number of Share

That is to say, it is the earnings per share that could be distributed to shareholders if the company chose not to retain any earnings. Clearly the higher the figure, the better it is; this measure enables companies of different sizes and industries to be compared. In our example:

- Earnings per Share (EPS) = Net Income/Average Number of Share

$$= 5000/10 = £500$$

- Price Earnings (P/E) Ratio = Share Price/Earnings per Share (EPS)

This ratio shows the price that investors are currently willing to pay for a certain share, in relation to its earnings, based on its most recent set of accounts (quarterly or annually). A high ratio implies market confidence in the company's current and future performance. However, the ratio is only of use in comparing companies in the same industry because average values vary widely between different types of business. Typically a high-risk/high-reward industry will have a high P/E ratio while utilities have lower P/E's. For example, technology companies might have P/E's above 20, while energy companies might be below 10 on average.

Our example company isn't quoted on the stock market, so the price earnings ratio doesn't apply.

4.3.5   Summary

It is possible to touch on just some of the key techniques involved in analysing a set of financial accounts and to explore some of the key ratios, generally in use. Nonetheless we hope that the key lessons are clear:

- A systematic review will cover all the key aspects of a company's performance.
- It is important to consider the past, current and future prospects.
- There are no black and white answers. Ratios and results will vary from industry to industry and from circumstances to circumstances. Nonetheless, they provide objective measures on which to base our judgement.

4.4   Business planning

4.4.1   The planning hierarchy

As we have seen, successful ICT companies require certain attributes. They require a clear understanding of their customers; they require good product management; they require access to reliable and technically advanced networks; and they require well-managed and efficient resources. Perhaps, more than anything, they need a clear vision of the future and the ability to create a plan for success.

In reality, they need several, interlinking plans. Telcos seek to finance and build a network, develop a portfolio of products based on that network and to generate revenue from the marketing, sale and operation of those products.

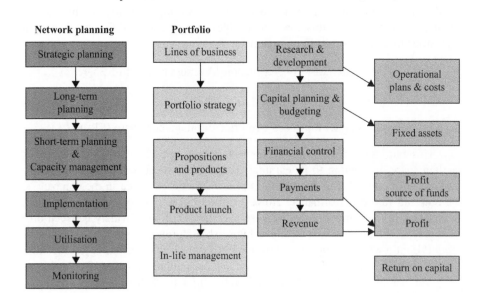

Figure 4.1   Hypothetical planning hierarchy

Therefore, as a minimum, they require a network plan, a portfolio plan (that combines products and marketing) and a financial and business plan. While each of these plans is separate, and developed by different parts of the company, they need to relate to one another. For example, the business plan will need to identify the capital funds required to build the network; the network will need to have the capabilities and capacity to support the products and to accommodate customer demand; and sale of products and services will provide the revenue to generate forecast profit. There are many more points of connection and alignment required in a successful business plan. Figure 4.1 shows a hypothetical example of how business planning might work in a typical Telco.

We discuss network planning in Chapter 6 (sections 6.4–6.6) and portfolio planning in Chapter 8 (section 8.3.2), so it is appropriate here to explore the business and financial plan in greater detail.

4.4.2   The planning year

4.4.2.1   The business plan

In most companies the planning is achieved through a linear process that runs to an annual cycle. If we imagine a company with a financial year that runs from April to March, the planning process is likely to commence in the summer with a group of senor strategists and financial planners developing a 3–5 year view of the future. They will take any previous plans into account and be partially mindful of external environmental influences, such as the macro-economic outlook, technological developments, future customer expectations, etc. The plan will be informed by the company's strategy and vision and is likely to rely on top-down analysis, rather than a bottom-up build. The output of such a plan will, ideally, paint a clear vision of the future financial and business shape of the company and will illuminate the strategy. It will identify investment needs; financing requirements; portfolio developments; any mergers or acquisitions that may be envisaged; and any gaps in the company's capabilities, such as particular people skills.

4.4.2.2 The operating plan

Once the long-term plan has been agreed and perhaps formally adopted by the company board, the planning process moves onto a far more detailed operational plan. This is likely to be a 2-year plan (perhaps with rather less detail included for the second year) and will take the big picture output from the long-term plan and put flesh on the bones. External environmental trends will remain important, but they will be matched by equally important internal issues, such as spare capacity in production processes; the need to develop better sales capability or new product developments. The central planning team is likely to consult divisional or business unit planners to ensure the overall plan is both deliverable and meets the aspirations of the company strategy. Ideally it will be a combination of a top-down/bottom-up approach. From the outputs of the plan the company should have clear targets of business growth; market share; new products to be launched; marketing programmes; resources and productivity improvement; profit and profitability.

4.4.2.3 The annual budget

Finally, in our hypothetical planning year, divisions and business units will be required to develop a bottom-up, detailed annual budget that delivers the overall profit requirements of the first year of the operating budget. They will do this between January and March with the hope of having the budget agreed and authorised before the new financial year. The purpose of the budgeting process is to provide the financial control that will ensure delivery of the planned profit and profitability. For this reason it may be that the overall annual budgets may need further devolution, down to individual cost centres, products or even specific managers. The focus of this stage of the process is almost entirely internal, with the aim of allocating revenue and resources in considerable detail. The outputs of the budget analysis will consist of many detailed spreadsheets, but in broad terms it will encompass product sales; target prices; revenues by product or line of business; headcount; pay rates and labour productivity; expenses and overhead cost.

The overall annual planning process is described in Figure 4.2.

4.4.3 Capital planning

We have discussed elsewhere the particular nature of capital expenditure (see e.g. Box 1.3), but it is perhaps worth reiterating the key points. In essence, capital expenditure is used to buy assets that have a value to the business beyond the year in which they are purchased. The expenditure would include, e.g., equipment provided under contract, the cost of installation and commissioning, and in some cases the workforce costs associated with the planning and positioning of the new

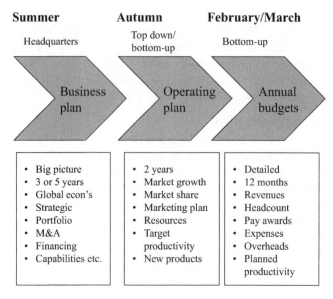

Figure 4.2   Indicative annual planning process

equipment known as *capitalised labour*. Capital expenditure, therefore, appears as part of the cash flow statement for the year in question but not in the profit and loss statement; rather the cost of the asset purchased is spread across the usable 'life of plant', in the form of depreciation.

However, there is another aspect of capital expenditure that marks it out as different from general expenditure. Capital tends to get spent on programmes and large projects that take several years to build e.g. the development of a next generation network (NGN). Once the programme or project or has been approved and the capital authorised the funds must be found for each year of the work. This doesn't sit easily with a planning process that ultimately aims to set an annual budget, but it is clearly essential that funds can be found for work that is already underway – maybe nearly complete – at the start of the financial year.

The special nature of capital expenditure, therefore, drives a series of particular business processes that we shall explore here:

* the planning process
* programme and project governance
* capital authorisation
* business cases

Figures 4.3 and 4.4 illustrate the key approach to planning for capital funds. At the start of the year in question a series of programmes and projects are already underway, authorised in previous years. The programmes will include projects still to be granted capital authority, but the presumption of the planning process would be that they will go ahead (though, clearly a balanced business decision would still

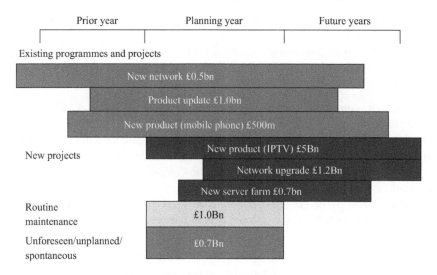

Figure 4.3   Capital planning inputs

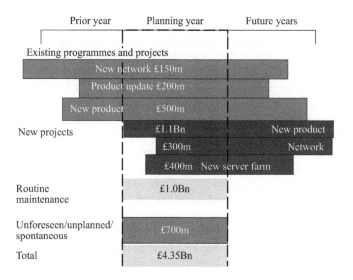

Figure 4.4   Annual capital budget

be needed to ensure they are the best use of any scarce capital funds). In our example we have:

- Pre-authorised programmes and projects for a new network (total programme cost estimated at £0.5 billion), product upgrades for existing products (£1.0 billion) and the launch of a new product, a mobile phone (£0.5 billion).
- Planned new projects that are likely to seek authorisation in year, such as another new product, IPTV (£5.0 billion), a network upgrade (£1.2 billion) and

a new server farm (£0.7Bn). None of these projects will be completed in the year in question.

- Planned routine maintenance (£1.0 billion).
- A contingency for unforeseen, unplanned and spontaneous capital projects (£0.7 billion).

When it comes to completing the annual budget, however, the finance department need to understand what funds they are likely to need to raise in the planning year in question. It is therefore necessary to calculate the capital funds appropriate to just the coming year. Often this information will be fixed in the supply contract, or clearly laid out in the authorisation document. Sometimes, however, it will be dependent on agreed metrics (e.g. a target penetration, successful technical trials, etc.) and therefore will have to be estimated. Figure 4.4 shows that, in our example, the relevant capital required might be £4.35 billion [2].

4.4.4   Capital programmes, projects and authorisation

As we have already seen, capital funds are essential to managing, developing and growing a business. In essence it is concerned with the future prospects of the company and is a fundamental aspect of new programmes. These may be concerned with new products; enhancing existing or building new networks; other infrastructure; buying companies, technology or even knowhow, as discussed in Chapter 6. Larger schemes, e.g. building an NGN, might initially be described as a programme – essentially a coherent group of component projects that achieve the overall programme objective. Of necessity a programme is a high-level view, with less detail that a project business case (see section 4.5). Nonetheless, as a minimum, a programme will have a clear statement of its aims and benefits; a clear strategic fit and position in the market; and an outline financial analysis, including estimated capital requirements. In addition, a programme will describe the component projects and give an indication of the order of implementation and the likely timescale for business cases to come forward for formal authorisation.

Capital authorisation processes differ from company to company but, certainly, good governance and financial disciple demand a clear policy and published rules. These are normally embodied in standing orders that need to be well promulgated and understood throughout the company. Fundamentally, the authorisation process relies on a certain degree of hierarchy in the company's structure (see Figure 4.5). Standing orders will require that the most important decisions (those that require a certain level of capital expenditure) be authorised at company board level. In fact, authorisation processes often include both capital and non-recurrent current expenditure i.e. one-off expenditure that is required to implement the project, but which isn't deemed to be capital. Non-recurrent current expenditure often includes training, software and the cost of the project implementation team.

The board might delegate authority for smaller projects to the operating committee, management board or to a divisional board. These organisations might delegate further, perhaps to individual senior managers. Quite often the detailed scrutiny of the business cases is completed by a financial concurrence group or

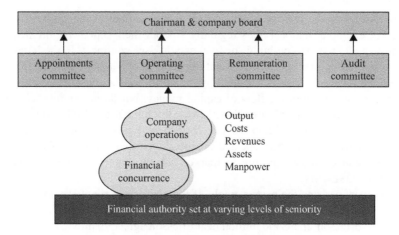

Figure 4.5   Capital authorisation

investment committee consisting of senior managers from all the key business disciplines e.g. marketing, engineering, financial, etc. This provides for a wide review by managers who are both expert in their area and likely to need to support the business idea and to *own* the implementation.

Some companies will require programmes to be authorised prior to giving authority for capital spent on any single project. Equally, boards may demand gateway reviews, or funding in stages, so that they have a chance to review progress (e.g. against the initial business target) before authorising the next tranche of funds. Finally, good practice would include a post-implementation review of projects so that lessons can be learnt and improvements made in future project planning.

4.4.5   A final thought on capital expenditure

Raising capital (or using retained profits) to invest in future opportunities is a risky business (as we will explore in the next section). It, therefore, often makes good business sense to avoid capital expenditure altogether. There are a number of routes open to the enterprising company. For example, rather than buying capital assets, a company might choose to lease the necessary equipment. This spreads expenditure over the revenue-earning life of the asset; however, it might cost more overall once financing costs are included. An extension of this approach would be to ask an equipment-manufacturer to 'build and operate' a network, thereby reducing both the risk and cost of managing the asset once it is in operation.

However, these approaches leave the risk and reward for the new venture with the Telco. In other approaches, market and product risk is shared. For example, an ICT company might persuade equipment-manufacturers to provide equipment free of charge initially, in return for a share of the revenues or profit as they materialise. Clearly this requires suppliers to be happy to share risk, or exceedingly keen to do business! Equally, it reduces the retailers' risk but also limits their potential benefits.

If we look at the real world we can see that a number of models have developed. For example, major Telcos regularly smooth their capital requirements by entering either 'pay-as-you-build' or 'pay-as-you-use' contracts. In the latter case the equipment-supplier takes the risk that the demand may not achieve the forecast; in the former case the risk is still held by the customer (i.e. the service provider). Both approaches smooth the flow of capital money, but do not entirely eliminate the need for capital expenditure. Nonetheless, throughout the 1990s and 2000s at least one large European Telco used both of these types of contractual arrangements for the roll-out of network and broadband services. In particular they significantly reduced their exposure to market risk by entering a 'pay-as-you-use' contract for DSLAMs.

However, perhaps the best example of the equipment-supplier taking the risk occurred when Bharti Tele-Ventures outsourced the majority of their Indian mobile network to Ericsson. In a $400 million deal Ericsson agreed to provide, manage and maintain the network, leaving Bharti Airtel as a purely sales and marketing organisation. This model proved so successful that by 2010 Ericsson had won 5 more outsourcing contracts (worth almost $5 billion) and Bharti was the most successful mobile service provider in India [3, 4].

Incidentally, risk sharing can also apply in the other direction. When a mobile phone company provides an expensive smartphone free of charge in return for the customer signing a 2-year contract, the mobile phone company is knowingly taking on extra risk, i.e. that the customer will default before the 2 years is up. On the other hand the mobile company expects to make extra margin for the risk they are taking.

We have already explored the issues of capital projects, authorisation and planning. We will now explore the preparation of business cases – a key management tool in any ICT business.

4.5   Business cases

4.5.1   The purpose and use of business cases

As we have seen, companies will typically require all new investment in networks, company acquisitions, buildings and new products to be authorised at an appropriate level. The usual vehicle for this is the business case, although the detailed form can vary widely between companies. Some will require a very closely specified written document, with pre-set appendices on issues such as the market, finances, technology, etc. Others will be rather more flexible in the form and content. Some companies might require no more than a presentation to the board.

The purpose of the business case is to demonstrate that the project is viable, an appropriate way to invest shareholder money, and the best use of limited capital funds. For these reasons there will be emphasis on strategic fit; the size and nature of the target market; customer needs; forecast income and costs; and detailed investment appraisal. There will normally be a formal request for investment authority to cover the necessary capital and non-recurring costs.

Formally, the purpose of a business plan is to demonstrate the viability of the proposed project, but it serves additional purposes. The project may be viable, but in a world of opportunities, is it the best use of shareholders' funds? It also provides a plan against which to monitor progress and to assist in holding managers to account. It may formally include targets that must be achieved before the project progresses to the next stage. It will form an input to the business planning process, enabling the finance department to:

- forecast the need for capital funds
- identify possible new revenues
- foresee resource requirements with more assurance

Finally, as we saw earlier, it will be the basis of a formal request for financial authority.

4.5.2 Business case content

While the output of the business case process is likely to be a paper, it is far from simple to write. Nor is there a standard template for success. Rather, the paper will form a high-level summary of a lot of detailed work by the project team, each member working on their specialist area but also acting together as a team. It will include detailed research and is likely to be data-rich. The work may take many months to complete or just a few days, depending on resources available and the complexity of the project. In practical terms, and to aid readability, the paper is likely to have a short, clear narrative free of detail but making the case strongly. It is likely to be supported by a number of more detailed appendices covering the technical description, market data, customer research, financial analysis, etc. The overall requirements of a business case are summarised in Figure 4.6.

While there is no standard template, a good case is likely to include what is described in the following subsections.

4.5.2.1 An executive summary and introduction

Ideally this will be short and to the point. It will include what is intended e.g. to launch an IPTV product; the fundamental business reason for the proposal e.g. to fulfil an emerging customer need; the forecast outturn in financial and business terms e.g. to achieve 50% market share in 3 years and to break-even in 15 months; and what is required of the board, e.g. to provide capital authority to progress.

4.5.2.2 A description of the market

This section will include an estimate of the current and forecast size of the market to be addressed, how that market relates to other relevant markets, and how well the market is currently served. Of necessity it will refer to other players addressing the market in question and their relative strength. For these reasons some of the models referred to in Chapter 7 may be employed (e.g. Porter's five forces). In the case of a new product launch or enhancement the case will go on to describe the customer need that is being addressed – ideally supported by extensive customer research (survey material, focus groups, etc.). Finally, the product will be described – not in technical

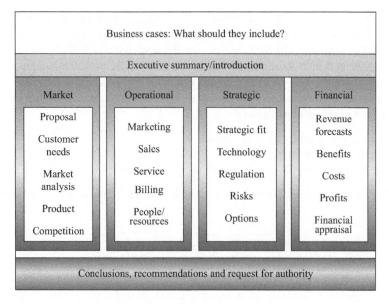

Figure 4.6   Business case content

or engineering terms but in relation to fulfilling the identified customer needs. A proposed target customer segment may be described together with an outline marketing plan, potentially based on the 4 P's (see section 7.5). It should be possible to estimate customer sales and target revenues to include in the financial appraisal.

4.5.2.3   The operational considerations

This section needs to answer a series of practical questions such as how will the product be marketed and sold; the form of customer service e.g. call centres or online; will field engineers be required; are new operational systems required; how will the product be billed and paid for? Significant resources are likely to be required. The case should identify the types of resource needed and how they will be obtained and the cost. Again, detail and data are likely to drive this analysis; the costs will form an important input to the financial appraisal.

4.5.2.4   The strategic fit

The business case should explain how the proposal fits with the stated strategy of the company; this might include the market being targeted, use of existing assets, and the product portfolio and technology strategy. There have been numerous examples where ICT companies have got this wrong:

- The Telco choosing to expand overseas with no knowledge of the regulatory regime or culture.
- The mobile company adding content to its portfolio without the matching technical strategy to support the service.
- The media company adding banking to its portfolio without the brand to make it acceptable to their customers.

Given the nature of the ICT industry, it is important that the case should cover regulatory considerations. For example, a proposal to acquire another ICT company might be seen as concentrating market power. Will this be of interest to the regulator and how might their concerns be mitigated? Similarly, many retail and wholesale telecommunications products are subject to regulatory price controls. This needs to be made clear and compliance demonstrated.

Regulation is just one form of risk that applies to a project. It is important that the business case identifies all the key risks and actions to mitigate them. Key risks often revolve around technology, market developments, customer response, competitor action and delays in implementation. Such risks need to be evaluated and minimised by remedial action. Sensitivity analysis is an excellent method of formally tackling risk identified in a business case. We cover risk and sensitivity analysis in more detail in section 4.7.

Finally, it is good practice to provide the authorising group with some options, in the form of alternative proposals that aim to address a similar business issue e.g. provide business growth, or fill a gap in the product portfolio. These options may be viable but less attractive than the preferred option, or might prove to be unprofitable once they are fully assessed. Nonetheless, it is important that the project team have been open-minded in considering the issue at hand and can show that they have looked widely for the most appropriate answer.

4.5.2.5 Financial appraisal

A key element of any business case is the financial appraisal. However, the importance of the financial results is often misinterpreted. Philosophically, it is not the case that financial outcomes are all that matter, it is the fact that investment appraisal is a way of creating a quantified answer to all of the inputs to the case. So, the quality of the market research; the assumptions about market take-up and price-levels; the planned technical solutions; the level of operational efficiency built into the case all contribute to the forecast financial outcomes. In section 4.6 we will look in detail at some of the methods of investment appraisal that can be applied, but it is useful here to mention the key measures that will assist in assessing any new project. Table 4.2 summaries some of the key measures and their purpose.

4.5.2.6 Conclusions, recommendations and a request for financial authorisation

Finally, the business case will conclude with a short paragraph drawing out conclusions from the analysis, making recommendations and requesting authority to invest a stated amount of capital and non-recurrent expenditure. There are two classes of expenditure that would normally require authorisation:

Capital expenditure consisting of the investment in assets with a future revenue-earning life, e.g. equipment, network plant, servers, buildings, vehicles, etc. It might also include some capitalised labour i.e. salaries for planners, project managers, etc., who are required to implement the assets.

Non-recurring current expenditure i.e. one-off expenditure that cannot be capitalised, e.g. certain software development, integration costs, testing and field trial, training, consultants and agents fees, etc.

Table 4.2   Investment appraisal measures

Measure	Definition	Use
Pay-back Period	The time taken to earn back the cash costs of the project.	A key determinant of whether to carry out a project as uncertainty increases as time passes.
Net Present Value (NPV) and Discounted Pay-back Period	The total net value of the year-by-year project cash-flows (income and costs) allowing for the *time-value* of money.	Must be positive to ensure a project will be profitable over its life, after allowing for cost of capital and risk.
Internal Rate of Return (IRR)	This is the rate of return that ensures that a project has an NPV of zero.	Useful to compare with the target rate of return set by the company (Test Discount Rate (TDR) or to compare other competing investment opportunities.
Return on Capital Employed (ROCE)	The profit over the life of the project in relation to the capital invested.	Must be greater than the cost of capital for the project; a useful way of comparing competing projects.

The form of the final paragraphs of a business cases is best shown as an example:

The board is asked to note the strong case to launch an IPTV product in the United Kingdom, thereby contributing to the strategy of enhanced brand recognition and increased average revenue per customer (ARPU). The product will be targeted at young professionals and working families and is expected to achieve 20% market penetration within 3 years and 50% within 10 years. Discounted pay-back is less than 3 years and NPV is positive over the life of the project.

The board is recommended to approve the project and to review progress in 12 months.

The board is asked to give financial authority to £3.6 billion capital expenditure over the next two years and £1.1 billion non-recurrent current expenditure, to be spent in years two and three.

A worked example of a full business case to support the launch of an IPTV service is shown in Box 4.2.

4.5.3   Selling the business case

It is said 'that people buy from people' and so it is with business proposals. Therefore, the success of a business case will depend on the team that has prepared it and their belief and passion to make the proposal work. It is therefore important for the sponsor of the idea to personally consult with key senior managers in the authorisation process. These would certainly include the finance director, the marketing (and sales) director, the chief engineer, possibly the head of R&D.

It is essential that when the case goes forward for authorisation it doesn't contain any surprises, that there is a clear fit with the wider company strategy and other existing programmes and doesn't make claims or rely on assertions

Box 4.2   An example business case: Internet protocol television (IPTV)

It is not possible to include a full business case here. However, an example can be outlined. The following provides the key points and headlines of an imaginary case of a major incumbent Telco seeking to gain financial authority to launch an IPTV service. Obviously, a real business case would deal with subjects in much greater depth and would therefore be significantly longer, though it is a good practice to ensure the final document is succinct and readable.

Introduction and executive summary

This business case proposes the launch of an IPTV service at a capital and recurrent current account cost of £4.7 billion. The strategic purpose is to replace declining telecommunications' profits, meet growing market needs and extend our product-reach with existing and new customers. The project exhibits strong financial results.

The market

Terrestrial and satellite TV are well-served markets, but extensive market research has identified new customer requirements; these customer needs revolve around the convenience of the service (e.g. time-shifting); the pricing (only paying for what individual customers are really interested in) and the quality and breadth of content (bundling, foreign channels, personalisation, etc.). Professional people and young families will represent out target customer segment.

The size and growth of the relevant market sector are ... (include data)

We face strong opposition in the market from both terrestrial and satellite TV operators. It will be essential to differentiate our service and to focus on unserved customer needs.

Subject to market testing the features of the proposed service might include:

- real-time and time-shifted TV
- video-on-demand
- high definition (HD) and 3D video
- an Innovative pricing model such as pay-on-demand, channel subscription, free services, etc.
- TV-on-the-go and location-based services
- personalisation services
- block-buster, premium and hard-to-find content

Operational considerations

There are significant operational considerations as follows:

- There needs to be widespread broadband access to provide a rich market.
- The network will require additional capacity in both backhaul and trunk transmission to enable IPTV streaming.
- Skills are required in buying, editing and presenting content. Securing these skills will be difficult and expensive as there is a global shortage and they are much in demand.
- Customer service will require a major call centre (both sales and service) together with field force capability. It is currently intended to outsource the call centre work, while the field force capability will be achieved through multi-skilling our existing field engineers.
- It is intended to deploy a proprietary, integrated stand-alone order-handling, fulfilment and billing system.

Strategic fit

The strategic rationale for this project is two-fold:

- to increase revenue and profit to replace declining telephony revenues
- to create an increased revenue stream from our investment in NGN (both access and core)

These aims underpin the company's objectives of customer retention and market share capture and growth.

This proposal also fits closely our technology strategy; the key features and capabilities of ITPV are fully consistent with the NGN strategy currently being pursued. IPTV architecture will deliver services using the Internet protocol suite over a packet-switched network e.g. the Internet. Access will be by broadband (initially ADSL but increasingly over fibre [of one version or another]) and will rely on Internet protocol multimedia subsystem (IMS) to configure services and mange functionality. See the appendix for detail.

A number of risks have been identified. Mitigation plans are in place. The three keys risks are:

- technology unproven in large-scale applications – mitigate by field trials and scaling-up
- access to compelling/premium content (and associated costs) – mitigate by understanding consumer requirements in advance through focus groups, market research, etc.
- strength and financial clout of competitors – mitigate by product differentiation, contingency pricing plans and good knowledge of cost structures

In addition there are a number of specific regulatory issues that apply to this type of service. We will need to obtain a broadcast licence. Also, we may have to make wholesale access available. However, there is an equivalent opportunity in that we may be able to gain wholesale access to e.g. SKY content. We are currently discussing these issues with the relevant regulators.

Despite the excellent strategic fit for this proposal, a number of alternatives have been considered. The main two are:

- launch an online gambling service
- launch a financial services business

Both these options have been discounted as they provide lesser financial returns, fit poorly with our technology strategy and are less attractive to our target market segment.

Financial appraisal

The proposal shows strong financial results. The overall project payback is 4 years (on a discounted basis) with a positive NPV of £xxxM over 10 years. The return on capital employed is an average of yy% over the life of the project. In regard to sensitives, an optimistic case would pay back in 3.5 years, with an NPV of £zzm; and a pessimistic case 6.5 years with a negative NPV. (See the appendix for detailed assumptions and financial appraisal.)

Conclusions, recommendations and a request for financial authorisation

The board is asked to note the strong case to launch an IPTV product in the United Kingdom, thereby contributing to the strategy of enhanced brand recognition and increased average revenue per customer (ARPU). The product will be targeted at young professionals and working families and is expected to achieve 20% market penetration within 3 years and 50% within 10 years. Discounted pay-back is 4 years and NPV is positive over the life of the project; there are acknowledged technical, operational and market risks, but mitigation plans are in place.

The board is recommended to approve the project and to review progress in 12 months.

The board is asked to give financial authority to £3.6 billion capital expenditure over the next two years and £1.1 billion non-recurrent current expenditure, to be spent in years two and three.

Appendices:

- Detailed Market Research
- Technology Considerations and Architectural Diagram
- Detailed Product Description
- Demand and Revenue Forecasts, Operational Requirements and Cost projections
- Marketing Plan
- Detailed Investment Appraisal incl' Assumptions, Risks and Sensitivities

that cannot be supported by reliable data. In particular, it is often the case that new products or businesses substitute or undermine existing revenue streams e.g. broadband services replacing existing leased-line services. It is essential that this level of *cannibalisation* is recognised and quantified in the business case. See also Chapter 8 on the risks of product cannibalisation and revenue substitution.

Finally, it is usual to include breakpoints or gateways in the project plan to enable a review of progress and modification if targets are not being met. All too often, achieving authorisation for the business case is seen as the goal! In fact the goal should be a successful, profitable outcome that achieves the stated aims of the business case.

4.5.4   Business cases: summary

In this section we have seen that:

- Companies will always require all *capital investment* and *non-recurrent* expenditure to be appropriately authorised. Requests for authorisation may take many forms but frequently a written business case will be required.
- Business cases will be required for a variety of projects e.g. technology upgrades, new property builds, acquisition of a business, or launch of a new product or group of products.
- Business cases should comply with the rules laid down locally by the company or division in question. They should have the support of all key members of the product team.
- A well-written business case is clear, persuasive, well backed by data, and makes good business and financial sense.

4.6   Investment appraisal

4.6.1   Introduction and background

As we have seen investment appraisal is a vital aspect of any business case or investment decision. Its aim is to determine if any particular proposal is likely to make an acceptable profit and return on the investment. It is therefore a support to strategic decision-making and aims to ensure that the likely returns on an investment are great enough to justify the risk. No individual would invest in a risky venture that didn't return as much as they could earn in the building society. Similarly, well-run companies don't invest in new ventures that return less than risk-free investment such as government bonds. They are also mindful that any new venture that earns less than the company's current cost of capital will diminish the company's value. But capital funds are also scarce and so investment appraisal can be a useful tool to set priories or help to choose between competing proposals. Clearly, investment appraisal is a forward-looking exercise and the quality of the answer is dependent on the quality of the input assumptions and the forecasts. The answer can therefore never be *right*, but the aim is to achieve the *most likely* or conservative estimate. We should also note that, because the future is not certain, forecasts become less certain the further out in time we go. This phenomenon is referred to as the time value of money and requires good decisions to be made based on discounted cash flow analysis.

Finally, we should note that not all decisions are purely monetary. There may be *non-financial* benefits attached to certain proposals. These might include regulatory requirements; opportunities to improve relations with strategically important customers or suppliers; as well as improving the strategic position of the company.

Such cases should be rare and, in any case, some attempt should always be made to associate a quasi-cash benefit to any proposal. (As an aside, a highly respected finance director (FD) in a major Telco used to be heard to say 'If the engineers say it's *essential* it's likely to cost the company millions; if the marketing director says it's *strategic* it will likely cost billions. It's neither essential, nor strategic to go bust!' While this is clearly a joke it does underline the need to rigorously question any unquantified benefits claimed for a project.)

4.6.2   Cash flow and simple payback

Unlike the published financial accounts, investment analysis is based on cash flow. It is a simple equation.

Cash In − Cash Out = Net Cash Flow

Cash in will, in most cases, arise from revenue but might also include sale of assets or licence contributions, etc. *Cash out* will certainly include the purchase of assets or any form of operational expenditure. In any period *net cash flow* might be negative or positive.

Let us consider a hypothetical project to launch a new mobile music service. It is a very simple proposal. On day 1 we purchase the necessary equipment, such as servers, etc., (£300 million) and launch the service. In year 1 we earn £250 million in revenue and incur operational expenditure – pay, equipment, running costs – of £150 million. We expect the product to have a viable life of 5 years before it's superseded. The cash flow statement for this simple proposal is shown in Figure 4.7.

Cash flow statement (£m)					
	Year 1	Year 2	Year 3	Year 4	Year 5
Cash in (revenue)	250	250	250	250	250
Capital expenditure	300				
Operational expenditure	150	150	150	150	150
Net cash flow	−200	100	100	100	100
Cumulative cash flow	−200	−100	0	100	200

Figure 4.7   Example project cash flow statement

We can now look at the payback period for our project. Cumulative *net cash flow* (the sum of the all the previous years' *net cash flow*) is zero after 3 years. In other words it takes exactly 3 years to earn back the initial cash outlay. This is shown graphically in Figure 4.8.

It is somewhat unrealistic to buy equipment and to launch a new product immediately. Similarly it is unrealistic to expect to earn a full year's revenue in year 1 and for revenues not to grow year on year. Nonetheless, our simple example serves to illustrate the main points.

There are more significant problems with simple payback analysis. First, the project with the shortest payback might not be best. Consider Figure 4.9. A project that pays back in less than 2 years has much less cumulative cash after 5 years than a project that takes longer to payback. It is a function of the rate of growth of the revenue and positive net cash.

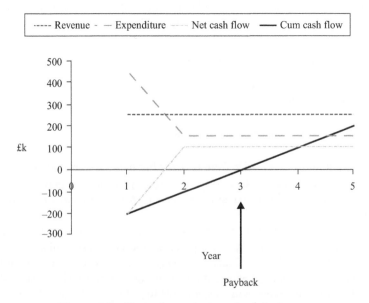

Figure 4.8   Example project payback in 3 years

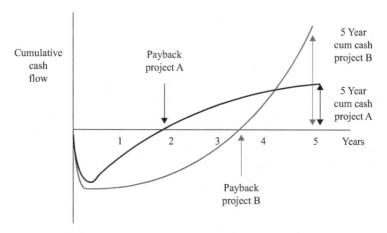

Figure 4.9 Comparative payback periods – the shortest payback may not be the best

The second reason that simple payback may be a misleading indicator is the time value of money. This is where we must consider applying discounted cash flow analysis.

4.6.3 *Discounted cash flow (DCF) and net present value (NPV)*

When we come to think about investment appraisal we find that money changes its worth over time! A pound today is worth more than a pound tomorrow. If we think about money in a deposit account, we would expect a pound invested today to be worth more in 10 year time. Similarly, given the choice, we will pay bills later, rather than earlier, because a pound is of less value in a year than it is today. One reason for this is that we can earn interest on our deposit (and if its compound interest we earn interest, not just on our initial sum, but also on the interest). It goes without saying that if we can invest £100 today and earn £105 by year 2, then £105 in 2 years' time is worth only £100 at today's value.

But there is a second reason and that is risk. The further out we look the greater the risk; and the more uncertain the project the less reliance we will put on the worth of our money by the end. Money put on deposit in a bank is at little risk, though even there some risk applies. But money invested in a commercial venture is certainly at risk; one just has to be aware that the majority of new start-ups fail.

The *time value of money* is therefore a function of the length of the project; what our money would earn in a risk-free investment; and risk involved in the project.

If we consider the typical investment project there is a major investment in year 1 (or even before the project starts (year0)), but the major revenues don't start flowing until years 2, 3 or later. Our cash expenditure will be at today's value, but our earnings will come later when they are of lower value. We, therefore, need to discount future cash flows to allow for the *time value of money*. We do this by applying an appropriate discount factor to each year's net cash flow where the discount factor is calculated as:

$$1/(1 + \text{TDR})^n$$

and where TDR = test discount rate and n = the year in question.

Clearly, the test discount rate is an important factor in this calculation and will represent a target rate of return that reflects both the cost of capital of the company in question and the company's appetite for risk. The TDR will normally be set by the finance department and should be applied to all project appraisals. It should aim to ensure a reasonable return after discounting for an acceptable level of risk, both the risk associated with the industry in question and the company's specific appetite for risk. It should, in any case, always exceed the weighted average cost of capital (WACC) for the company (see Box 2.5).

There are two further considerations in appraising a proposal – the life of the project and any terminal value that it may have. In considering project lives we must assess market needs (a mobile phone or tablet loses market attractiveness in 18 months or 2 years at most) and also technical obsolescence (will Wi-Fi become obsolescent, once 4G is launched?). But we should also bear in mind the difficulty of forecasting costs and revenues in 10 or 20 years. In any case, discounting ensures future years have less and less impact. Overall, therefore, it is prudent to keep lives relatively short. Similarly, a project may have some value once the product is withdrawn from the market (either as an asset to sell to another company or as scrap value). Nonetheless, such *terminal values* are difficult to assess and will be relatively small once discounted, so prudence would suggest disregarding them or setting them very low.

We are now in a position to calculate the value of the whole project, expressed in today's values. We do this by progressively discounting each year's cash flow (positive and negative) and accumulate then into a single value expressed in today's value of money. This is the net present value or NPV of the project. This calculation is shown diagrammatically in Figure 4.10, which shows the reduced values of equal annual incomes during the costing period.

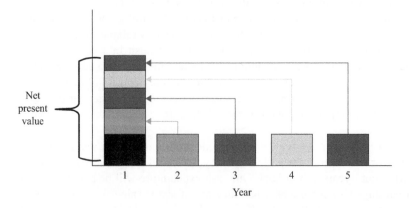

Figure 4.10   Calculation of net present value

Discounted cash flow statement (£m)					
	Year 1	Year 2	Year 3	Year 4	Year 5
Cash in (revenue)	250	250	250	250	250
Capital expenditure	300				
Operational expenditure	150	150	150	150	150
Net cash flow	–200	100	100	100	100
Test discount rate (TDR)	10%				
Discount factor	1.00	0.91	0.83	0.75	0.68
Net DCF each year	–200	91	83	75	68
Cumulative DCF: year end	–200	–109	–26	49	117
Net present value (NPV)	117				

Figure 4.11   Example project NPV = £117 million

We can now return to our example project (Figure 4.11) and note that we have a net cash outflow of £200 million in year 1. This is at today's value. However, in every subsequent year we have forecast a positive net cash flow of £100 million. In terms of the time value of money, each £100 million becomes less valuable as time goes on. We, therefore, need to discount that positive cash inflow to allow for time, interest and risk. We do this by applying the discount factor described earlier. In our example we have used a project life of 5 years, a TDR of 10% and we have disregarded any terminal value. So we see that in our example project £100 million positive cash flow in year 2 is worth only £91 million at today's prices and £68 million by the final year of the project. Therefore, to find the true worth of our project we have to find the sum of the discounted cash flows, over the whole life of the project. This sum is the net present value (NPV) of the project and must be positive to make the project worth investing in. If NPV turns out negative it implies that the risk of the project is too great and that returns will not cover the cost of borrowing to fund the project (or the notional cost of borrowing if we are financing it from company reserves). In the case of our example the NPV is positive (£117 million) and therefore we are likely to go ahead unless there is a project that achieves similar aims with a higher NPV (see Internal Rate of Return (IRR) analysis below) [5].

4.6.4   Discounted payback

Now that we understand the *time value of money*, we can also understand the limitations of calculating a simple payback period (see section 4.6.2). We should always discount future earnings for the *time value of money*; when we do the payback period will invariably be longer because we are placing less reliance on future earnings (see Figure 4.12).

4.6.5   Internal rate of return (IRR)

The IRR of a project is the rate of return that ensures that the project has an NPV of exactly zero. As we have seen previously, NPV is the total worth of a project expressed at today's value and allowing for a certain amount of risk. The amount

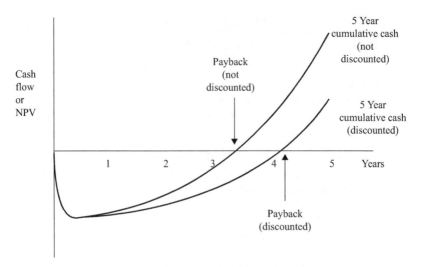

Figure 4.12   Discounted pay-back period

of risk considered acceptable is reflected in the level of test discount rate chosen. A high TDR allows for a lot of risk; a low TDR assumes little risk.

But there are circumstances when we want to know the *maximum* level of risk that we can anticipate and still cash remains positive. This is particularly true when comparing a number of projects that achieve similar aims but with different investment profiles. Some projects require high upfront investment but start to earn revenues early. Others might require less early investment but then earn modest returns, but for many years.

To find the IRR of each comparison project we keep increasing the TDR and re-calculating the NPV until it equals zero. The IRR is therefore the highest value of TDR that can provide positive cash flow (strictly it is neutral cash-flow) over the whole life of the particular project; and it represents the highest level of risk that that project will bear. In a comparison, then, one would choose the project with the highest IRR, so long as it exceeded the company cost of capital.

On a practical note, many computer spreadsheets will calculate the IRR of any set of cash flow numbers. Alternatively, if you wish to calculate the IRR yourself, trial and error is laborious but probably the easiest way. In Figure 4.13 we compare our initial example project with another approach to achieving the same aims. As we have seen our initial example project has an NPV of £117 million if we use 10% TDR. However, if we recalculate using 35% TDR our NPV becomes zero. We can compare those results with an alternative project that has NPV of £200 million at TDR of 10% but a zero NPV at only 19% TDR. Therefore, although the alternative project appears to have a higher NPV it will bear less risk as time goes on. All things being equal then we would choose the first project [6].

4.6.6   Return on Capital Employed (ROCE)

Finally, we should consider a measure that relates to profit, rather that discounted cash flow. While return on capital employed (ROCE) doesn't allow for the *time*

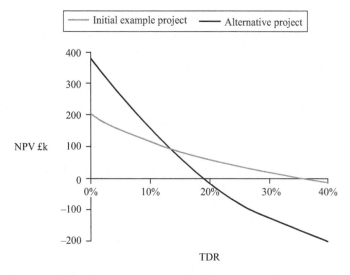

Figure 4.13   Comparing alternative projects using internal rate of return (IRR)

value of money it has other advantages. It is directly related to the profit that we hope to make and that will contribute to the future profitability of the whole company; and it takes into account the level of assets employed to earn that forecast profit. The formula for calculating ROCE is therefore:

$$\text{ROCE} = \frac{\text{Average Operating Profit}^*}{\text{Capital Employed}^{\times}}$$

*Average per year over the life of the project
× Total Assets less Current Liabilities

Once again it is essential that the project ROCE is greater than the company's cost of capital and that comparison is made with the industry average. Low-risk industries would expect a relatively low forecast ROCE; a high-risk undertaking would seek an appropriately higher ROCE.

4.6.7   Investment appraisal – summary

We have seen the importance of rigorous investment appraisal for any project or business case. Well executed investment appraisal increases the likelihood of investing in projects that offer acceptable returns and low levels of risk. It avoids investing in projects that are likely to fail financially and it provides a means of choosing between competing projects. Investment appraisal is an indispensable part of any business case.

4.7   Risk management and sensitivity analysis

A further vital part of project management (see also the appendix) and the development of business cases is the identification and management of risk. No project is without risk; the key is to identify, quantify and mitigate the likely risks – as far as that is possible. ICT is especially subject to risk given the cutting-edge nature of the techno-logy; the significant up-front investment involved; and the speed of change in the marketplace. Nonetheless, risks that are identified can normally be mitigated. A list of possible risks would include the following, though it is by no means comprehensive:

- The technology doesn't work, runs late or doesn't deliver the expected benefits.
- The project runs late for other reasons, such as shortage of skilled resources, etc.
- The macroeconomic environment moves unfavourably.
- Costs exceed forecast.
- Market requirements/expectations change.
- Demand for the service doesn't meet forecast or takes time to develop.
- Competitive response (e.g. by reducing price or adding functionality).
- Government policy changes e.g. in terms of regulatory requirement.

Each of these risks can be mitigated to some degree and the mitigation should be reflected in the project plan and business case. For example, the risk to costs can be mitigated by setting fixed-price contracts, but they may be more expensive. The risk to technology not working as specified can be mitigated by greater resources being devoted to pre-launch testing, but the extra time and costs will impact the overall project finances. The biggest risk with the deployment of new technology is the potential delay in it being available (new equipment launch dates from the suppliers can be unreliable). Any delay in equipment being delivered by the supplier will cause slippage in the start date of the product, and hence a reduction of revenues falling within the costing period. Unfortunately, mitigation of this factor is difficult, though a strategy of using more than one supplier of similar equipment can help.

Given the wide range of risks, which may occur singly or in combination, there are numerous possible outcomes for any project plan. *Sensitivity analysis* is one useful technique to summarise the financial effects of all these possible outcomes. We have seen the importance of NPV as a way of judging the value of a project after discounting a certain level of risk. We can therefore calculate the degree of change that will take our project NPV to zero (i.e. to the point where the project return is likely to just cover the cost of capital after allowing for risk).

For example, if we consider our example project, we can bear an overall reduction in forecast revenue of 11% before the NPV becomes negative. This revenue reduction might occur through changed market expectations, competitive response or a delay in launching the service. Similarly we might bear a 39% increase in capital costs or a 19% increase in current account expenditure. Obviously, a number of these risks might occur in combination; we must choose a representative set of assumptions to test. This approach to sensitivity analyis is shown diagrammatically in Figure 4.14.

An alternative method of assessing the overall risk of a project is to compare the best (most likely) forecast with possible alternatives. It would be usual to include both optimistic and pessimistic variants. Each scenario would include a set of assumptions about all the key risks; a range of investment appraisal measures would be calculated and assessed. Figure 4.15 shows the approach dia-grammatically, using discounted payback as the most appropriate comparator.

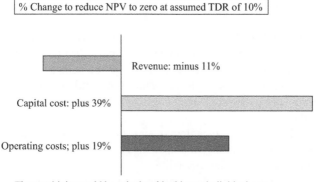

% Change to reduce NPV to zero at assumed TDR of 10%

Revenue: minus 11%

Capital cost: plus 39%

Operating costs; plus 19%

The sensitivity could be calculated looking at individual cost
or revenue assumptions or combinations

Figure 4.14   Project sensitivity to changing assumptions

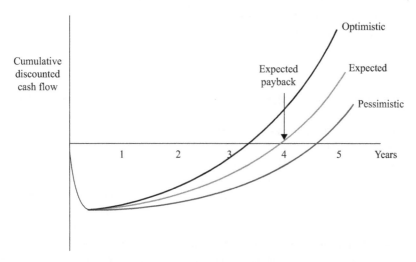

Figure 4.15   Overall project sensitivity

Despite these approaches to risk and project sensitivity, there is no certainty in this world. These approaches can only ever give a little additional confidence about adopting a certain course of action – or the contrary of course. If forecast revenue can be halved and the NPV still remains positive, albeit at a lower value, then we can feel confident to go ahead. If a few minor changes to our assumptions (perhaps reflecting a less optimistic view of market take-up and technical trials) double the payback period perhaps we shouldn't go ahead. In any case we will want to plan mitigation action to ameliorate any risk that we encounter [7].

4.8   Summary

In this chapter we first looked at the formal, external financial accounts of a business – how they are prepared and what they mean. We then looked at how the accounts might be analysed to determine how well the company is managed today and what its future prospects might be. Many stakeholders are interested in such matters, but they are particularly relevant to current and potential investors i.e. to the stock market.

The focus then shifted to the internal company aspects of finance, starting with financial and business planning; the challenges associated with capital planning; and the authorisation of capital expenditure.

Finally we explored the development and purpose of business cases and gave particular attention to investment appraisal and the assessment and management of risk.

Our aim here was not to provide a comprehensive textbook explanation of financial accounts, or to replace your friendly accountant! It was to give a level of financial understanding and awareness appropriate to help non-financial managers in the ICT industry operate more successfully in whatever field of endeavour they choose to pursue. Hopefully we have succeeded.

References

[1]   Myddelton, D.R. 'Managing Business Finance' [Paperback], Pearson Education, 2000, Chapter 1.

[2]   Myddelton, D.R. 'Managing Business Finance' [Paperback], Pearson Education, 2000, pp. 77–79.

[3]   *Business Standard*. 'Bharti, Ericsson in $400m network outsourcing deal', New Delhi, 6 August 2015, http://www.business-standard.com/article/companies/bharti-ericsson-in-400m-network-outsourcing-deal-104021001110_1.html

[4]   *The Economic Times*. 'Ericsson wins Airtel's $1.3Bn Outsourcing Project', New Delhi, 6 August 2015, http://articles.economictimes.indiatimes.com/2010-04-01/news/28469946_1_network-outsourcing-maintenance-and-management-contract-ericsson

[5]   Myddelton, D.R. 'Managing Business Finance' [Paperback], Pearson Education, 2000, pp. 59–62.

[6]   Myddelton, D.R. 'Managing Business Finance' [Paperback], Pearson Education, 2000, p. 63.

[7]   Myddelton, D.R. 'Managing Business Finance' [Paperback], Pearson Education, 2000, Chapter 8.

Bibliography

Jeter, L.W. 'Disconnected: Deceit and Betrayal at WorldCom', Wiley, 2003.

Chapter 5

Network economics

5.1   Introduction

Network operators have to invest significant amount of capital expenditure (CapEx) to build and extend their networks. For example, BT – the UK's incumbent operator – invested £2.4 Bn in its fixed network in 2013/14 [1] (see also Box 9.3). The challenge for the Telco is to build just sufficient network capacity to meet the expected demand for service by the customers in terms of volume and quality. Overprovision results in wasted investment by the Telco, while underprovision results in reduced quality or the inability to carry the offered traffic, with the consequential loss of potential revenue. In this chapter we consider the economics of the network in terms of its size, structure and the technology used, while the network strategy and planning activity necessary to achieve this optimum network are examined in Chapter 6.

   We begin by examining a simple model which shows the network as an economic system with inputs and outputs. This gives a basis for deriving a set of fundamental economic principles that are applicable to all types of tele-communications networks (and in some cases they even apply to other kinds of networks, e.g. the railways). Having developed an appreciation of the economic principles involved in the size and structure of networks, we explore the other major cost variable, namely the choice of technology. In particular, the current shift by most Telcos towards converting their existing networks to IP-based (packet switched) technology – often referred to as 'next generation networks' (NGN), is examined in detail. Finally, we draw together the economic components of a business case for NGNs.

5.2   The network as an economic system

5.2.1   The commercial model

The networks – fixed and mobile – represent the major asset of a Telco, being the basic infrastructure that enables it to provide telecommunications services, i.e., its main business. Our model in Figure 5.1 illustrates how a set of customer-demand drivers influences the coverage and capacity of the network which, subjected to the cost drivers, determine the size and structure of the network. The output of the system is a set of revenues and costs, which directly feed into the P&L of the company. This model is really an elaboration of the commercial model introduced in Chapter 1 (Figure 1.9), but it does enable us to examine the cost dynamics of the network in

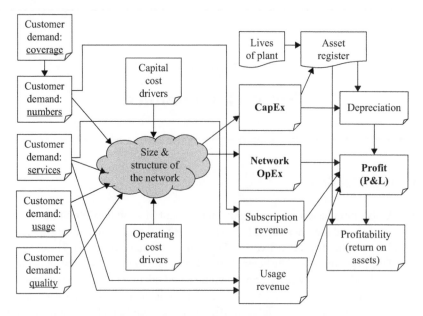

Figure 5.1   The network considered as an economic system

detail. First, we will work our way through all of the components of this system and examine the interactions between them.

Input factors that determine the size and structure of the network:

(a) *Customer demand: coverage.* The geographical extent of the network, i.e. its coverage, is a fundamental parameter for any Telco since it impacts on the customer population that can be served, and hence the revenue, and also effects the costs of the network build. In many instances the national regulator will impose a requirement for particular areas of the country to be served – e.g. with the awarding of the 3G and 4G mobile licences [2] – as discussed in Chapter 2. For the mobile operators the cost of coverage is determined by the deployment of the cell sites and the choice and siting of the antennas, whilst for the fixed network it is determined by the choice of the topology and type of technology used for the subscriber access lines. (Chapter 6 covers the planning of both mobile and fixed networks.) As the model shows, the coverage feeds into the number of customer (or 'subscribers') to be served.

(b) *Customer demand: numbers.* Both mobile and fixed networks need to be dimensioned to meet the expected customer connections – either as capacity in the cell sites or as numbers of subscriber access lines. In the case of the fixed network there are relatively high costs involved in terminating the subscribers' PSTN lines or the broadband lines at the Local Exchange. Thus, the numbers of subscribers served in an area affects the transmission and switching costs in the network.

(c) *Customer demand: services.* The design of capacity and functionality of the transmission and switching equipment in the network is directly influenced by the range of services that customers subscribe to. Examples of the services include various speeds of broadband, switched voice calls, leased lines, and virtual private networks (VPNs), all of which will determine the functionality and capacity required in the network.

(d) *Customer demand: usage.* The amount of usage of services by the subscribers directly determines the capacity and structure of the parts of the network that are sensitive to the volume of traffic flowing between network nodes. For example, the average number of calls during the busy hour period and the duration of those calls will set the required capacity of the route switches in the local exchanges of the PSTN, as well as the backhaul network and all the transmission and switching in the core network.

(e) *Customer demand: quality.* The quality that the customers require of their services will affect the network size and structure in two ways. First, it limits the extent to which cost economies can be achieved in the design of the network, e.g. the extent of contention and concentration; and second, it sets the level of robustness in the network and the degree of resilience in its design using duplication, meshing, alternative routeings and protection systems. (These factors are considered in Chapters 6 and 9.) In addition, the quality required for particular types of customers and services, particularly in terms of acceptable latency, will influence the network design and technology choice. On the other hand, increased or guaranteed levels of quality can attract customers in a competitive environment.

(f) *Capital cost drivers.* We will see further how the structure of the network is directly determined by the relative capital costs incurred in purchasing the transmission and switching equipment used to build, expand or enhance the network. These costs vary with the type of technology involved, quantities of equipment purchased (varying by volume discount and contractual arrangements) and the capacity of the equipment deployed (due to economies of scale). Other capital cost drivers include the costs of software needed to control the equipment and manpower costs involved in its planning and installation.

(g) *Operating cost drivers.* These cost drivers cover the manpower costs for the Telco's staff involved in running the network on a day-to-day basis, provision of service to customers, and may also include the charges for using third parties to maintain the network. Operating costs also include powering and cooling the equipment, lighting, maintenance of support infrastructure and buildings, running the vehicle fleet, etc. Chapter 9 looks in detail at the operations of a network.

Output factors from building and operating the network:

(h) *CapEx.* This covers all of the annual capital expenditures on building or enhancing the network – culminating in the CapEx programme, as described in Chapter 6. Although this capital has to be paid when incurred, for the purposes of the P&L account these expenditures are depreciated over the economic lives of the equipment or plant (known as 'lives of plant' or 'asset lives'), so that only a proportion of the capital costs are accounted for each year, shown as the depreciation cost in Figure 5.1. As described further, all the assets acquired with the capital expenditure are detailed in the company's asset register.

(i) *Network OpEx.* This represents the annual total of the operating costs incurred in running the network and providing customer service, as examined in Chapter 9 and Box 9.3.

(j) *Subscription and usage revenue.* For the purposes of this model, all the service revenues are decomposed into the subscription and usage categories. The subscription component covers all fixed rate payments for the service, including any quality options. The usage charge covers all the volume dependent income, including fixed price, threshold limited throughput or simple data volume and call charges. Of course, in practice the actual tariff format will vary greatly, depending on the price package being applied, including discounts, etc.

(k) *Profit and profitability.* The model in Figure 5.1 assumes that the P&L assessment of profit (or loss), excluding the non-network items, is derived from the sum of the revenues less the sum of the depreciation and Network OpEx. The profitability, in the form of 'return on assets', is calculated using the sum of the net assets, as taken from the asset register.

5.2.2   Network assets

In accountancy terms, the general definition of an asset is an entity that has some future revenue-earning potential. This definition is important because it enables the often considerable cost of an acquired asset to be spread over its lifetime, rather than having to show the outlay all attributed in one year's P&L accounts. Given the high cost of investment in networks and the long-term nature of the telecommunications business, it is not surprising that a great deal of care is taken by Telcos in how they assess the worth of their network assets.

Interestingly, although there are some (national and international) mandatory rules that define how assets should be shown in the accounts assumptions made for the lives of network equipment and plant are deemed a management decision for the Telco. It is therefore necessary for the Telco to state clearly in its published annual report and accounts what assumptions are made concerning the lives of their network equipment, and hence the logic for the particular level of depreciation charges for that year. Failure to adhere to clearly-stated depreciation policy can lead to prosecution of the company's directors! (The CEO, CFO, and some other officers of WorldCom, a US Telco, went to prison because they authorised the shifting of operational expenditure into the capital account (spreading it over several years) in order to improve the (apparent) profitability, and so protect the company's share price – see also Box 4.1 [3].)

By definition, the asset life should be equal to the economic life of the telecommunications equipment or plant – i.e. the assumed time that the equipment can continue to provide service economically. The economic life will be limited by:

(a) The longevity of the design of the equipment and the technology used.
(b) The onset of obsolescence, due to better technologies becoming available.

This means that the economic life of a piece of equipment is usually less than its physical life. Therefore, the equipment may continue to be used in the network (and earning revenue) long after its economic life expires – in which case it is deemed to be fully depreciated (staying in the asset register with zero value) , and no more depreciation payments are made for that equipment in the P&L accounts. A network operator, therefore, has to decide what lives it should apply to particular types of equipment and plant being purchased and used in the network. Usually, these decisions are made by a committee of engineers and accountants. Of course, the committee's views on suitable lives will change with time, as technology progresses, and the replacement or upgrade plans by the Telco will inevitably influence their decision on the onset of obsolescence. Some typical examples of asset lives of network equipment and plant are given in Table 5.1.

As indicated earlier, the original policy on the appropriate life to assign a type of equipment may need to be changed later if there are unexpected developments that materially bring forward its obsolescence. So, e.g., if equipment type A costing £20,000 had been given a 10-year economic life the Telco would expect to allocate a depreciation payment of £2,000 in the P&L account for each of 10 years. However, if equipment type B using more cost-effective technology becomes available in year 6, the Telco may wish to replace the existing Type A equipment with the better new alternative. There are then two accounting options to follow: either keep the type A equipment and continue to pay the £2,000 depreciation charge for each of the remaining 4 years, as well as the depreciation for newly purchased Type B

Table 5.1   Some typical telecommunications lives of plant

Type	Equipment or plant	Economic life
Digital TDM circuit-switched fixed exchanges, as used in the PSTN, Centrex and other business and corporate networks	Equipment	Originally 10 years
Digital TDM circuit-switched mobile exchanges, as used in GSM, 3G, 4G (with circuit-switched fallback) networks	Equipment	Originally 10 years
Digital packet-switched systems, as used in GPRS (2½G), 3G, 4G, mobile networks	Equipment	10 years
Digital packet-switched systems, as used in NGN fixed networks	Equipment	10 years
Copper cable, as used in the Access Network of the PSTN	Plant	Originally 20 years
Optical fibre, as used in the Access, Backhaul and Core Transmission Networks	Plant	10 to 15 years
Optical fibre, as sued in submarine cable systems	Plant	20 years
Duct in the ground, as used in the Access, Junction, Backhaul and Core Transmission Networks	Plant	30 years
Wireless masts	Plant	10 to 15 years
Operational buildings	Plant	50 years
SDH multiplexors	Equipment	7 years
IP routers	Equipment	2 years
Computer servers	Equipment	2 years

equipment; or discharge the remaining £8,000 of depreciation charges for Type A equipment in one year's P&L accounts – known in accounting terms as a 'write off'. The effect of a write-off is to have a once-only hit on the profits for the company for that year. Usually, the management of the Telco will make a strong public statement to the investment community explaining that the need to recognise the advantages of the new technology, etc., so that the adverse investor reaction to the loss in profits that year can be pre-empted. Box 5.1 gives an example of the effect of the right off for copper cable investment due to replacement by optical fibres.

Given the range of economic lives shown in Table 5.1 and amount of capital spent on the network, it is interesting to consider composition of the asset base of a Telco. Figure 5.3 illustrates the relative proportions of the network asset base for a typical PSTN (public switched telephone network). We can see that some 35% of the total asset base is in the links between the subscribers' premises and the Local Exchange (LE), of which 33% is in the actual copper cables, and 40% is in the underground cable ducts. However, 40% of the total is in the biggest asset component – the LE, of which 54% is invested in the concentrator switches terminating the subscribers' lines. The rest of the LE asset base includes just 7% for the group switches and 12% for the exchange-control processors. It is important to note that over three quarters of the investment in LEs relates to subscriber-line connections, as shown by the breakdowns in Figure 5.3. In fact, it is generally true that about 75% of the cost of any telecommunications electronic equipment is due to line termination (usually known as 'line cards') – whether it be multiplexors, circuit switches, data routers, broadband access line equipment, etc.

Also in Figure 5.3 it is interesting to note that just 15% of the total asset base is in the transmission links between the local and trunk exchanges (known as the Junction or Backhaul Network), and a tiny 3% of the assets are in the trunk exchanges. The conclusion of this analysis is that three quarters of the capital investment in a PSTN is in the subscribers' line and LE, and that the trunk portion of a national network, despite the fact that it is long distance, represents a small proportion of the total asset base. We look at why the presence of subscriber's lines makes the LE so much more expensive than the (subscriber-free) trunk exchanges later in this chapter.

5.3   Network structure

In this section we consider how economics influences the size and structure of any type of telecommunications network, whether it be a fixed PSTN, mobile, data or Cable TV network. We begin by looking at the basic network structure of access, points of aggregation and the core network.

5.3.1   Access, points of aggregation and the core network

The basic economic premise of a telecommunication network is that the capacity provided by the operator is shared among many users, thus enabling the economies of scale to be exploited and the benefits made available to all the network's subscribers. This means that there are not normally direct transmission links

Box 5.1   Example of write-off of the copper cable depreciation

Figure 5.2, covering several years, shows the depreciation charge on copper cable as the dotted area is equal to the area under the copper cable capital spent profile. Copper cable was originally assumed to have a life in the

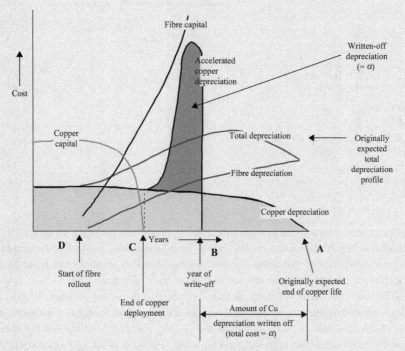

Figure 5.2   Example of write-off of copper cable depreciation [Reproduced with permission from K.E. Ward]

network up until point A. Within the economic lifetime of the purchased copper cable the Telco decided to deploy optical fibre instead, starting at point D. The capital spent profile and its depreciation are shown in the figure, as is the originally expected total depreciation charge of the copper plus the optical fibre. Because the copper cable had then become obsolescent it was decided to cease its deployment at point C and write-off the chunk of copper cable depreciation covering period A-B, total cost $= \alpha$), as shown by the vertical shaded area at time point B.

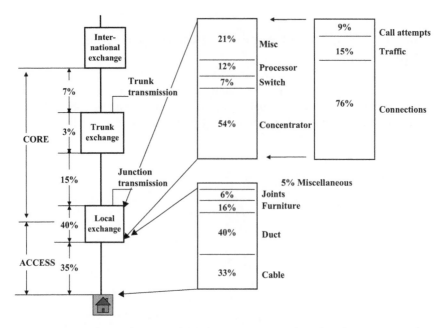

Figure 5.3   Analysis of a typical PSTN assets [Reproduced with permission from K.E. Ward]

between users, but rather they are linked to a parent hub from which connectivity is available to all other users. In the case of fixed networks this link is dedicated to each susbscriber/user, while for the mobile case the link is made available by the allocation of free radio channels for the duration of the call.

The interconnection of all these hubs is provided by a core network. Within the hubs is an important boundary – known as the aggregation point – between the dedicated transmission capacity of the user's (or subscriber's) link and the shared capacity leaving the hub onto the core network. The role of an aggregation node is to provide a saving in the capacity of the core network (backhaul, core switching and transmission) by exploiting the fact that the users are active for only a small proportion of the time. This is true not only for fixed or mobile telephony users, but also for people accessing websites, which tends to be in short bursts of activity. Therefore, by collecting – i.e. aggregating – just the active traffic from a population of users served by the access network, the core part of the network needs to have just enough capacity to carry the maximum aggregated traffic flow. The aggregation ratio, Ar, is the proportion of total access capacity entering the aggregation node divided by the core network capacity leaving it. The result of aggregation is that the loading on the core network channels is around 60–70%, compared to the typical average of 5% occupancy of each access line. This improvement in loading significantly improves the network economics, given the long distances between nodes in the Core Network (which usually spreads across the whole country).

The larger the Ar, the greater is the saving in the cost of the Core Network. However, since the output capacity of the aggregation node is less than the input

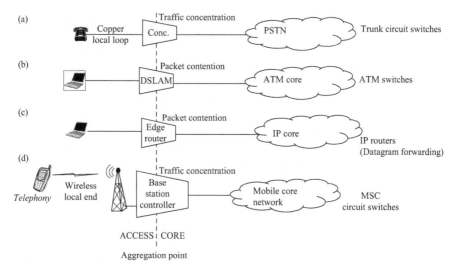

*Figure 5.4   Aggregation points for telecommunications networks: (a) Telephony;
(b) ADSL broadband; (c) IP network and (d) Mobile (e.g. GSM)*

capacity, larger values of *Ar* will have increasing likelihood of peak levels of traffic from the users being blocked (in the cases of circuit switching) or delayed (in the cases of packet switching). Thus, in setting the value of *Ar* the Telco has to balance the network cost savings on the one hand against the quality of service offered to the users on the other. It is a constant challenge for Telcos to balance the necessary network equipment and operational cost savings versus the quality offered to the customers. In this chapter we consider mainly the capital cost of building the networks, while Chapter 9 addresses the operational costs and the current account for the Telco.

The aggregation nodes for the different types of network are shown in Figure 5.4.

(a) For the telephony PSTN case, the aggregation is provided by the concentrator switch in the LE. Typical values of *Ar* for telephony circuit switching, also known as the concentration ratio, are around 10:1. This means that at any one time only 10% of the access lines can send or receive calls. Any subscribers attempting to set up a call when all outlets from the concentrator switch block are occupied will not receive dial tone – fortunately, this is now a rare occurrence for subscribers in the United Kingdom.

(b) In a broadband network using ADSL over copper pairs, the aggregation is provided by the DSLAM (digital subscriber line access module) which is located in the local exchange. Typical values of *Ar* for packet aggregation, also known as the contention ratio, are between 50:1 and 20:1. Unlike the circuit-switched case, any broadband lines generating data traffic during the busy period beyond the capacity of the outlet from the DSLAM will incur a delay and slowing of the packet rate – a common experience for many broadband users in the United Kingdom.

(c)   For an IP network the user lines terminate on the edge IP router, which is the aggregation node – with less output than input capacity. Typical values of *Ar*, the packet contention ratio, are around 20:1. As with the broadband case, the packets are delayed (and in the extreme situation, lost) in the event of the outlet capacity from the edge router being full. However, people's usage of data services is rapidly changing. Of particular importance to network operators is the increasing use of IP networks for music and TV programme streaming with the consequential increase in occupancy times – necessitating the operators to use much lower values of Ar in order to provide adequate quality of service.

(d)   For the mobile case, the aggregation is provided by the base station controller, BTC, (or equivalent), which allocates the send /receive radio channels to the appropriate cell for the duration of the call or session – there is no concentrator switch as in the case of the PSTN. However, since the mobile voice calls are circuit switched at the MSC, any call attempts when all radio channels in the cell are occupied will be denied. Moreover, if a mobile handset engaged in an existing call moves into the fully occupied cell the call cannot be handed over from the neighbouring cell and it will consequently be dropped. The values of Ar (concentration ratios) for voice in mobile networks are usually higher than for the PSTN. The packet traffic over mobile networks also experiences delay and packet loss in the event of any traffic overload due to aggregation, as described for the (c) IP network earlier.

5.3.2   Links, nodes and hierarchies

All types of networks are made of links and nodes. In the case of telecommunications networks (fixed and mobile) the links provide transmission across their length, and the nodes provide all the service functions like the switching for a voice call, routeing of packets for broadband web access and other data services. Nodes also provide service functions like computer servers for web hosting, back-up and cloud services. The optimum structure of a telecommunications network is set by achieving the minimum capital cost, within the constraints of any technical limitations of the equipment used. We can identify two general rules for determining this economic optimum network.

> *Rule No. 1*: The optimum network design is driven by the economics of balancing transmission costs against nodal (e.g. switching or routeing) costs.
> *Rule No. 2*: The direct interconnection of *n* nodes requires $n/2(n-1)$ links between the nodes – a relationship known as 'Metcalfe's Law'.

The application of these two rules determines whether the necessary connectivity between any two points in the network should be achieved by direct links or indirectly via intermediate links and nodes. This is true for all forms of telecommunications networks (voice and data) including the Internet. As an example, we will now apply these rules to determine the structure of a PSTN, noting that all other networks follow similar principles.

As described earlier, telephone users in the PSTN are connected by subscriber lines over the access network to the LE, which is the edge node of the switched network. The number and location of the LEs required to serve a community of subscriber lines is determined by the economics of the disposition of these lines and topography of the area on the one hand, and the transmission limitation on the maximum length of lines on the other. (A similar limit is set in the cell size of a mobile network.) Thus, a PSTN is essentially a collection of catchment areas of subscriber lines each containing an LE – the economic optimum location of which is at the centre of gravity of the subscribers, so that the total cost of the all the lines (which is normally proportional to line lengths) is minimised. We look at the planning implications of this economic objective in Chapter 6, while we concentrate below on the economics determining the network structure.

(a) *Number and location of local exchanges*: Let's imagine a small town. We could serve all the telephones with a single large local exchange. The optimum location for this exchange is such that the sum of the access line costs (which are proportional to length) is a minimum – i.e. the exchange is located at the centre of gravity of the population of phones within the town. Not surprisingly, this means that telephone exchanges are usually located in the town centre. In practice, the exchange may need to be sited away from the optimum centre due to a lack of suitable building space (or even the presence of a pond just where you want to place the exchange) – the extra CapEx incurred is known as the 'out-of-centre costs'.

For larger towns, the distance – and hence the cost – of access lines to the outlying subscribers will become high enough to justify the deployment of two or more LEs to serve the town's population. As before, the optimum number of LEs is determined from the trade-off of switching versus transmission costs (Rule 1). However, in addition to the total costs of subscriber access lines, which will decrease with the number of exchanges, there is the new cost of providing the $n/2(n-1)$ links (known as 'Junction routes') between the n exchanges necessary to give full connectivity between all telephone lines (Rule 2). The optimum location of each the local exchanges will be based on minimising the total cost of the subscriber' access lines, as described above.

(b) *Number and location of trunk exchanges*: We can extend this logic to a region of a country comprising a large number of local exchanges (there are a total of about 6,500 in the United Kingdom). Again the network structure is determined by the trade-off between the transmission cost of the $n/2(n-1)$ junction links, which interconnect the n local exchanges, against the cost of the provision of one or more 'trunk' exchanges. If trunk exchanges can be justified within the region, each will act as a parent to the local exchanges within its catchment area. Each local exchange then needs only one link to its parent trunk exchange, since all other exchanges within the catchment area are accessible via the trunk exchange. Clearly, each of the T trunk exchanges needs to be connected to every other trunk exchange via a total of $T/2(T-1)$ trunk routes to provide full connectivity – and this cost needs to be included in the trade-off calculation. The optimum location of each the trunk exchanges will be based on minimising the total cost of the links between the local exchanges and their parent trunk exchanges.

(c) *Number and location of trunk transit exchanges*: Taking the logic one step
     further we can consider a network of trunk and local exchanges extending
     across the whole country. Again, network cost can be optimised by deploying a
     further level of switching – 'trunk transit' exchanges to replace the $T/2(T-1)$
     links between the trunk exchanges. The M trunk transit exchanges need a mesh
     of $M/2(M-1)$ links to provide full connectivity between LEs.

This process can continue with increasing number of switching layers until the
structure offering the optimum total network cost is reached. A simplified picture
of the resulting network structure is shown in Figure 5.5. This shows a network of
26 LEs parented onto six trunk exchanges and one trunk transit exchange. The
physical structure of Figure 5.5(a) can be redrawn as a traffic routeing hierarchy, as
shown in Figure 5.5(b). Our example has two levels of trunk switching hierarchy.
Typically, national PSTN networks have between one and four levels of trunk
switching within their traffic routeing hierarchy, depending on the size and geography of the country. Of course, due to Rule 1, whenever there is a major technology change in the network the number of nodes in the network and the number
of levels in the traffic routeing hierarchy will change in line with the new balance
between transmission and nodal costs. An example of this was in the 1980s when
BT converted its predominantly analogue PSTN with a core network of 399 nodes
and 4 hierarchical levels to one using digital switching and transmission with just
80 trunk exchanges and a 2-level hierarchy (see Box 5.2). A similar restructuring
can be expected when the existing digital TDM network is replaced by the IP
packet technology of an NGN. We consider the economic aspects of choice of
technology later in this chapter.

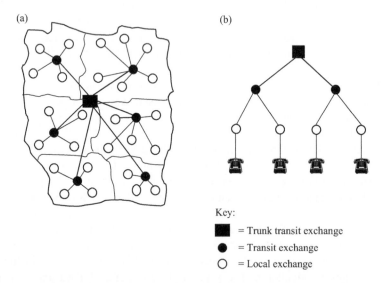

Key:
■ = Trunk transit exchange
● = Transit exchange
○ = Local exchange

*Figure 5.5   The PSTN structure in physical and routeing hierarchy form:
            (a) physical network structure and (b) equivalent traffic routeing
            hierarchy*

Box 5.2 The change of the UK trunk network structure due to the shift from analogue to digital (TDM) technology

Figure 5.6 gives a simplified view of the change in numbers of exchanges and the number of levels in the routeing hierarchy as a result of the complete replacement of the analogue technology by digital. By 1976 BT's PSTN network comprised a 3-tiered trunk network with a primary level of some 360 analogue (mainly Strowger) 2-wire switched Group Switching Centres (GSC) and an overlay Transit Network of 39 (crossbar) 4-wire switched exchanges – District and Main Switching Centres (DSC and MSC). In addition, there was a special arrangement for London (L/MSC and Special Purpose Unit). The transmission used in the trunk network was predominantly analogue (FDM systems) using 10pps 1VF and DC signalling in the GSC network, and high speed multi-frequency signalling in the Transit Network.

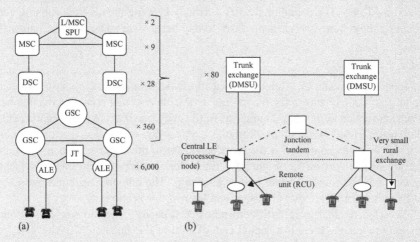

Figure 5.6   A simplified view of the two versions of BT's PSTN: (a) BT's analogue PSTN and (b) BT's digital (TDM) PSTN

In contrast the all-digital replacement PSTN – using TDM digital transmission and TDM digital switching – comprises just a single trunk level of about 80 fully interconnected digital main switching units (DMSU). There were also changes in the structure of the local switching network to take advantage of the introduction of large processors for call control. All local and trunk exchanges were replaced by the new digital systems (System X and AXE10) using SS7 inter-processor signalling by 1998.

For completeness, Figure 5.6 also shows the use of Junction Tandems (JT) in both the networks. This provides transit-switched connectivity between the many LEs within large urban areas (e.g. London and Birmingham) and so reduces the need for direct links.

Whilst the links forming the traffic routeing hierarchy are necessary to ensure that there is connectivity across the network between any two subscribers, further network economies can be obtained by setting up direct links between nodes between which there are high levels of traffic flow. These optional routes that by-pass the hierarchy are provided only where economically justified. We look at this and other planning aspects resulting from the application of Rule 1 and Rule 2 in Chapter 6.

5.4   Economics of network design

In this section we look at the economic factors affecting the design of a tele-communications network in terms of its structure and types of equipment, taking in account the required capacity and set of services, the coverage and terrain served and the disposition of its users. This view concentrates on the factors that determine capital costs, since these will represent the level of investment in building this vital asset for a Telco. The operational costs – namely keeping the network running and providing the required services on a day-to-day basis – are addressed in Chapter 9.

5.4.1   Principles

An economist's view of a telecommunications network is of a system with two capital cost components: a fixed cost, and a variable cost that relates to the number of subscribers or to the total amount of traffic they generate. We can capture this relationship in following simple formula: $C = A + Bx$, where C is the total capital cost; A is a fixed quantity; B is the cost coefficient of the variable cost; and x is the quantity of users or traffic. Figure 5.7 illustrates this formula in graphical form, showing the fixed and variable cost relationship. The components contributing to the fixed and variable costs are examined later in this section.

We can see that even with zero network capacity – i.e. no subscribers and hence no revenue – there is a capital cost of A. It is also apparent that the slope of the total cost line determines the cost of each added subscriber or unit of network capacity, x, which is known as the 'marginal cost'. This is an important parameter because it signifies the cost of adding one more subscriber or one more call, or

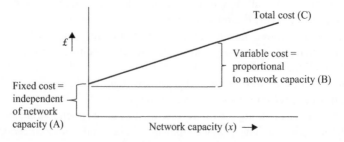

Figure 5.7   Fixed and variable costs of a telecommunications network

carrying one more megabyte, as appropriate. The key challenge for a Telco is to build a network which has a low fixed-to-variable cost ratio, so that the total unit cost (i.e. C/x) can approach the incremental cost. The network will then be in the optimal position of having a cost base that is proportional to volume of user/subscribers, i.e. its revenue-earning capacity. In short, the fixed costs of a telecommunications network should be minimised.

The next major economic principle for consideration is that of network, or platform (the combination of a network and its support systems) integration. Traditionally, individual networks support only a few services, in many cases only one. Therefore, a Telco will typically have several platforms each supporting just one or two services. Although this situation is recognised by all as being economically inefficient, since each platform has its own fixed cost (A), the limitations of the network technology and designs usually mean that it is easier and quicker for Telco's to introduce new services by deploying new platforms specifically for that service. There have been several attempts since the 1980s to overcome this problem by developing so-called 'integrated networks' (and platforms) that can replace the plethora of platforms to support a wide range of services. The most notable was the introduction of the integrated digital services network (ISDN) form of digital access in the 1980s, as described in Box 5.3.

Box 5.3   Integrated digital services network (ISDN)

ISDN is an internationally standardised form of integrated access from data terminals and analogue phones at the subscriber's premises carried over digital access lines (long before the current 'broadband' access systems were developed) to the digital (TDM) local exchange. This service, which was first introduced in the 1980s, provided 2 or 30 channels which could be used for data or voice (with signalling). The line rates are 144 kbit/s or 2 Mbit/s for the 2- or 30-channel system, respectively [5]. ISDN was marketed to subscribers as being 'integrated', with the benefit of reducing costs for the collection of services carried, and offering extra capacity with a degree of flexibility in the way it was used.

The principle of platform integration was also exploited in the core of the network with the introduction of the multi-service platform (MSP) concept by several Telcos in the 1990s [6]. The idea was that rather than build a new platform for each new service a single core platform supports all the services, with separate aggregation nodes and service-specific interfaces presented to the users at the edge of the MSP, as shown simplistically in Figure 5.8. It is interesting to note that the cost advantages of next generation networks, which we examine later in section 5.5.2, are primarily driven by the MSP nature of their architecture.

MSPs have been used to provide a range of leased-line (private circuits) and data services. Such platforms offer several important economic advantages for the

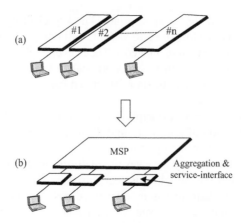

Figure 5.8   The multi-service platform (MSP) concept: (a) n separate dedicated platforms and (b) single multi = service platform equivalent

Telco. The first is that because the MSP is essentially a high-capacity simple data platform it can be constructed cheaply since the expensive interfacing and intelligence features are in the aggregation and service-interface units. Therefore, the MSP tends to have a relatively low fixed cost, with the total cost determined mainly by the volume variable component. Furthermore, the more services that the MSP supports the lower the apportioned cost is for each service – allowing the operator the chance of competitively pricing.

The second important economic characteristic of an MSP is that the system is relatively immune to fluctuations in the volumes of the individual services being carried. This is due to the aggregating nature of the MSP, where an unexpected increase in the traffics volumes in some of the services can be offset by a drop in some of the other services – often referred to as the 'swings and roundabouts' nature of an MSP. Given the difficulties in accurate forecasts of traffic, particularly for new services that do not have a historical trend to follow, there are significant cost advantages in having an MSP rather than individual platforms, each dimensioned to meet the forecast level of traffic for its specific service.

The third economic characteristic of an MSP is that it provides a ready platform, usually nationwide, that will allow the rapid introduction of new services. This can offer a big competitive advantage for the operator – in terms of cost and, importantly, time-to-market – compared to a new entrant who needs to rollout a dedicated network for that new service, with the consequent extra costs and deployment time involved.

Fourth, there should be significant economies in the Telco's operational costs in running a single MSP rather than a set of dedicated service platforms. These savings stem from the efficiencies of having just one set of maintenance procedures, tools, training programmes for the operational staff, a smaller range of spare parts and savings in levels of stores. There can also be significant capital savings from having a unified set of operational support systems, as described in Chapter 9.

We finish this principles' review by introducing one of the important measures of economic efficiency of a network, namely: the fill factor. In building a network the operator's objective is to establish sufficient capacity to be able to support the levels of telecommunications traffic. However, to ensure sufficiency when the forecasts of required capacity will undoubtable be uncertain operators need to include some spare margins in the dimensioning of the network build. In addition, the network plant and equipment tend to be constructed in specific sizes – e.g., optical fibre cables come in fixed sizes of 4, 8, 16, 32, 48, 96 fibres within the cable sheath – which means there is invariably inherent unused capacity in the network. The effect of both lower actual traffic from subscribers and the step-nature of equipment throughout the operators' networks creates an aggregated proportion of spare capacity, which can be viewed as capital overspend – sometimes referred to as the 'burden of spare plant (BSP)'. A measure of the economic efficiency of the total network asset is its fill factor, usually known as the network utilisation factor (NUF). Of course, the optimum NUF for a Telco's network is a trade-off between minimising the BSP versus the ability to meet customer demand from installed capacity. A typical NUF that provides the required comprise is around 75%.

5.4.2 Access network

The defining characteristic of the fixed Access Network is that a dedicated link is provided for all subscribers to the first node of the network. This means that irrespective of the amount of traffic generated or received by a subscriber a link must be provided. Unlike in the Core Network, there is no opportunity for cost saving by traffic aggregation of the access link, although some degree of transmission aggregation is often possible. Therefore, the fixed Access Network is always the most expensive part of a Telco's networks. It is for this reason that new entrants into a telecommunications market usually opt for the choice of interconnecting with the incumbent Telco to gain access to the subscribers, while providing their own Core Network. This situation also explains why there is a big debate around the World and with Telcos and national regulators involved in how to best provide the rollout of broadband access links for the consumer and small business market, as well as the sparsely populated rural areas of the country (see Chapter 2). A different situation applies in the case of a mobile (cellular) access network, which we discuss separately later in this subsection.

The Formula 1 introduced above in the macro context of the total network cost also applies in a modified form at the micro level to the individual transmission or switching systems used in a network. In the case of the transmission systems used in fixed Access Networks the formula becomes:

Line system cost, $C = A + B.d$, where A is the terminal cost; B is the cost per unit length, and d is the distance between terminals.

In the case of the copper local loop, as used to provide a fixed line telephony access circuit in the PSTN and cable TV networks, the terminal cost is low and so the Bd product dominates, making the overall cost essentially proportional to the length.

(However, in practice the local loop is usually made up of several lengths of different gauge copper cables, each having their own B factor, so Bd is equal to $B1d1 + B2d2$, and so on.) There are two further important cost components affecting the subscriber lines, namely: the external plant providing the infrastructure for the subscriber's line and the provision of multiplexing transmission capability. The range of cost components for the local line is summarised simply in Figure 5.9.

Considering Figure 5.9(a), the easiest example is that of the copper local loop mentioned earlier. In this case the terminal equipment at the subscriber's premises is simply a line socket box usually located close to the point where the external cable enters, and into which the subscriber plugs their own telephones. (This is the network terminating and test point, (NTTP) representing the end of the Telco's network [7].) There is a similarly simple low-cost termination at the exchange end where the line is terminated on a set of metal tags at the main distribution frame (MDF). However, of course, there is also some form of infrastructure to physically support the copper cable, known generically as 'external plant', the extent of which is shown in Figure 5.10, and this can add significant extra cost.

The type of external plant used depends on many cost factors, including: the physical terrain, the topography and the disposition of the population of served subscribers, as well as non-cost considerations such as the geo-type of the neighbourhood. Generally, the cheapest form of external plant is overhead suspension from 'telegraph' (sic) poles. Historically, all access cables were carried this way. However, the density of subscribers in urban areas and the centre of towns make this method unwieldy, as well as unsightly. In such cases, the copper cables are carried in buried ducts under the streets. This results in the situation shown in Figure 5.10, where underground external plant is used for the entire length of lines in the centre of town and urban areas, while in all other areas the more cost-optimal hybrid arrangement of overhead provision from premises to the serving point close to the street cabinets and thence underground to the exchange. Rural areas,

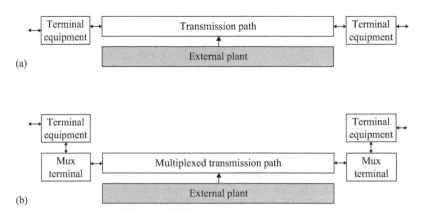

Figure 5.9   Cost components for a transmission system: (a) simple transmission system and (b) multiplexed transmission system

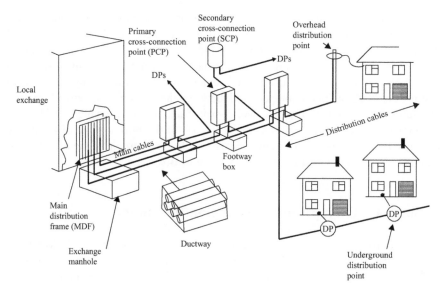

Figure 5.10   The external plant in a fixed access network

particularly if remote, have most of the length back to the exchange carried over-head. The final section of the local loop closest to the subscriber's premises is known as the distribution point (DP). In the overhead case the DP is a terminal block at the top of the pole serving 10 or so subscribers' drop wires, while the underground DP comprises a run of breakout joints in a multi-pair cable, each buried underground in a small tube serving the premises (see Figure 5.10).

The cost of digging up the streets and laying duct routes is considerable, due to the high level of civil engineering activity involved, therefore the ducts represent an important long-term investment for the Telco (typically 40% of the local line asset base, as shown in Figure 5.3 above). Normally, duct-ways with several bores are used so that many cables can be supported and that once laid they are available for the addition of new cables throughout their economic life of some 50 years or more.

The external plant also includes the items known as 'street furniture' – the cabinets and the footway boxes at the side of the road. These form an important economic role in the cost management of the copper access network. The main challenge is being able to meet demand for new subscriber circuits throughout the service area from pre-provided capacity, rather than reacting urgently to install new cables. Of course, this objective has to be balanced against not incurring too great a BSP. There are different approaches to this challenge around the World [8]. In Europe the problem is usually managed through the use of flexibility points – known in the United Kingdom as primary and secondary connection points (PCP and SCP), as shown in Figure 5.10. Sufficient copper cables are provided from the PCPs to serve all premises, including those not yet connected, within their catchment area (typically within a few streets) because of the short distances involved this over-provision is not too onerous, with NUFs of around 60%. Each PCP is connected to its local exchange by one or two large capacity copper cables (typically 100–4,800 pairs in one sheath), sufficient to meet the forecast aggregated capacity of the working connected pairs at the PCP cabinet – this capacity tends to be smooth and predictable and can be expanded in economic instalments, with a typical NUF of over 75%. A continuous copper circuit is provided by a technician visiting the PCP, cabinet to jumper a link between the 'distribution network' pair to the subscriber and the 'main network' pair back to the exchange. Additional economies are achieved in rural areas by the use of the further stage of flexibility provided by the SCPs, as shown in Figure 5.10 [9]. An example of the inventory of external plant in the United Kingdom is given in Box 5.4.

Box 5.4   Example of external plant inventory

The external plant in the Access Networks of most operators around the World is extensive and represents a large asset base. The following figures give an idea of the scale of external plant in the PSTN of the United Kingdom [9]:

- 48,000,000 pair km, 495,000 sheath Km of copper cable
- 3,000,000 manholes and footway joint boxes
- 186,000 bore km, 124,000 route km of duct
- 65,000 flexibility cabinets (primary and secondary)
- 2,500,000 distribution points (DP)
- 4,500,000 telegraph poles

Now considering Figure 5.10(b), which shows the costs components for a multiplexed line system. This category in the fixed access network includes ISDN, leased lines and the range of copper-and-optical fibre broadband access systems. The costs for this category are complicated by the fact that two or more access circuits are carried as 'channels' on the multiplexed system. Looking at the cost for one circuit, we have the terminal cost for a channel (A_c) and the share of the multiplex system cost for distance d_m, with its terminal and line equipment ($A_m + B_m.d_m$). Thus, the total cost for the access line is $A_c + A_m + B_md_m$, plus the share of the external line plant cost. As we have discussed earlier, the external line plant cost can be significant in the case where new underground duct or an overhead pole route needs to be installed specifically for this new circuit. However, it is important to note that the cost of the channel-termination equipment (typically manifested as a 'line card') is usually the predominant cost component in the total circuit cost – a factor we will return to later in this section.

Although the copper Access Network was originally constructed as part of the PSTN, dimensioned and structured on the economic basis of providing telephony service, it has since the 1980s increasingly taken on the additional roles of supporting digital leased line, data and broadband Internet access services. For convenience, Figure 5.11 summarises the various options for providing broadband

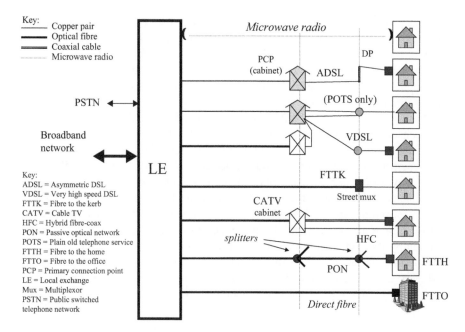

Figure 5.11   Summary of options for delivering broadband access

access [4]. Although potentially all these options are available to a Telco, in practice only a few are adopted by any individual operator. For example, Cable TV operators use only the hybrid fibre coax (HFC) system with cable modem (the latter not shown in Figure 5.11), most Telcos use ADSL, VDSL, FTTH and PON, some Telcos also use point-to-point microwave radio (e.g. WiMAX), particularly in the rural areas. The choice of which system to deploy is primarily based on their economic characteristics, the geo-type of the served area, whether there is existing equipment and external plant, etc., as well as the specification of the services being delivered. The value of this service, in terms of revenue generated, may also be a factor.

From Figure 5.11 it will be noticed that the architecture of the broadband systems ranges from all coper (e.g. ADSL) through varying combinations of optical fibre and copper to all fibre – with the cost and speed capability increasing correspondingly. However, the economics of the broadband access systems is complex and not simply dependent on the proportions of copper and fibre they contain. The unit costs characteristics of these systems show rapid reduction from increasing penetration, i.e. the proportion of premises passed that are connected to the network. Acceptable unit costs are not reached until the penetration reaches around 30%. Thus, many alternate operators, e.g. the cable TV companies, will deploy their HFC networks only in the streets where they expect to gain customers from at least 30% of the premises passed – hence creating patchy service-availability areas around the country.

The penetration-dependency is driven by the costs of the channel terminal equipment at the customer end (see Figure 5.9[b]). It is therefore the number of potential connections accessible within the reach of the channel termination, which is set by the location of the channel equipment – in the middle of the street, at the curb edge or in the basement of office blocks or in single occupation premises – that determines the unit-cost characteristics.

A study made by Analysis Mason for the Broadband UK Organisation shows in a dramatic way how the geo-type of the territory affects the deployment cost of broadband access systems [10]. According to this study the cost of deploying fibre to the cabinet (i.e. VDSL) broadband access starts to increase more rapidly as the rollout moves into more sparsely populated areas, with longer line lengths back to the exchange. For example, the cost of covering the 10% of the population located in the most remote areas of the United Kingdom would cost as much as covering the first 35% of the population (who are in the urban areas). This cost characteristic is driven by the dominant cost of external plant necessary to carry the broadband transmission system, as well as the influence of the spare number of premises that can be served from each cabinet. The regulatory and political issue associated with the steep cost curve are discussed in Chapter 2.

5.4.3   Circuit-switching traffic efficiency

We continue our review of the economics of network design by examining the efficiencies of circuit switching. These are important not only for the cost optimisation of the fixed PSTN and other networks, but also for the radio-channel allocation and voice switching within the mobile networks.

Let's begin by considering how to determine the required capacity of the switch-blocks in the local exchange to serve a given population of users. The switch block can be considered as a system that offers a number of servers to handle the telephone calls. Since calls originate or terminate at different times and the duration of the calls is variable, a statistical formula devised by A.K. Erlang (a Danish mathematician) is used to dimension the circuit-switched networks. His formula (known as Erlang B) defines the relationship between the number of servers (e.g. channels through a switchblock), the amount of traffic carried and the probability of calls not being handled (i.e. lost), known as the 'grade of service' (GOS). The amount of traffic switched is measured in units of Erlangs (E), where 1 Erlang is defined as one circuit occupied for one hour. The important characteristic of this formula is that there is a non-linear relationship between the traffic carried and the number of switch channels (servers). This means that it is more cost efficient to have a network of a few large exchanges rather than one of many small exchanges.

Table 5.2 shows the application of Erlang's formula for different values of GOS. The non-linearity is clearly apparent, e.g. for a GOS of 1 lost call per 1,000, 5 circuits can carry 0.76 E (i.e., 0.152 E per circuit), whereas 50 circuits can carry 32.51 E (i.e., 0.65 E per circuit) – so the loading on the large switch-block is 65% compared to the 15% loading on the smaller switch-block. The other characteristic

Table 5.2   Circuit switching traffic efficiency

No. of circuits	Erlang capacity with GOS				
	1:1000	1:500	1:200	1:100	1:50
1	0.001	0.002	0.005	0.01	0.02
5	0.76	0.90	1.13	1.36	1.66
10	3.09	3.43	3.96	4.46	5.08
20	10.11	10.79	11.86	12.8	14.0
50	32.51	33.88	35.98	37.9	40.3
100	75.24	77.47	80.91	84.1	87.6

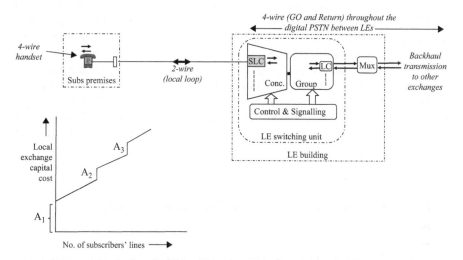

Figure 5.12   PSTN subscriber's line and the main economic elements at the local exchange, together with the LE capital cost characteristic

illustrated in Table 5.2 is that higher loadings per circuit are possible if higher probabilities of congestion (i.e. worsening GOS) can be tolerated. For example at a GOS of 1 lost call in 50, the 5-circuit switch can carry 1.66 E, i.e. a loading of 0.332 E per circuit, over double that for a GOS of 1:1,000. This trade-off between quality of service for the user and the cost to the operator is a constant theme in managing the network business, and is a major determinant in the design and dimensioning of all telecommunications networks.

We next need to consider the cost elements of a circuit-switched exchange, starting with the local exchange (LE). A simplified view of the subscriber's line and the local exchange showing the main cost components is given in Figure 5.12. Operators are continually trying to find ways of reducing the capital and operational costs of this part of the network, since even a small improvement will be magnified by the huge numbers of subscribers to create a worthwhile impact on its cost base. One of the earliest and probably the greatest savings was the use of a device to convert the inherently 4-wire nature of a telephone circuit to 2-wires – i.e. a pair – for transmission to the local exchange. As Figure 5.15 shows, the telephone handset in the subscriber's premises has 4 wires: one pair carrying the transmission in the Go direction from the microphone, the other carrying the received (Return) transmission to the earpiece. The 4-to-2 wire conversion device (known as a 'hybrid transformer') in the body of the telephone enables a single pair to be plugged in to the network termination box of the 2-wire local line (known as the local loop). This halving of the cost of copper cable in the PSTN, which is used throughout the World, is at the expense of the conversion device at both ends of the copper loop and some performance limitations, particularly the need to prevent echoes occurring.

Each subscriber's line terminates (via the MDF) on a subscriber's line card (SLC) at the concentrator switch-block, where the 2-to-4 wire conversion is provided. From here onwards throughout the digital (TDM) circuit-switched networks (including mobile networks) the transmission is 4-wire, with separate Go and Return channels. The SLC performs all the functions of terminating a copper pair of wires from a telephone instrument onto the digital electronic equipment in the exchange (see Box 5.5). Several of these functions require physically robust relays, resulting in the SLC being relatively expensive. The summation of the cost for SLCs, one per subscriber's line, amounts to about 70% of the total cost of a local exchange (as also shown in the asset breakdown in Figure 5.3).

Box 5.5   The BORSCHT functions of a subscriber's line card

BORSCHT is an acronym describing the set of functions a SLC must perform to terminate a subscriber's line at a digital LE [5].

B = Battery. The provision of 50VDC line power.
O = Overload protection for the delicate electronic equipment in the event of accidental high voltages on the line (e.g. due to lightening or other induced high voltages).
R = the application of Ringing current.
S = Supervision (the detection of 'off-hook' line condition).
C = Codec providing analogue-to-digital (and vice versa) conversion of the speech channel.
H = Hybrid transformer, providing 4-to-2 wire (and vice versa) conversion.
T = Test of the end of the line's metallic path at the exchange.

Figure 5.12 also shows the cost characteristic for a digital (TDM) local exchange. This shows the classical $A + BN$ cost relationship: the fixed cost component A is made up of several steps when extra equipment is added, and B is the cost-per-added subscriber (essentially the cost of an SLC), and N is the number of subscriber lines terminating on the LE. The main fixed cost, A_1 covers such items as the signalling and control processor equipment, the concentrator and group switch-blocks, racks, power supply, testing equipment, element managers, etc. As each subscriber is added the cost rises evenly until the traffic generated by the subscribers requires extra switching equipment, control processor equipment or line cards (LC) for the 2 Mbit/s trunk links leaving the exchange, causes a step increase in the total cost curve.

We can now easily consider the cost characteristics for all the other types of digital (TDM) circuit-switched exchanges, i.e. those without subscriber lines, namely in the fixed network: junction tandems, trunk exchanges and international exchanges in the PSTN, various types of business services exchanges (e.g. Centrex, VPN), and in the mobile networks the voice switching provided by mobile switching centres (MSC). The basic cost structures of all these exchange systems is similar and can be deduced from Figure 5.12 by eliminating the subscriber lines, SLCs and concentrator switches. They will all follow the basic formula in the form of $A + BE$, where A is the fixed costs (which will vary with the type of exchanges, with the highest value being with the MSCs due to the complexity of the mobility control equipment), B is the average cost per Erlang, and E is the amount of traffic in Erlangs being switched. In practice, these cost characteristics will also have step functions due to the granularity of switch-blocks and the need to add 2 Mbit/s (or 1.5 Mbit/s) line cards for terminating additional external links as the traffic levels increase. Of course, the absence of subscribers' lines makes the capital cost of all these exchanges significantly less than that of the LEs.

The conclusion to this subsection is that Telcos set the design GOS for their networks at a level that gives acceptable quality to their users and is cost-effective. In practice the GOS for the local exchanges in the PSTN, the most costly part of the network because of the need to provide traffic concentration and terminate subscribers' lines, is usually set at 1 in 200, while in the highly-loaded cost efficient core of the network a GOS objective of better than 1 in 1,000 is typical. The second point to note is that wherever possible the network is structured and the traffic routed in order to contain the fewest and largest exchanges as possible. Finally, the cost of LEs is dominated by the SLCs, amounting to some 70% of the total. In fact, it seems to be a general empirical law that for all telecommunications equipment – including IP routers and other data equipment – the fully provided cost of terminating line cards is about 65% to 70% of the total cost of the equipment!

5.4.4 Mobile networks

Mobile networks differ from the fixed PSTN through the use of wireless rather than wirelines for the subscriber access links. In most other respects the two networks are similar. However, due to the widespread use of wireless by many other bodies the availability of suitable radio spectrum for the mobile public services is highly restricted and carefully allocated by the national regulators or governments, all within strictly defined bands agreed internationally at the ITU assembly in Geneva. Since the nominated chunks of spectrum are allocated to operators within a country – either through a 'beauty contest' or auction, often at high price, – for a specific services or type of network, it represents a valuable asset. Today's mobile network operators, with their huge customer base, are economically viable only because they deploy cellular systems that enable the relatively small chunks of spectrum to be re-used extensively throughout the mobile service area. This, plus the rapid advance in capability, and decline in cost, of semiconductor electronics used in both the network and the handsets, has led to the ubiquitous worldwide adoption of mobile services for voice and data.

A distinguishing feature of mobile networks is that the subscribers' perception of them is centred on the features and attractiveness of their handset, which is very much a personal device. The majority of handsets are used as data terminals supporting applications ('apps') interacting with service providers other than the mobile operators (i.e. for 'over-the-top services') as well as GPS satellites. Therefore, the economic challenge for a mobile network operator is to create an infrastructure that gives high throughput data transport to an ISP – at 'broadband' speeds – as well as good quality telephony and text messaging.

There has been rapid development of the technology used in mobile networks since their introduction in the mid-1980s, manifest as a series of generations [11]. Therefore, Figure 5.13 gives a simplified view of a generalised mobile network structure, applicable to 2.5G, 3G and initial versions of 4G (i.e. LTE) in order to identify the economic characteristics. In mobile cellular systems the serving area is covered by a set of closely packed, slightly overlapping, wireless cells. The size of the cell is determined by the wireless signal strength transmitted from the antenna at the base station at the centre of the cell. Therefore, the cell boundary is set by the

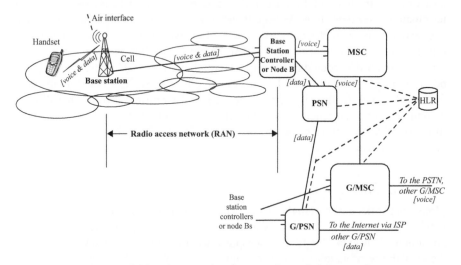

Figure 5.13   A generalised view of a mobile network

threshold of detectable signal by the handset. Beyond the boundary the signals will be lost in background noise, so the operator's set of frequencies can be used to carry the mobile calls and data sessions simultaneously in any distant cells without mutual interference. In practice, the spectrum is split into, typically, seven groups, each allocated to a cell so that the service area is covered by clusters of seven cells.

The use of the frequencies designated for each cell are managed dynamically by a base station controller (BSC), or Node B, which oversees a catchment area of cells creating the radio access network (RAN). By allocating a carrier frequency on demand to a base station when an incoming or outgoing call to a handset within the cell is initiated, the BSC is performing the role of traffic concentrator – as shown earlier in Figure 5.5. Therefore, the traffic aggregation point is at the BSC; whenever all carrier frequencies are in use within the cell additional handsets will not be able to send calls or data. The first economic challenge for mobile operators is setting the locations of the cell areas so as to provide the required mobile service coverage on the one hand, while ensuring that there is sufficient radio carrier capacity available to meet the expected customer demand within each cell. We look at the planning process to meet this challenge in Chapter 6.

Mobile networks have the important economic characteristic that no subscribers' lines are terminated, with consequent saving in SLC costs. Of course, most of the SLC functions are provided but they are not part of the mobile network infrastructure; instead, they are provided within the handset, and so it is the subscriber and not the operator that pays directly for them! This arrangement is possible because there is a well specified 'Air interface' (see Figure 5.13) that all handsets adhere to in order to work on that type of network.

The subscriber's voice and data traffic are jointly carried over the RAN, but are separated at the BSC so that the voice traffic can be switched at the digital (TDM) circuit-switched MSC, while the packetised data is handled by the IP router in the Packet Service Node (PSN). A further economy is gained by reducing the cost of interworking to other networks (e.g. the PSTN) by limiting the ingress and egress of the mobile network to a few gateway nodes (G/MSC and G/PSN), as shown in Figure 5.13. Finally, the management of the mobility for the subscribers is provided by the home location register (HLR), which is a remotely centralised database, holding real-time information on the active network's registered subscribers.

An increasingly important economic challenge for the mobile network operators is minimising the cost of backhaul and core transmission between BSC and MSC/PSN and the MSC/PSNs and the gateway nodes. These transmission links have to have sufficient capacity to carry all the aggregated data and voice traffic generated by the subscribers on the networks (including traffic from roaming users). The data and increasingly video traffic, with ever increasing session durations, creates the need for large backhaul and core transmission capacity to be provided economically.

5.4.5   Core transmission network

The Core Transmission Network of a Telco is a common resource for the transmission of all telecommunications services between its network nodes. Since the

Telcos tend to locate all their equipment in just one or two operational buildings (usually referred to as 'telephone exchanges') in each town, the transmission station in the basement of the buildings can be used to carry services to all of the units there – e.g. PSTN Local exchange, trunk exchange, mobile switching centre, leased line units, and several data nodes. Most large Telcos, including the mobile network operators build their own Core Transmission Network to the extent that it is cheaper than leasing capacity from the incumbent Telco. Invariably, the incumbent Telco operates the most wide-spread Core Transmission Network, and this tends to be the 'network of last resort' in the event of overload or breakdown in the smaller networks.

The challenge for a Telco is provide all the required set of paths between the nodal points (exchanges and cell sites) in the most economical manner. Since these networks are built using high capacity transmission systems over optical fibre or microwave radio, both of which introduce only low propagation delays, there is flexibility in the way that the paths can be routed, so they need not necessarily go directly over the shortest distance. This allows the Telcos to take advantage of the pronounced economy-of-scale characteristic of the high-capacity transmission systems. For example, the unit cost per 2 Mbit/s tributary stream carried over a 1 Gbit/s transmission system is 100 times the price of being carried over a 10 Gbit/s system.

The cost model shown in Figure 5.9(b) applies, following the $A + Bd$ characteristic, so that once the external plant is established the cost of high capacity transmission has a low distance-related cost (optical fibre plus repeaters), B, and it is the terminal cost, A, that dominate. Obviously, the cost advantage of using higher capacity transmission systems only applies if there is a good fill of the capacity by the tributaries. The economic conclusion to this analysis is that it is always cost effective to use a small number of high-capacity transmission systems (known as 'fat pipes') rather than a larger number of 'thin pipes'. Consequently, the Core Transmission Networks comprise a set of fat pipes linking the major nodes, with many feeder links from the remaining nodes; where possible tributary streams will be extracted from the fat pipes at flexibility points and inserted into other fat pipes, even if this causes roundabout routeing, in order to gain good fills on the fat pipes [4].

5.4.6   Network resilience

As stated at the beginning of this section, the objective of network design is to create an infrastructure that is cost optimised to provide the just sufficient capacity, coverage and capability to meet the expected customer demand. However, as we have seen, the minimum cost invariably involves a structure with the smallest number of large capacity links, consistent with full connectivity across the network. The problem for a Telco is that a cost-optimised network is usually also vulnerable to any traffic overloads due to unexpected surges in demand or the breakdown of one or more of the high-capacity links or nodes. A public telecommunications network needs to have a reasonable degree of resilience to such events. Therefore, the cost-optimised network design will always have redundant capacity added to provide a degree of resilience – examples include: standby alternative links, automatic re-routeing of traffic and use of standby service-protection network alongside the most vulnerable high capacity links. The economic choice for a Telco is what proportion of capacity redundancy should it add, or looked at another way: how much excess capital spent is justified? Unfortunately, predicting the level of resilience added to a network is difficult to quantify, so simple rules of thumb are usually followed. Typical excess capital spent for resilience is around 8–10% of the total investment meeting the optimum design.

5.5 Choice of technology

The choice of technology used by the equipment and plant of a network has a significant impact on its economics – influencing the cost of buildings, extending, upgrading and operating the network, as well as its optimum size and structure (as shown in Box 5.2). Of course, a Telco may have to use a particular type of technology if they are to provide certain service features. However, even in that situation there may be a number of equipment suppliers offering design variants. In this section we first examine the various economic parameters associated with the technology used in a network. We then look at the major technology transition affecting the telecommunications industry – i.e. the move from circuit-switched to packet-based networking, characterised as an NGN.

5.5.1 The economic parameters of a technology

Simplistically, it would seem that the choice of technology for a Telco, given that the functionality is much the same between the various offerings, would be directed to the lowest cost equipment. But the true cost of equipment for a Telco is much more than just the purchase price! This is because once purchased and installed the equipment will need to be operated over its useful lifetime – so it is the so-called 'whole-life cost' (WLC) of the equipment that needs to be considered. Let us examine the components of WLC and the economic parameters of the cost of technology.

(a) *Capital cost.* This is primarily determined by the equipment manufacturer. However, it is notoriously difficult to generalise about the purchase prices of equipment because they are invariably masked by particular commercial arrangements between supplier and the Telco. For example, there may be volume discounts relating to the quantity purchased, or long-term relationship pricing, or special loss-leading introductory pricing. Nonetheless, the unit capital cost (as considered in sub-section 5.4.2) does form an important determinant of the economics of a particular technology and equipment type.

(b) *Extendibility.* Invariably, equipment installed in a network will need to be extended some time later due to the growth in traffic carried. A key economic parameter is therefore the ease with which the capacity of the equipment can be extended. In practice, this might require the simple addition of slide-in modules on one hand, or the establishment of a new standalone unit (with all the associated costs) on the other hand. The main requirement for a Telco is that the equipment design has a reasonable unit cost within the expected range of capacities. For example, in the case of a mobile switching centre (MSC) the range may extend from 2,000 to 20,000 Erlangs.

Another aspect affecting the extendibility is the 'volume granularity' of the equipment, i.e. the size of the building blocks making up the equipment. For example, SLCs in an LE are typically contained within an eight-line plug-in card; therefore, provision of the minimum capacity in order to serve one more line would introduce seven lines of spare plant (i.e. a BSP).

(c) *Operational cost.* The cost of operating particular equipment can be a major contribution to the whole life cost, especially when radically new technology or design is involved. So, even if a supplier offers the new equipment at an attractive capital cost compared to the existing types being used, the Telco may find that it cannot afford to introduce the cheaper equipment! The operational costs cover activities normally provided through the use of element-manager equipment, as described in Chapter 9. Such activities include:

- Assigning capacity on the equipment.
- Bringing a new subscriber into service, usually referred to a 'provisioning'.
- Alarm management.
- Routine and remedial maintenance including diagnostics, usually referred to as providing 'assurance' of the required quality of service.

(d) *Operational support systems cost.* A new suit of operational support systems (OSS) may be required in order to perform the operational functions (Chapter 9), covering customer service activities – such as order taking, billing, fault management – as well as element managers. In addition, there may be integration costs resulting from the provision of extra support-system equipment needed for the new equipment to integrate into the Telco's infrastructure. Examples of such integration equipment are mediation devices to interwork between the new network equipment and its element managers and the established support systems of the Telco.

(e) *Running costs.* The annual costs of electricity, including the provision of fuel for standby power generators, can be a significant factor in the choice of technology. This is particularly important in the case of computer servers in data centres, but it may also be a cost penalty in moving to a packet-based technology in the network, which tends to be less power-efficient than the older TDM equipment.

(f) *Accommodation costs.* Although the trend since the 1980s has been for the amount of accommodation, in terms of floor area, to reduce with successive generations of technology in network and computer equipment, there may be additional requirements for cooling, ventilating and air-quality control. Many of the electronic modules used contain ever more dense arrays of computer-integrated circuits and devices which generate sometimes intense amounts of heat. There may also be requirements for the physical construction of the accommodation to be modified, e.g. the provision of false raised floors to channel the inter-rack and power-distribution wiring.

(g) *Economic life.* As discussed earlier in this chapter, the economic life of equipment directly determines their depreciation contribution to the company's P&L account. Particular technology and equipment designs, depending on their maturity, will vary in their time to obsolescence, and hence their economic life.

(h) *Update and enhancement costs.* These costs relate to the periodic software updates that are usually associated with modern electronic telecommunications and computer equipment – including the ease with which such updates can be made, whether there is a break to service, and so on. There may also be a cost difference between the choice of technologies and types of equipment incurred in enhancing the equipment to bring in new features, or make some operational improvements.

The conclusion to the above discussion is that a Telco needs to consider the range of whole-life cost factors when making a choice of technology and equipment design – noting that the lowest purchase price may not result in the most economical solution. A further important conclusion is that a Telco needs to plan carefully the approach to introducing new technology to the network so as to minimise the additional costs involved, noting the time taken to fully deploy new equipment across a widespread network may take several years. We look at the strategies for the roll-out of new network equipment in Chapter 6.

5.5.2   The case of NGN

At the turn of the new millennium the telecommunications industry began the process of specifying a new generation of networks that would address the major changes in customer's expectations and their acceptance of the Internet, by taking advantage of the increasingly capable data networking technologies. The concept of what is now widely called the next-generation network (more usually known as 'NGN') has firmed up over the last few years into a radical new architecture and a set of new technologies and standards, which is to be the target for all future networking – supporting fixed, mobile, voice, data and video applications.

(a) *The drivers for NGN*: There have been several drivers towards the NGN concept, including:
   • The rise of the Internet.
   • Consumers increasing acceptance of web-based services for commercial activity, especially online shopping.
   • The roll out of high-speed broadband access for residential customers.
   • Developments (increasing capability and decreasing costs) in the technology of the enterprise market – e.g. IP routers, LANs, extension of ethernet speeds and distance.

- Widespread adoption of computer-based activity as a way of life (e.g. emails, text, social networking, web-based TV). This shift results from the rapid developments in consumer electronics – e.g. laptops, PDAs, mobile handsets (which now contain huge computing power).
- Development of good-quality voice-over-internet-protocol (VOIP) technology.
- Widespread adoption of VOIP based services – e.g. Skype and several of the cheap international calls services.
- TV broadcast as well as catch-up, together with a variety of video services delivered over the web.
- Increasing cost of planning and operating existing networks.

The commercial effects of the above shifts include:
- Increasing obsolescence of the existing networks (PSTN, private circuit networks, ATM-based networks, TDM technology, etc.)
- Consumer expectations of low (or free) costs of services
- Increasing difficulty for Telcos to forecasting the demand for services – and hence the increasing cost penalties in planning the long-lead time investments for the existing service-specific networks.

(b) *The definition of an NGN*: In the context of the above drivers, we can appreciate that the characteristics of an NGN may be summarised as follows [12]:
- Packet-based multi-service platform supporting a wide range of services and applications (voice and non-voice). In practice this means that the NGN is based on a single IP network, which supports all services.
- Broadband capabilities together with the end-to-end management of the quality of service (QOS).
- Comprises fixed (wireline) and mobile (wireless) customer access.
- Ubiquitous mobility and converged services. This means that the features of mobility – i.e. location and service management – are available to users throughout the NGN over fixed and mobile access. This feature may also be included within a set of converged services.
- Layered architecture with the service control separated (functionally and physically) from the switching and transmission (often collectively known as 'transport') platforms.
- Provision of special interfaces to service providers, so that third parties can provide services over a Telco's network.
- Comprises a services creation capability in which a set of 'service building blocks', comprising hardware, software and processes, may be used to assemble new services and features.

This internationally agreed definition above forms the basis of new network designs and the progressive upgrades of networks around the World.

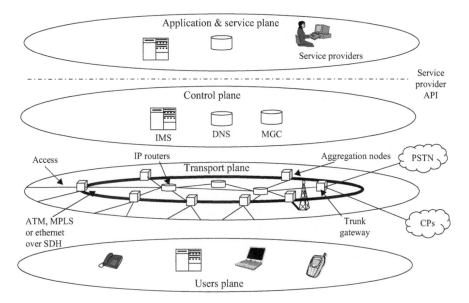

Figure 5.14   The logical conceptual architecture of an NGN

In practice there are two strands of NGN being developed: one for fixed networks which introduces a strong data capability and potentially offers a replacement for the PSTN – usually referred to an NGN; and other is the 4G mobile network. Both versions are based on an-IP multi-services platform. The expectation is that eventually both strands will come together, so that there is a single core network with both fixed and mobile access, offering a combined set of services.

Figure 5.14 gives a simplified view of the logical architecture of an NGN with its distinctive set of layers. The purpose of the layering is to allow the design and construction of the transport activity within the network – namely, transmission and switching and routeing in the Access and Core Networks – to be undertaken independently of the control of the services provided, which reside in the control plane and application-and-service plane. It is interesting to note that this idea of separating transport from control and services has long been an economic objective of the tele-communications industry, but now the developments in software and IP technology make this now a realistic objective.

The transport plane represents the common-service platform, described in section 5.4.1, which is realised using IP packet routing (at OSI Layer 3), supported with virtual path management (at Layer 2) and high-speed radio and cable transmission (at Layer 1). Links to the subscribers/users, together with the various service interfaces, are provided by the range of Access Networks shown. Therefore, all services over the NGN are handled by the single IP-based common services platform. The transport plane also includes the trunk gateway function providing interconnection from the NGN to the PSTN and the networks of other communication providers (CP).

The control plane contains all the relevant control elements (such as IP multimedia subsystem (IMS), media gateway controller (MGC) and associated domain-name server [DNS]). The idea is that this control plane provides a set of resources that can be used by a number of services. The key feature of the NGN concept is that there is the potential for a number of service providers other than the Telco running and owning the NGN to gain access to the control plane to build and run their own services, which would be run on the Telco's network. This facility could be provided via the service provider API (application programming interface), as shown in Figure 5.14. However, such opening-up of the control of the network by a Telco to other service providers, of course, presents complex commercial, reliability and security issues for the Telco and national regulator.

As mentioned earlier, 4G is considered to be the mobile network form of NGN. We can see the concept of this by considering Figure 5.13, in which the MSCs are eliminated and voice as well as data switching is provided by the multi-service platform of PSNs.

(c) *The business case for introducing an NGN*: We can now pull together our views on network economics to identify the main arguments to be presented in a business case for a Telco to introduce an NGN. (We consider the general purpose and use of business cases in section 4.5.1.) As with any business case seeking a large commitment of capital spent to deploy a new network which will rapidly replace the existing infrastructure, the arguments fall into four main categories: the rationale, revenue drivers, cost drivers and risks.

The scene setting introduction to the business case presents the set of drivers given in (a) above is clearly strong enough to indicate that the time is right for Telcos to make a radical change to their network infrastructures [13]. However, the next part of the case – the revenue drivers – is more difficult to muster. Normally, deploying new network infrastructure is justified by a new set of revenues from the services introduced, but today's complex highly competitive marketplace, with the dominance of over-the-top services, price declines and changing customer habits, makes incremental revenue opportunities difficult to identify. Therefore, the primary drive for the NGN business case is the expected savings in future capital investment due to the economies of using an IP common-services platform as a replacement for the existing networks, especially the ageing PSTN. Operational cost savings, leading to lower OpEx can also be identified. Finally, there are risks in deploying new technology to replace existing platforms, particularly when these (like the PSTN and leased lines) may be generating much of the Telco's revenue. Box 5.6 outlines some examples of the cost and revenue drivers for the NGN business case.

Box 5.6   The NGN business case

Cost drivers

It is generally agreed that the potential cost savings provide the main inputs into an NGN business case. The main cost drivers are as follows:

- *Multi-service platform.* The use of a single multi-service platform provided by the NGN offers potentially significant cost savings compared to the alternative of building and operating several service-specific platforms, as described earlier. The extent of this saving largely depends on the strategy chosen for implementing the NGN – specifically how soon the NGN replaces the existing platforms.
- *VOIP technology.* There are potentially CapEx savings in the use of VOIP technology as a replacement for the circuit-switched PSTN and other networks.
- *Broadband access.* CapEx cost savings can be achieved through the association of the rollout of fibre-based access broadband with the installation of the NGN aggregation nodes due to equipment integration.
- *Simplified network.* The NGN architecture represents a more simplified structure than the existing networks it will replace. Such simplification offers savings in equipment (i.e. CapEx), as well as in the planning and operation costs (OpEx).
- *Layered architecture.* As described earlier, the NGN's layered architecture offers the potential for separate development of the control and service capabilities from that of the transport infrastructure. It is expected that this separation will enable cost optimisation of the various layers, giving future cost savings.

Revenue drivers

New revenues resulting from the introduction of an NGN, although important and potentially exciting, are likely to be relatively small due the increasingly competitive and uncertain telecommunications market. Nonetheless, we can identify the following broad categories where new revenues could be expected.

- *IP multimedia control.* The NGN is expected to enable many new forms of multimedia services and features.
- *Broadband access.* As mentioned earlier when considering the cost advantages, the association of high-speed broadband rollout with the deployment of NGN should also enable revenue opportunities.
- *Converged services.* The NGN should enable a range of new converged services.

Examples of services enabled by the above three sets of features taken together include fixed-mobile voice, TV over the NGN (IP TV), location and presence and web-voice services.

Such a business case, by its nature, presents an idealised view of the argument and what can be achieved. In practice, the identified costs savings attributed to the deployment of an NGN will be achieved only if the old platforms are withdrawn within the specified timescale and the services transferred safely. Given the scale of the task – the number of subscribers' lines and network nodes to be converted, the physical limitations on what work can be done simultaneously within buildings, the availability of the worksforce, etc. – the deployment will need to be period will be several years. During this period the old networks and new network side by side across the country, both carrying live traffic and valuable customers' connections – i.e. the revenue source! Thus, a major determinant of the economically successful implementation of the NGN business case is the pursuance of an optimised implementation plan. We look in more detail at the planning for network conversion in Chapter 6.

5.5.3   The big network-economics debate: circuit or packet switching?

Many people are surprised to learn that circuit switching still played an important role in providing voice services for fixed and mobile networks long after the advent of voice over IP (VOIP) technology. However, there is relentless pressure to move voice switching onto an IP platform, particularly as a result of the adoption of NGNs. It is interesting to look at the economics of switching voice over IP rather than circuit switches.

First, what are the arguments made for putting voice over IP, either over the public Internet or over an operator's IP network, rather than keeping them on a circuit-switched network? Typically, they are as follows:

- It's cheaper.
- More efficient in that network links are able to be highly loaded with interleaved packets.
- Circuit switching is yesterday's technology and everything is moving to IP anyway.

Let's examine these arguments briefly.

(a) *It's Cheaper*: Actually, it is not! Most people's perception is based on the price they pay for the service rather than its cost to produce. The prices of most VOIP services, like Skype, Vonage, etc., and the special cheap international calls (which use VOIP), are indeed very low price or even free of charge! However, there is no inherent cost difference between the two voice switching technologies. Indeed, as Box 5.7 shows there is even very little difference in the actual technologies. However, there are cost savings that accrue instead from an economic important change in network architecture.

We saw in section 5.4.3 that some 70% of the capital cost of fixed local exchanges is due to the subscribers' line cards, which provide the interface equipment. However, we also saw that mobile exchanges do not bear this cost because most of the functions are provided by the handset – and a similar situation occurs for the VOIP equivalent of a subscriber's line.

Box 5.7   No inherent cost difference between a digital TDM Circuit Switch-block and an IP Router

Digital circuit switching systems (as used in the PSTN and mobile networks) are based on a T-S-T switch-block comprising a space switch (S) sandwiched between inlet and outlet time switches (T), as shown in Figure 5.15. Circuit-switched call connections are established as 64 kbit/s Go and Return paths through the TST assembly held for the duration of the call. The address information relating to the call is carried in a separate message-based (SSNo7) links between the control systems of each switch-block in the call connection. Once the required connection across the switch-block is established the control system is free to deal with other calls.

An IP router (as used in the Internet, data networks, data service nodes in mobile networks, NGNs) is a digital space switch sandwiched between input and output buffers (which are essentially time switches), as shown in Figure 5.16. The routeing of IP packets is achieved by storing each packet in the input buffer until a free path is available through the space switch and hence into the output buffer, where it is held until the output link becomes free. The address information for each packet is held in its header, so that each packet in a stream has to be interrogated by the router control system before it is individually routed.

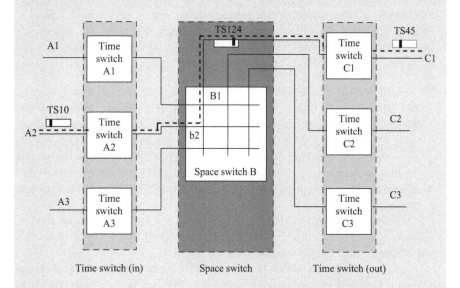

TDM circuit established between A2/TS10 and C1/TS45 via b2:B1/TS124

Figure 5.15   A TDM T-S-T digital switch-block

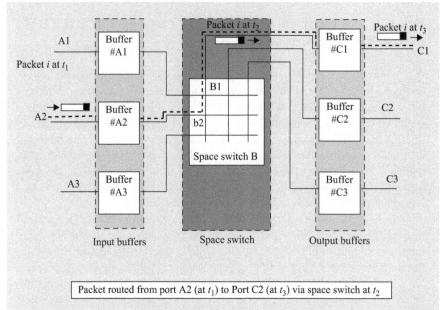

Packet routed from port A2 (at t_1) to Port C2 (at t_3) via space switch at t_2

Figure 5.16   An IP router

Both systems described above use similar very large scale integrated (VLSI) printed-circuit semiconductor technology and, apart from the differences in the control arrangements, there is a remarkable similarity between digital circuit switches and IP routers.

The different locations for this interface equipment, providing the BORSCHT functions, are shown in Figure 5.17. It can be seen that the BORSCHT functions are provided in the VOIP handset or a VOIP-application run on a lap top or home computer with headphones and microphone; so it is the user rather than the operator who buys the interface equipment. This shift of interworking costs gives a real cost benefit to VOIP where this architecture applies.

Finally, of course, we have seen in section 5.4.1 that the use of a single multi-service platform offers cost saving compared to the set of individual platforms – this is true whether or not IP is used.

So, the conclusion is that the use of VOIP technology in place of digital circuit-switched technology is cheaper only if the interface costs are shifted to the user and/or it is embedded in a form of multi-service platform.

(b) *Greater network efficiency*: It is true that because packet switching is a delay system the outgoing links from a router can be highly loaded with closely packed interleaved IP packets. However, this efficiency is achieved through

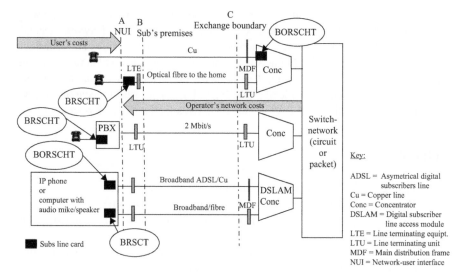

Figure 5.17   Shift of user-interface costs

causing variable delays to the packets. This is a problem for real-time services which do not tolerate variable or long delays to the IP stream. Indeed, the use of VOIP generates a high-volume stream of IP packets, which can cause severe disruption to the other data packets being carried on the network. In practice, acceptable quality of service is achieved either by partitioning the IP network (which negates the unified platform advantage) or running the links at much reduced loadings, typically 45% (which is worse than the TDM trunk network loadings of near 70%). So, there is not generally an improvement to network efficiency by using VOIP rather than circuit switching.

(c)   *The trend is towards IP*: This is indeed true. It is fair to say that the trend towards IP networks is driven by the rapid advances in data networking equipment, which itself is following the steady improvement in terms of cost declines and bit capacity improvements being achieved by the computer industry. This rapid rate of improvement in VLSI, battery, and screen technology is also driving the rapid advances in handset and device equipment (smart phones, tablets, laptops, TV sets, cameras, etc.) – and the consequent user's hunger for applications and services.

5.6   Summary

In this chapter we began the examination of the wide subject of network economics by first considering a network as an economic system at a macro level. We then considered the importance of the network as a set of assets which drive the depreciation costs in the company's P&L account. The next two sections looked at the economics driving the size and structure of the optimum network design, first noting the importance of aggregation points, access and core parts of any network. Second, we looked at the economics of the various parts of the networks, fixed and mobile in more detail. Our consideration of network economics concluded with a review of the factors affecting the choice of technology and how that affects the capital and current accounts of a Telco. The example of NGN/4G with its major shift in technology gave an important perspective of the economic challenges facing network operators.

References

1. BT Group plc. "Annual Report and Form 20-F 2014".
2. Ofcom. Statement: "3G rollout Obligations", www.Ofcom.org
3. Jeter, L.W. *Disconnected: Deceit and Betrayal at Worldcom (Business)*, John Wiley & Sons, New Jersey, 2003.
4. Valdar, A. "Circuit Switching Evolution to 2012', *The Journal of the Institute of Telecommunications Professionals*, Vol. 6, Part 4, 2012, 27–31.
5. Valdar, A. *Understanding Telecommunications Networks*. IET Telecommunications Series 52. Institution of Engineering and Technology, London, 2006, Chapter 6.
6. Valdar, A. *Understanding Telecommunications Networks*. IET Telecommunications Series 52. Institution of Engineering and Technology, London, 2006, Chapter 8.
7. Valdar, A. *Understanding Telecommunications Networks*. IET Telecommunications Series 52. Institution of Engineering and Technology, London, 2006, Chapter 5.
8. Skey, P. "Overseas Practices", *Local Telecommunications*. IEE Telecommunications Series No. 10 (edited by Griffiths, J.M.), Peter Peregrinus, Ltd, London 1983, Chapter 15.
9. Kingett, D. "Local Line Plant Planning and Practices", *Local Telecommunications*. IEE Telecommunications Series No. 10 (edited by Griffiths, J.M.), Peter Peregrinus, Ltd, London, 1983, Chapter 4.
10. Analysis Mason, www.analysismason.com.
11. Hearnden, S. "The History of Mobile Communications", *The Journal of the Institute of Telecommunications Professionals*, Vol. 6, Part 4, 2012, 42–45.
12. ITU-T. "General Overview of NGN", Recommendation Y2001 (2004).
13. Martucci, J., Elvidge, A. and Villez, J. "On the Factors Influencing the Business Case for Next Generation Networks", *The Journal of the Institute of Telecommunications Professionals*, Vol. 1, Part 1, 2007, 35–40.

Chapter 6

Network strategy and planning

6.1 Introduction

Building a telecommunications network incurs significant capital investment which takes many years of running revenue-earning services to recover. We saw in Chapter 5 how the infrastructure assets – optical fibre cables, exchanges, cell sites, routers, etc. – have economic lives ranging from 2 to 25 years. The challenge for a Telco is to build the network with sufficient capacity and geographical coverage to carry the predicted levels of traffic at the appropriate quality of service, recognising the cost and longevity of the infrastructure and the possible difficulties in adjusting the configuration to meet changing requirements. This challenge has become progressively more difficult since the 1990s with the opening up of competition, the vast expansion of data networking and the Internet, and the huge explosion of application providers (e.g. OTT), leading to less predictability in the market and more rapid technology churn. The purpose of this chapter is to examine the principles of planning a telecommunications network in the most efficient way. We also consider how to determine the strategy for the networks enhancement and growth that meets the challenge.

We begin our consideration of network strategy and planning by identifying its links to the various levels of business planning within a Telco. In the next section we see how the business objectives and strategy set the framework for the network strategy, and how this directs the planning of the network. (Of course, this relationship may at times be somewhat circular in that the network capabilities can also influence the business objectives.) The topic of network planning is first described as a set of key principles and then the planning approach, based on the economic rules introduced in Chapter 5, for the main parts of a telecommunications network are discussed. We then look at the planning principles for creating a major transformation of the network technology and structure. Finally, we consider one of the most important outputs from the network planning process in a Telco, which is the list of equipment that needs to be purchased and installed – i.e. the capital investment programme for the network.

6.2 Links to business planning

Since the network is the main capital cost component for a Telco it is not surprising that the planning process – which in effect results in a CapEx shopping list each year – needs to be closely linked to the business planning activity within the company.

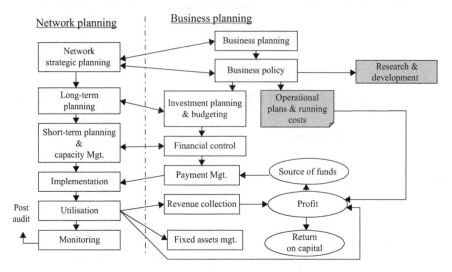

*Figure 6.1   The link between the network planning and business planning
processes [Reproduced with permission from K.E. Ward]*

Figure 6.1 shows the linkages between the hierarchy of functions within the realms of
network planning and business planning within a Telco. The relationship shown in
this model is also considered in the context of corporate governance in Chapter 4.
Clearly, as described in Chapter 3, the overarching positioning of the Telco's business
is set by the business planning activity, which identifies the markets to be addressed
and the commercial model for the company. For example, to address the consumer
mobile market a Telco may be a full mobile network operator (MNO) with its own
network infrastructure, or take the role of a virtual operator (MVNO) having its own
customer service support systems, but no network infrastructure, instead leasing
capacity from an MNO. For convenience, we will describe the network strategy and
planning processes for full network operators since this requires the management of a
major CapEx programme, hence bringing out all the key issues for discussion.

The business planning activity and the business policy form the main input to
the development of the company strategy, as described in Chapter 3. It is at this
level that the objectives for the company's R&D activity (see Chapter 8) and for the
strategic network planning are set. The business policy gives a framework for the
all-important set of plans for the company, namely: the operational plan and budget
covering the running costs for the next one or two years – essentially setting the
OpEx budget (see Chapters 4 and 9) on the one hand; and the investment plan and
budget covering the CapEx budget, which is the focus of this chapter.

It is useful at this stage to briefly describe the various levels of network
planning:

Network strategic planning. This sets the framework for the future planning of
the network, covering expected changes of technology, structure and functionality. In particular, it covers the range of new products to be supported

and any expected changes to the commercial model, and is closely linked to the company's business strategies.

Long-term planning. These plans cover the long-lead time items in the network (e.g. buildings, major transmission routes and radio towers), typically identifying the need for land acquisition, building work, etc. Generally, such planning covers 3–5 years ahead.

Short-term planning and capacity management. This planning activity covers 1–2 years ahead, detailing network capacity, links and routeing for transmission, switching, data routeing, etc. In addition, the actual utilisation of the network needs to be assigned – a process known as 'capacity management'. For example, the allocation of the traffic route between a cell site transmitter and its control node to a particular transmission link and optical fibre.

Referring again to Figure 6.1 we can see that the long-term plans are linked to the investment planning process, which is particularly important if major investment in new network equipment is required in the future, for which the company needs to allocate funds and seek external funding where necessary. At the short-term plan level, however, there is the need for financial control on the commitments made for the purchase of equipment and plant. Later, once the plans are implemented and equipment is installed and paid for, the capacity can be utilised. Control of payments to equipment suppliers is applied at this level (financial management) because this directly affects the cash flow of the company. Finally, of course, it is the utilisation of the network that drives the revenue for the Telco, as shown in Figure 6.1. Also, at this point the asset utilisation factor (AUF; see Chapter 5), namely the proportion of network capacity that carries live – hence revenue earning – traffic, can be calculated. The AUF then provides a measure of planning efficiency, since it identifies the extent of over-provision of capacity within the network.

We may now summarise the differing objectives of network strategy and network planning. Network strategy aims to set an overall framework covering some 5 or more years for the planning of the network, which aligns with the business objectives of the company. In contrast, the objectives of network planning are to design and dimension a network such that its geographical coverage and traffic capacity is sufficient to carry all required services at the appropriate quality levels, without undue over-provision, and within the annual CapEx budgets set by the business planning process of the company.

6.3   Network strategy

6.3.1   *The role of network strategy*

An interesting way of describing how network strategy directs the development of the network is shown diagrammatically in Figure 6.2. This shows the functionality of the existing network (measured by performance, connectivity, service and operational features, etc.) being enhanced over time, culminating in some vision for

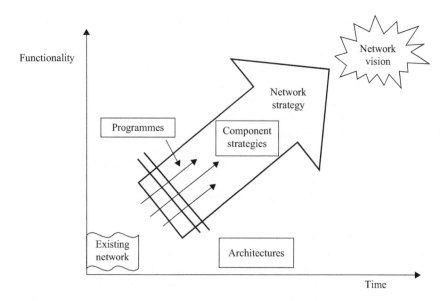

Figure 6.2   The role of network strategy

the future network. The big arrow, therefore, represents the strategy that will take the network towards the vision. Within the overall network strategy is a set of component strategies covering the range of network elements and the operational support systems, all of which are to be developed in the direction of the vision. The delivery of the component strategies depends on suitable programmes, with resources and money allocated. Finally, we can imagine network architectures as being an orthogonal view across the set of strategies and programmes, showing how they all fit together to make a coherent set of plans at various times throughout the conversion period. We will briefly examine each of these concepts:

Network vision. The vision is a view of the target network that would best meet the commercial and technical environment at some point in the future. Typically, the vison is set 5–10 years ahead. However, with the increasingly shorter technology generation cycles (particularly in terminal devices) and the rapidly changing social setting for consumers and business customers affecting their usage and expectation for ICT services, the visions need to be ever closer in time. Developing the vision within a Telco involves activities similar to those described for scenario forecasting, as described in Chapter 8. However, the network vision needs to be more specific in terms of characterising the features and performance of future technologies, the societal settings that determine customer behaviours and, most importantly, the commercial models that will exist within the ICT environment. Many Telcos have 'futurologists' within their R&D teams who specialise in the technological and societal futures, and work within the company to develop the appropriate network visions [1].

Component strategies. Each of the main elements of the network and the operational support systems – e.g. the fixed access network, cell sites, switching centres (fixed and mobile telephone exchanges), backhaul, core transmission networks, data networks, signalling and control networks, and element and network management systems – needs to be progressively enhanced and upgraded to make best use of new technology and achieve functional, operational or cost improvements. Therefore, a strategy is required for each main element of the Telco's networks, and these need to fit within the overall network strategy, as illustrated in Figure 6.2. For example, the strategy for converting the fixed access network to superfast broadband – targeted at the network vision where all subscribers have unrestricted broadband access to IP-based services – will cover the necessary equipment and plant upgrades, the choice of equipment and architecture of the broadband access network.

Network programmes. The key message is that nothing happens without a programme being in place. Therefore, no matter how elegant and attractive a network strategy might be, there will be no changes in the network until a business case is produced within the company to seek authority for committing capital and current account budget for specific deliverables, e.g. installation of 200,000 high-speed broadband lines by a certain date. The programmes will also need the allocation of staff resources and a manager responsible for its execution. At any one time, a large Telco may have several hundred network programmes in place. Many of the programmes will be continuing the installation of the same technologies to meet capacity and coverage growth requirements, for which only a perfunctory business case is needed. A more thorough business case is needed, however, when the programme is installing new technology or a large-scale increase in coverage or capacity. The business cases will be the responsibility of the relevant product manager (see Chapter 8), where the network or support system enhancements are due to a product launch or enhancement, otherwise it would be a manager in the Network Planning Department that is responsible. It is important to note that many enhancements will be to just the customer service management or network management systems (see Chapter 9), perhaps only software upgrades, and not involve any network elements.

Architectures. Architectural views show how the various network and support system elements all work together to provide the infrastructure for all the Telco's services (products). There are four different types of architectural views, as shown in Figure 6.3 [2].

Commercial model view: Showing the commercial relationships between the various players. For example, in the case of an IPTV network, the view would cover the carrier providing the broadband IP distribution network, the video service provider and the various video content providers (some or all of these roles could be taken by the one Telco or spread among several players).

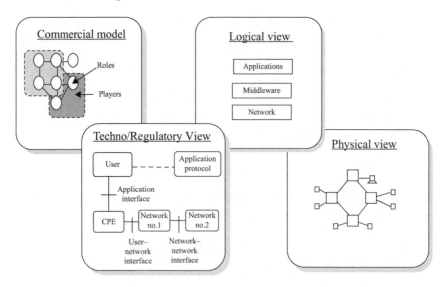

Figure 6.3   The four architectural views

Techno-regulatory view: Showing the network boundaries and interfaces to
   end users (UNI) and other networks (NNI), clearly demarcating the extent
   of the Telco's network service responsibility.
Logical view: This conceptual view allocates the network to a set of functions, usually organised on a layered basis. Such views can be useful in
   the way that complex physical networks can be easily understood, particularly by non-experts and senior management. Examples of a logical
   architectural view are the standard OSI 7-Layer model for data networks,
   and the 3-layered ICT model shown in Figure 6.3.
Physical view: This view, which shows the disposition of each network
   element and their linkages, is probably the most commonly understood
   meaning of network architecture. It has the advantage of showing a bird's
   eye view of all the network elements, but in a complex case it can be
   difficult for the functions of all the boxes and links to be readily apparent.

Following on from the discussion of architectural views above, we need to
consider the set of networks, and resources that constitute the telecommunications and ICT infrastructure supporting today's homogeneous market of services
and players. Figure 6.4 attempts to capture the complex interrelationships
involved using a simple pictorial model. This shows the user connected via either
fixed or mobile access networks via their respective core networks. Calls can be
handled entirely within the originating networks or through interconnection links
to other network, as shown by the black links between the network boxes. Of
course, in practice there can be many fixed and mobile networks within a
country. In addition, there will be alternate network operators, often based on

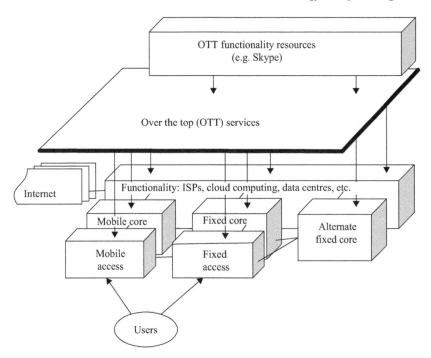

Figure 6.4　ICT: telecommunications networks, services and applications

just core networks using the fixed or mobile networks for access to the users through call interconnection, or some form of leasing capacity (e.g. local loop unbundling [see Chapter 2]).

Users gain access to network resources beyond simple calls or IP routeing, via the networks to what is shown as the 'Functionality' box in the model. This covers the wide range of resources such as ISPs, for access to the Internet as shown in Figure 6.4, Cloud computing data centres, IPTV service nodes, etc. This base layer of networks, including the Internet, and functionality resources, forms the basic platform which supports a wide range of applications run over the top – often referred to as 'Over the Top (OTT) services'. These so-called OTT services (actually, they are technically applications being run over the network services!), such as Skype, WhatsApp and Facebook, have their own functionality resources, as indicated by the top box in the model.

The commercial relationships between the various network operators, the service providers running the functionalities in the base and over the top layers, in many complex permutations, are best shown using a commercial model architectural view for a particular situation. Such models in turn can be related to the value chain analysis introduced in Chapter 1. A further dimension in the modelling of networks, functional resources and OTT services is the mapping of players to the roles.

Typically, a national Telco owns and runs a PSTN or NGN (Access and Core network), a mobile network, possibly other networks like cable TV (shown as Fixed access and Core in Figure 6.4), as well as cloud computing resources, IPTV resources and is also a major ISP for the country. Although there are mobile-only operators, nowadays many Telcos have both mobile and fixed broad and access and core networks, as discussed in Chapter 1. Our consideration of network strategy and planning below, therefore, relates to the totality of networks and resources owned and operated by the Telco.

6.3.2   The network strategy process

The process for producing the network strategy to cover all of the networks of a Telco is illustrated in Figure 6.5. As we saw earlier, the network strategy must be closely linked to the company's business planning process, and obviously it must follow the company's policy. In determining what needs to be done to the network and support systems, however, the Telco also needs to consider the changes expected in the target markets and the future products being formulated by the company. This commercial environment then forms the background to the development of the strategy for the network, focusing on what needs to change, and the opportunities offered by new and expected future technologies. In many cases the radical nature of the technological changes, and the predicted societal changes and customers' expectations, can lead to the strategy process generating ideas for future products which are not related to the marketing or product plans of the company.

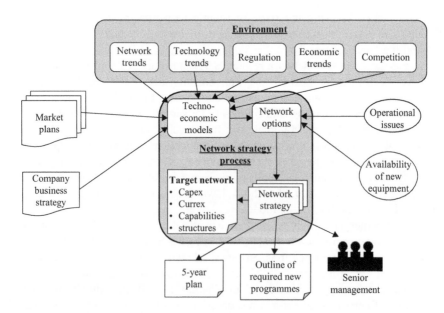

Figure 6.5   The network strategy process

This leads to a stimulating debate within the company which should bring fresh insight to what can be expected in the future.

An important input to the strategy-making process is that of operational experience from the network, e.g. fault problems with certain network elements, unexpected operational difficulties leading to poor performance or extra costs. These may lead to remedial actions – to improve quality or reduce operating costs – being incorporated into the strategy.

As indicated in Figure 6.5 the main method of generating network strategy is through testing a set of options using the commercial environment and technical opportunities. Where possible, the testing might be based on techno-economic models, using spreadsheet, scenario planning and other forms of evaluation. It is important that a full range of options are examined and evaluated against the agreed set of criteria for the company. Typically, the criteria for evaluating the strategy options include:

• meeting the company's strategic objectives
• any capital constraints
• limitations of the company's other resources and capabilities
• the current state of the networks
• future technology capabilities

Finally, the network strategy, with its component strategies, is constructed from the results of the option testing through a series of managed debates by a group of experts in workshop-style sessions, typically spread over several weeks [3]. This workshop process enables the conflicting views within the company, ranging from the conservative to the progressive factions, to be resolved and consensus formed. The emerging strategy can then be socialised with key players in the company before being finalised. Ideally, the network strategy should be approved at senior management level.

The network strategy process (see Figure 6.5) is usually completed annually, and synchronised to the business planning cycle within the company. Therefore, one of the important outputs of the network strategy is the forecast data for the later years (usually years 3–5) in the company's 5-year plan – typically in the form of capital investment for each year. Other outputs from the strategy include:

• Outline of required new network programmes to reduce operating costs or to improve quality.
• Guidance on the phasing out of old technology and the introduction of new technology or new network structures.
• Predicted changes to equipment economic lives and impending obsolescence.
• Details of the target architectures for the future.
• Outline of the major transformational programmes (e.g. introduction of NGN to the fixed network).

Box 6.1 presents examples of new network programmes that the network strategy for an incumbent Telo (with a PSTN) might have identified to address the four individual drivers ascertained during the strategy process.

Box 6.1   Examples of drivers and component strategies

Examples	Drivers	Component Strategies
1.	Extend superfast broadband nationwide	• Deploy VDSL • Deploy access optical fibre
2.	Reduce current account spend on the access network	• Attack copper black spots • Deploy NGN
3.	Reduce number of network platforms, and simplify traffic and transmission routeings	• Deploy simplified high-speed data core network • Deploy NGN
4.	Exploit new technology to cope with rapid growth of bandwidth & need for lower unit costs	• Deploy ethernet • Deploy NGN

The introduction of new generations of equipment or technology to an established network needs to be undertaken carefully, following an implementation plan that minimises disruption to the normal operations of the network but still enabling rapid advantage to be gained from the new elements. Thus, e.g., the introduction of SDH transmission systems into the established PDH networks during the 1990s required careful separation of the old and new networks, with a managed set of interworking points between them. In addition, the new SDH links were mainly used to carry the critical leased-line services, in order to exploit the quality of service features of SDH. Other examples include the introduction of IPv6 into a predominantly IPv4 world, and the introduction of next generation networks (NGN) for fixed and mobile network operators. This challenge of managing the build-up of the new and the ramp-down of the old network while still providing network services is a continuing and inevitable feature of modern telecommunications networks. We look at the planning aspects of such network transformations in section 6.6.

Overall, the network strategy should be a credible story of what needs to be done to the network (and support systems) and why. A good test of the network strategy output is how well it is received by the management boards within the company – i.e. can it be described in a jargon-free manner with rationale and consequences clearly articulated? It is vitally important that senior management of a Telco understand why developments are required in the network, and what this means for operating costs, quality and ability to offer new products competitively in the future. If this has been achieved successfully the board members will be receptive to the individual business cases for the component strategies, plans and programmes that will be seeking authority over the next year or so (see Figure 6.5) as a result of following the network strategy.

6.4 Principles of network planning

6.4.1 Overview of the full network planning process within a Telco

We can now summarise the role of network planning for a Telco as taking as inputs the set of requirements from the product lines and the network strategy to create a set of plans for all parts of the network over the next 1 or 2 years. The resultant plans lead to expanding the network's coverage and capacity and, where appropriate, upgrades of its functionality; of course, the plans may also include extracting equipment due to obsolescence or product withdrawal. The overall end-to-end process within a typical Telco is shown in Figure 6.6.

The first part of the process is the analysis of the various inputs, namely:

- *Product forecasts* detailing the capacities of each of the network products, usually for years 0–2 (i.e. the current actual levels). Usually, the planning process needs the capacity quantities to be on a detailed geographical basis, thus broad-brush forecasts will have to be modified in order to give the required level of granularity. For example, a top-level product forecast of '3% growth over the next two years for Southern England' would need to be appropriately applied to all relevant routes between towns in that region.
- *Existing traffic matrix* provides the detailed total network description for year 0, on which the planning will be based.
- *Reported shortages* come from the operations part of the Telco, detailing hot spots or other under-provisions of capacity within the existing network that need addressing.

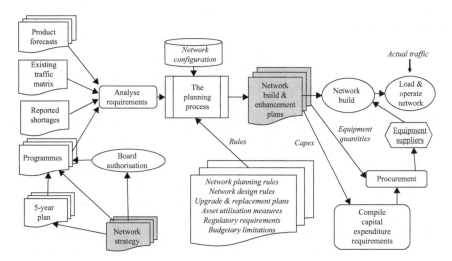

Figure 6.6   The network planning process

- *Programmes*, as described above the network programmes will detail the authorised network extensions in capacity or coverage, the deployment of new equipment, or the withdrawal of obsolete equipment, all of which need to be incorporated into the network plans. The need for new programmes is outlined in the network strategy.
- *Five-year plan*, which details the top-level figures for capacity and coverage, for new and existing products for years 0–5. As described above, usually the figures for years 3–5 are provided by the network strategy.
- *Network strategy*, as mentioned above, provides to the *board* the rationale for the new programmes and the capex budget, which assists in the authorisation of programme business cases.

The above listed inputs are combined to create a single list of requirements for the networks for years 1 and 2 in detail, with broad-brush requirements for the later years (usually 3–5). The planning process is essentially based around the application of a set of planning rules pertaining to the particular parts of the network(s) – we look at these in more detail later in this chapter. This activity relies on the availability of a database detailing the configuration of the networks being planned. It is important to realise that the planning process may be centralised to one HQ department or dispersed to a number of regional offices around the country, making the dissemination of planning data a crucial factor in the process. Traditionally, national Telcos had local offices responsible for the planning of the Access Network and local junction routes, with the long-distant network planned by regional and HQ offices.

Completing our view of the process, the set of plans is then passed for implementation in the field. Also, at this stage the planning process has generated the list of network items to be purchased, which forms the input to the CapEx budgeting process for the company, as discussed later in this chapter. The network build involves equipment suppliers and the Telco's technician force to install and commission equipment within operational buildings, or on external plant, as described in Chapter 9. Once installed in the network the capacity is used for live traffic – shown as 'load and operate' in Figure 6.6. A key factor at this stage is that the network is now carrying the actual levels of traffic generated by the users, which is usually different to the forecasts of traffic levels made 1 or 2 years earlier! Clearly, a serious under-forecast will lead to exhaustion of the network capacity earlier than expected, while an over-forecast will generate greater spare capacity (hence worse NUF) than planned.

The output from the network planning process each year is the set of plans for each of the programmes, covering equipment to be purchased, estimated capital costs, quantities and any special arrangements required. Usually, the plans include a proposed split between installation work that needs to be done by the network operator's technicians (covering accommodation, new racks, power supply, ventilation and cooling, cabling, etc.) and that to be undertaken by the equipment supplier. The planners, therefore, produce two main outputs – the capital requirements which feed into the company's business planning process and the programme requirements which feed into the construction and operational side of the Telco. In addition, the procurement teams in the company will need to manage the equipment supply for the programmes through the letting of new contracts, as necessary (see Figure 6.6).

6.4.2 Planning lead times

Despite the difficulties in producing accurate forecasts of customer demand and traffic flows several years in advance, the nature of telecommunications networks is that their design and build is a lengthy process requiring planning and dimensioning some time before completion. In the case of installing software-only upgrades to bring in a new feature or product, the timescale may be just a few months, but more generally when network equipment or plant needs to be installed, it may be as much as 18 months. The period between a need being identified and the 'brought-into-service' (BIS) date of the new product or network is known as the 'Planning lead time'.

Figure 6.7 gives a simplified flowchart of the components of the planning lead time for network provision. Usually the process starts with a period of evaluating alternative designs, together with finalising the specification of the network product (service) to be provided. It is at this point that the need for new equipment or a software upgrade is identified. In the case where there are several possible solutions from different equipment suppliers the Telco will need to implement an open tendering process to allow all relevant suppliers to bid for the project. The steps in the tender process, and the interaction with the Telco's product manager are described in Chapter 8, Box 8.1. There may then be

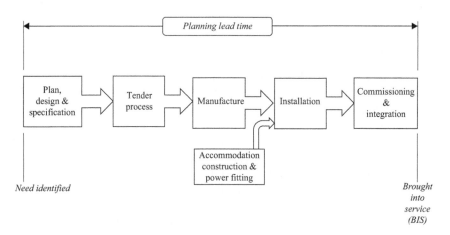

Figure 6.7   Planning lead time

some time needed for the manufacture of the selected equipment. However, the planning activity will also include preparing the accommodation for the equipment in an operational building (e.g. exchange or Core-transmission station). This will include the preparation of space within the building, which together with installation of power supply and ventilation/cooling equipment is necessary. In the case of some small rural exchange buildings, there may be quite a tight fit to install the new equipment before withdrawing the old – necessary in order to maintain continuous no-break service to the subscribers. Finally, once the equipment is installed time is required for its commissioning – i.e. setting-up and testing. In the case where new technology or a different manufacturer's equipment is installed there will be a need to test that it integrates adequately with other equipment and the support systems.

As mentioned earlier, the planning lead times will differ depending on the nature of the equipment or plant being installed, and whether the degree of new technology involved. Certainly, the implementation of a new generation of switching equipment – e.g. 4G mobile systems – will incur all the lead time components shown in Figure 6.6 causing a lead time of some 18 months. On the other hand, in the cases involving just adding more of existing-types of equipment, all but the installation functions may be null and the lead time will be far shorter.

Obviously, the presence of a long lead time can be frustrating for a Telco trying to react to a fast changing telecommunications marketplace, and there is a constant desire to find ways of reducing it – either through improved planning processes or by using network equipment that, once installed, can be easily flexed during its lifetime to meet changing requirements. In addition, there are usually 'fixes' that can be applied to installed network equipment to extend capacity as a matter of urgency. However, such reactive network provision is invariably more expensive than the installation of the same capacity following a *planned* approach, and therefore needs to be kept to a minimum.

6.4.3   Planning rules

The principle of network planning is that the design and dimensioning of network elements is governed by the appropriate economics for the particular equipment and network structure (as discussed in Chapter 5). However, in practice the planning of the extension of capacity and coverage using existing equipment and plant types is a continual process undertaken at all the planning offices within the Telco. Therefore, undertaking economic studies each time would be an unnecessarily burden, given the repetitive nature and similarity of such projects. Instead, the Telcos establish a set of economics-based planning rules which are then applied within the planning offices to all capacity and coverage extension projects. The rules cover which equipment should be used and the size of increments, based on the type of project, as we will see later in this chapter.

The size of equipment increment that should be planned is an important parameter for the planner, since if they are too large for the predicted rate of traffic growth there will be a large burden of spare plant (BSP, see Chapter 5), while too small an increment will mean more capacity will be required soon. Normally, the size of increment is set by the rate of demand such that sufficient capacity is installed to cover for a set time of growth (known as the 'design period') – e.g. enough capacity to last for 2 years of growth [4]. The design periods vary for different types of equipment due to their cost and the complexity of installing them, so as to optimise the BSP. In addition, of course, the size of increment will be dependent on the modular design of the equipment or plant. Take as an example: optical fibre cables which are available only in sizes of 2, 4, 8, 16, 32, 48 and 96 fibre pairs; so a forecast growth in demand of 24 fibre pairs over the design period would have to be met by either using a 32-pair cable incurring extra spare capacity (and BSP), or the installation of the cheaper 16-pair cable which will exhaust before the normal design period is up. Box 6.2 presents the economic drivers for design periods and gives some common examples.

Box 6.2   Design periods

Figure 6.8(a) shows the relationship between the curve showing forecast growth or historical (i.e. actual) demand and how long the increments of capacity will last before all spare capacity is used up. The optimum design period for each type of equipment/technology is derived from the cost trade-off between the burden of spare capacity and the planning and installation cost (Figure 6.8(b)). The other driver for design period – the economic order size – is shown in Figure 6.8(c), in which the stock holding cost is traded-off with the procurement cost, plotted against batch size.

Examples of design periods are:

Building sites: 20–40 years
Buildings: 10–20 years
Switching and transmission plant: 2–3 years
Cables: 5–10 years
Duct: 20 years

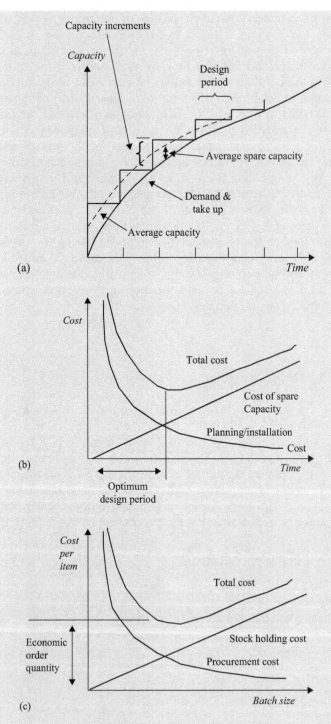

Figure 6.8   Design periods and their drivers [5]

We can group the planning rules into three sets:

(a) *Extending capacity* – i.e. Increasing capacity in existing networks at existing sites using the same technology, e.g.:
   • additional line cards in Subs Concentrator of PSTN Local Exchange

(b) *Extending coverage* – i.e. Increasing geographical coverage of an existing service through the provision of new sites or routes using the same technology, e.g.:
   • installing new cell sites to extend coverage of mobile 3G service

For both these cases, dimensioning and choice of technology depends on application of existing planning rules.

(c) *Deploying new network or technology* – i.e. Rollout of new network (new sites and links) using new technology, e.g.:
   • Deployment of 4G cell sites
   Deployment of NGN nodes and links

In these cases, dimensioning and choice of technology depends on the application of new planning rules that need to be developed once experience with the early deployments has been achieved.

6.5   Network planning: capacity and coverage

6.5.1   Introduction

This section attempts to summarise the salient points in the planning of the range of telecommunications networks to a level that is consistent with a review of telecommunications business. More detail can be found on this huge subject in the various books and journals cited. As a scene setter, Box 6.3 and Figure 6.9 present a comprehensive but simplified logical architectural view of the range of networks (often referred to as 'platforms') that may be owned by a network operator. (A physical architecture view, showing all the different network elements and how they are connected, in shown in [2].) Figure 6.4 shows how the mobile network, PSTN and the other platforms all rely on the access and core transmission networks for the communication conveyance and on higher functional layers for signalling and control, with the OTT network-function resources platforms discussed earlier. All the network functional layers are supported by the operations and maintenance platform, shown as a virtual plane in Figure 6.3 (described in Chapter 9). Network planning within a Telco embraces all these networks and platforms, as described further.

Box 6.3   The typical range of a Telco's networks

Figure 6.9 shows how the Telco's networks can be viewed as a functional architecture of a set of horizontal planes, each supporting those above. At the lowest level are the transmission networks (fixed and mobile Access, Backhaul and Core Transmission) providing transport for the traffic of all the higher planes. At the next level there are potentially several layers – e.g. ATM, MPLS and Ethernet – with IP and circuit switched exchanges above providing data routeing and voice switching, respectively. The traffic routes between these nodes are actually conveyed as digital bits in the transmission systems of the bottom plane. At the next level there are potentially several layers – the ATM, MPLS and Ethernet, with the level above, IP in the case of data services, and the circuit-switched telephony layer providing the PSTN and other services. The traffic routes between the exchanges in this level are actually conveyed as digital bits in the transmission systems of the transmission networks discussed further. A similar arrangement applies to the MPLS (multi-protocol label switching) label-switched paths and ATM (asynchronous transfer mode) virtual paths. The SS7 signalling for the circuit-switched exchanges, and the SIP for the IP network, resides in the next logical layer – again the physical conveyance of the bits for both forms of signalling are carried in the bottom transmission layer (e.g. TS16 in a 2Mbit/s block for SS7 signals). Above this level, there is the set of control nodes – IN for telephony, DNS, for IP-based data networks, etc. – with the physical link between nodes carried in the lower layers. Finally, the top layer comprises the various functional nodes (data centres, IPTV servers, cloud servers, etc.) – with links physically in the lower layers.

To complete the picture the operations and maintenance (O&M) functions are applied to all the layers – hence the vertical positioning of this platform in the architectural model. (However, purists might feel that the O&M plain is actually a horizontal at the very top of our model, since its communication links are also conveyed by the lower layers.)

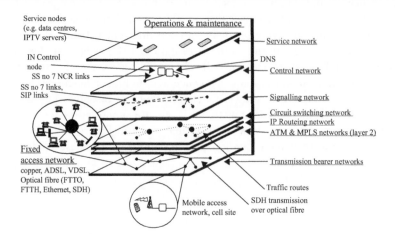

Figure 6.9   A Simplified view of the range of a typical Telco's networks

6.5.2   Fixed Access Network planning

All fixed Access Networks have the important characteristics of providing a permanent connection from each subscribers' premises to a serving exchange building. Such
networks for incumbent Telos, many of which have been in existence since the 1940s,
were constructed to support telephony service on the PSTN and are predominantly
based on copper cables, as described in Chapter 5, and shown in Figure 5.11a.
A substantial infrastructure of overhead poles, street furniture (connection boxes, etc.)
and underground duct-ways supports these copper networks, covering nearly all areas
of subscriber locations. The planning objective in such circumstances is predominantly constrained by the need to continue the exploitation of this infrastructure
in growing the capacity, and to a lesser extent increasing the geographical coverage of
the network. Although the copper network is designed to meet the narrow-band
transmission requirements for telephony, it has proved remarkably versatile since the
1990s in supporting progressively higher-speed transmission rates for ISDN, leased
lines and broadband (ADSL and VDSL) services.

However, fixed access networks introduced to provide essentially non-voice
services, such as TV distribution (i.e. CATV operators) and high-speed broadband
are instead based on coaxial cable and optical fibre, and in some cases microwave
radio links. (Although it is interesting to note that the CATV networks in the United
Kingdom also include copper pairs alongside the coaxial feeds in the final drop
from street multiplexor to provide their telephony service [6].)

All types of fixed access networks present the planning challenge of creating
sufficient spare capacity exactly where it will be needed in the future – e.g. extending
along the road in question – without incurring an unacceptably low asset utilisation
factor (AUF) (see Chapter 5). As a general rule, the highest proportion of the total
cost of an access circuit is in the part closest to the subscriber (where there is no
spreading of costs among other subscribers).

6.5.2.1   Capacity increase

Copper network

Normally, the addition of capacity separates into the E-side (i.e., from exchange to
primary connection point, PCP) and the D-side (i.e., Distribution from PCP to
subscriber's premises), as shown in Figure 6.10. The D cables serve specific roads,
so the planner's local knowledge of likely new buildings, conversion of buildings
to apartments, etc., enables realistic forecasts at the micro level to be made, and
hence the need for any new cables. The need for extra cables on the E-side is
assessed by summing all the forecasts on the D-side of each PCP. As Figure 6.10
shows, the copper Access Network is made up of a series of cables on the E and D
side containing a range of numbers of pairs and of varying gauges. These are
jointed at the footway boxes and PCP cabinets to create the continuous subscribers'
lines. The planner's role is to ensure that there are sufficient spare pairs to meet the
target probability that a subscriber line can be created on demand – normally this
would be set at about 95% [7].

Unfortunately, the Access Network with its tree and branch topology does have
inherent problem with network resilience, unlike the core network which is generally

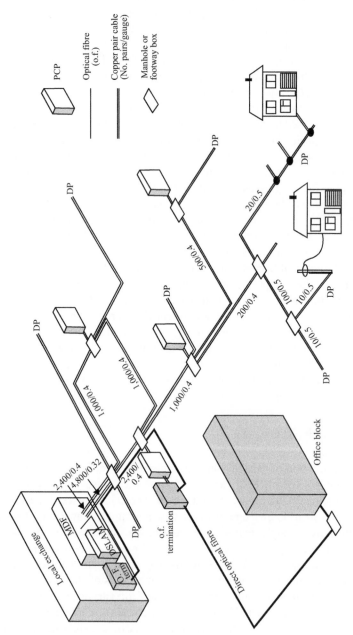

Figure 6.10 The fixed access network

highly meshed. Although in town centres, with close proximity of customers and adjacent cables, the resilience is better, this become a problem in sub-urban and rural areas where telephony and data services increasingly support businesses.

The planning rules itemise the appropriate cable sizes and copper-pair gauges, and the design period, to apply. Normally the extension of capacity is based on using the existing technology. In those cases where the need for extra capacity is urgent and normal planned provision would not be enough, there are expedient measures that can be taken. These include the use of pair-gain systems – i.e., multiplexers applied at the ends of existing pairs to double their capacity – which are easily deployed, but do have the disadvantage of not allowing the use of broadband systems (e.g. ADSL). Also, as mentioned earlier, such reactive solutions are more costly than the standard planned expansions.

Coaxial and optical fibre networks

Capacity expansion of such networks is normally done by the addition of line cards in the terminal units at the cable ends, or bringing in to use spare optical fibres or coaxial tubes in the existing cables.

The planning rules will also itemise the appropriate sizes of optical fibre or coaxial cables where new cable provision is needed and the design periods that should be applied.

6.5.2.2 Expansion of coverage

In extending the coverage of the Access Network the Telco has the option of using new technology rather than deploying more of the same, as in the case of adding capacity (described earlier). The planning rules cover the choice of technology – e.g. use of passive optical networks (PON) to cover large new sites, such as a new shopping centre away from the existing copper network [8], as well as the design periods.

Similarly, the planning rules for extending the coverage of the optical fibre network will indicate the size of the cables, the size and design of optical terminal units, as well as design periods. Such networks may be PON or point-to-point for fibre-to-the-home (FTTH) deployment. Similarly for optical fibre deployment to offices and other business premises using fibre-to-the-office (FTTO) systems.

A major feature of coverage expansion for any type of cable in the Access Network is the extent (and cost) of required new supporting infrastructure – ductways, pole routes, footway boxes, etc. Planning rules cover the specification of any new build needed, all of which will incur additional planning lead times.

6.5.2.3 Additional services

The rollout of new services, such as broadband ISP-access, is either provided using electronics to carry high-speed data over the existing copper telephony cables or it relies on the deployment of optical fibre for all or some of the link from user to exchange. In the case of using the existing copper pairs the electronic systems – known as asymmetrical digital subscriber line (ADSL) – are deployed at the subscriber's premises and terminate on access modules (DSLAMs) in the serving exchanges. The ADSL broadband systems are sufficiently adaptable to cope with a wide range of local line conditions, and so avoid the need to test to find suitable spare copper pairs from a cable – an important planning and cost consideration. Similarly, the rollout of ultra-fast broadband uses a hybrid fibre-copper solution – very high speed digital subscriber line (VDSL) – based on the use of existing non-selected pairs from the subscriber's premises to a near-by street cabinet which terminates an optical fibre link back to the termination at the serving local exchange (LE).

In the case of expanding the service area, new provision of broadband terminating equipment will be required at the various LEs for both the ADSL and VDSL systems. In addition, new optical fibre cables need to be installed from the serving LEs to each of the street cabinets, as shown in Figure 6.10. This new fibre provision may be able to use spare bores in existing duct-ways, or else expensive new duct provision will be required. All of these aspects, together with dimensioning, lead time and design periods, etc., are covered by the planning rules.

6.5.3   Mobile access networks planning

Mobile networks contain two distinct domains: the wireless cells providing the access and concentration functions, and the fixed core of the network providing the switching and distribution of the traffic [9], as described in Chapter 5 and shown in Figure 5.16. We will cover the latter in the following generic sections on switching, core transmission and data networks, since these elements are common for all sources of traffic.

The planning of mobile Access Networks is undertaken first to build the desired coverage through identifying the cell sites, and then to dimension of the capacities of each cell. The use of radio to provide the temporary access links from subscribers in a particular area on the one hand gives mobile networks their unique capability to provide personal communications on the move, while on the other hand radio propagation does impose complex and often unpredictable impairments to the performance of the service. Mobile access planning, therefore, involves a mixture of prediction using the laws of physics, together with technicians going into the cell area to measure the actual achieved signal strengths in the various locations.

6.5.3.1   Coverage design

Planning the rollout of new mobile service begins by examining the geographical areas involved. These target service areas are specified by the marketing and product management activities in the Telco, based on customer demand, product plans or pressure from the national regulator. Radio signal propagation is affected by the nature of the terrain traversed (e.g. hills, woodland, rivers and lakes,) as well as any man-made features, such as high buildings, bridges, moving vehicles, etc. These factors make the radio propagation both frequency dependant and time variant. The main planning and design steps are as follows.

(a)   The design work starts by gaining a detailed digital map of the areas to be served. This is then populated with the estimated densities of mobile users per unit area – these are often plotted as 'coverage polygons' showing the dense urban, urban, and suburban coverage levels, which will tend to form concentric circles around the town centres (see Figure 6.11). In the case of main

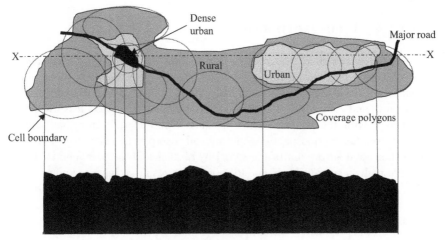

Terrain profile at section X-X

Figure 6.11   Coverage polygons and the nominal cells plan

roads, it can be expected that there will be demand for coverage along their length, creating corridors of users. As part of this process, clutter, land usage and presence of man-made obstacles is added to the map to present detail of the mobile-demand densities and the potential impairments to the radio propagation.

(b) Applying an average traffic-per-user figure to the coverage polygons enables a picture of demand to be plotted on the terrain map, so that a first attempt at defining cells is made – known as the 'nominal cell plan'. It is important to note that, although the cells are shown in Figure 6.11 as essentially circular, in practice due to the terrain and wireless propagation impediments the effective shape of the cells is much more irregular, and has variable levels of over-lapping with neighbouring cells.

(c) Suitable antenna locations are then identified for each cell, based on the nature of the terrain, and the coverage polygon of user-demand density. Within each cell, probabilities of the mobile signal strengths radiating out from the antenna are calculated at the main points of interest – outdoors, indoors, in car, etc. going up to the periphery of the cells – using the propa-gation model. Box 6.4 describes the propagation model based on the standard radio link-budget formula. It should be noted that the cells are not always circular: in the case of hot spots or where the coverage is needed along a road, highly directional antenna are used to focus the available wireless signal appropriately.

(d) The practical aspects of creating antenna sites are then examined with con-sideration given to road access, electrical power access, local authority acceptance of radio masts in such locations, and other factors, such as: local residence opposition, possible building developments in the area, etc.

Box 6.4 Propagation model

A simplified model of the wireless propagation between cell antenna and mobile handsets is shown in Figure 6.12. The path loss of the two wireless channels (Go and Return) – 'downlink' and 'uplink' (or in the US standards: 'forward' and 'reverse') – is determined by the free space loss, which decreases with the square of the distance travelled (d^2), together with attenuation due to atmospheric scattering and absorption. Other losses due to reflections and absorption of the signal by obstacles in the path, and by the several paths a signal travels due to reflections, result in, what is known as, the 'multipath channel'. This makes the power received a random variable, which can be predicted approximately by models developed either through measurements or through reflection- and refraction-based physics.

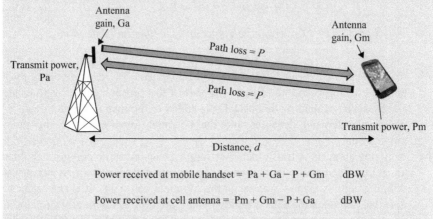

Power received at mobile handset = Pa + Ga − P + Gm   dBW

Power received at cell antenna = Pm + Gm − P + Ga   dBW

Figure 6.12   Propagation model for mobile network planning

The actual power received by the handset depends on the transmitted power at the cell antenna, modified by the directional gain introduced by the antenna itself, the path loss to the user (which decreases with d^x, where x ranges from 2 (free space) to 4 depending on the nature of the channel), and the directional gain provided by the user's handset antenna. The role of the propagation model is to create a set of probabilities of the signal strength at key points within the proposed cell areas so that the planners can determine their suitability [10].

(e)   Finally, the transmit power of the antenna at each of the cells needs to be set so that there is sufficient power to give an acceptable signal strength at mobile handsets on the periphery of the cells on the one hand, while not introducing interference to other cells in the vicinity on the other hand. The appropriate power levels can be predicted using Monte Carlo simulation tools covering a region or the whole mobile network [11]. Since all the antennas act as both transmitter and receivers, this technique uses iteration to allow for the mutual interdependence of the transmit powers of all the antennas on the amount of interference incurred at each receivers. For example, the transmit power needs to be increased to improve the strength of signal in relation to interference at the mobile handset, but this then increases the strength of interference by this transmitter at all neighbouring antennas, who in turn need to increase their transmission powers to improve the reception of their mobile handsets – thus worsening the signal noise level [10].

6.5.3.2 Capacity design

Once the set of nominal cells has been defined, which then sets the coverage area, the necessary capacities within each cell can be determined.

(a) An estimate of the generated traffic is made for each cell, based on market intelligence about the distribution of people within the coverage polygons and their activities. For mobile systems with data – i.e. GPRS, 3G and 4G – there is a particular need to predict the amount and type of data usage, and the associated quality-of-service requirements. It is important at this stage for hot spots, such as clusters of shops, to be identified. The number of radio channels required for each cell is then estimated based on the grade of service (GOS) to be offered using traffic tables of Erlang's formula (see Chapter 5).

(b) In the situation where an established cell site has experienced more growth in mobile traffic than planned, there are a number of ways in which extra traffic capacity can be created to alleviate the call congestion and improve the GOS for users. One method is cell splitting, in which the cell is divided into a number of sub cells or sectors, thus providing a proportion of the available channels in higher concentration where required for users, at the expense of reducing the available channels in the rest of the cell. Another method is to introduce to the congested area another cell with its separate set of channels – this might be on the basis of a microcell using highly directional antenna set to provide capacity in a small focused area, say within a shopping mall [10].

6.5.4 *Switching networks planning*

Planning and dimensioning of circuit-switching units ('exchanges') – in both mobile and fixed networks – is determined by two main parameters, namely: the amount of traffic (measured in Erlangs) to be switched [12] which sets the size of the switch-block, and the number of traffic routes terminating at the unit which sets the number of ports on the switch-block. Each traffic route terminates on a separate set of ports at the switch-block – these are at 2 Mbit/s (30 channel) or 1.5 Mbit/s (24 channel) depending on the standard used; thus large capacity routes will require several ports. The planning process for circuit-switched networks is, therefore, first to construct the traffic routes between the exchanges and then to calculate the total amount of traffic at each exchange that needs to be switched during the design period.

6.5.4.1   Number of traffic routes terminating at switch-block

The traffic routeing of a circuit-switched network is organised in a hierarchy, as described in Chapter 5 and shown in Figure 5.6. The hierarchy is established by a set of mandatory traffic routes between exchanges and their designated parent exchange. At the top of the hierarchy all the exchanges are fully interconnected by mandatory routes. This concept extends in the international network, as shown in Box 6.5.

Box 6.5   Standard national and international switching hierarchies

There is an internationally agreed set of designations for the various levels of exchange within national PSTN and international networks, as shown in Figure 6.13 which also gives the designations used in the United States and United Kingdom.

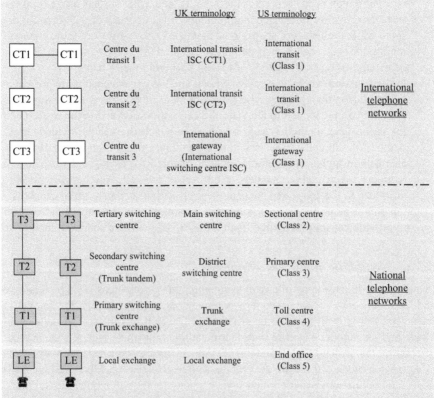

			UK terminology	US terminology	
CT1	CT1	Centre du transit 1	International transit ISC (CT1)	International transit (Class 1)	
CT2	CT2	Centre du transit 2	International transit ISC (CT2)	International transit (Class 1)	International telephone networks
CT3	CT3	Centre du transit 3	International gateway (International switching centre ISC)	International gateway (Class 1)	
T3	T3	Tertiary switching centre	Main switching centre	Sectional centre (Class 2)	
T2	T2	Secondary switching centre (Trunk tandem)	District switching centre	Primary centre (Class 3)	National telephone networks
T1	T1	Primary switching centre (Trunk exchange)	Trunk exchange	Toll centre (Class 4)	
LE	LE	Local exchange	Local exchange	End office (Class 5)	

Figure 6.13   Standard designation of national and international switching hierarchies

In the case of 2G, GPRS and 3G mobile networks, the mandatory routes are between cell sites and base-station controllers (or equivalent) and parent MSC (mobile exchange), together with links from each MSC to a gateway MSC. However, in addition, the 4G network has the new requirement for optional and mandatory routes between the cell base stations, giving more efficient traffic routeing between adjacent cells.

The key point is that all telecommunications networks have a set of mandatory traffic routes between their network nodes – necessary to ensure a minimum level of connectivity, usually forming a traffic-routeing hierarchy.

In addition, so-called auxiliary or optional traffic routes are provided between exchanges where there is sufficient co-terminal traffic to warrant a direct link (the direct versus tandem cost calculation) [4]. The threshold for warranting such a route depends on the distance between exchanges, and any administrative, regulatory or charging impediments to their direct connection. Typical examples of auxiliary traffic routes include LE to LE, LE to non-parent trunk exchange, and MSC to non-gateway MSC. These routes provide an overall network economy where there are high levels of traffic between two nodes through bypassing the backbone mandatory network and perhaps several stages of switching. As well as reducing costs, the inclusion of auxiliary routes introduces meshing within the network structure, which significantly increases its level of resilience, as described in Chapter 5, section 5.4.6 (see also Chapter 9).

Traffic routeings between networks are usually strictly limited to the mandatory routes between the appropriate gateway exchanges, although there are possibilities for auxiliary routes in exceptional cases – e.g. between LEs either side of a land border between countries.

Figure 6.14 shows the set of mandatory and optional auxiliary routes for a fixed PSTN and a mobile network. It is interesting to note, that although the mandatory routes are necessary to form the hierarchy, in practice the majority of traffic tends to flow over the various auxiliary (optional) routes, which consequently can be quite large (since these are provided only where high levels of traffic are expected to flow).

6.5.4.2 Switch-block traffic

The amount of traffic to be switched by a switch-block in an exchange depends on its position and role within the network.

In the case of a local exchange the originating traffic level is set by the number and type of subscribers in its catchment area of co-sited and remote concentrators. In addition, the switch-block needs to switch incoming – i.e. terminating – traffic destined for it subscribers, plus any transit traffic handled by the exchange.

For exchanges without subscribers, i.e. exchanges at higher levels of the hierarchy (e.g. junction tandems, trunk, international, mobile, and VPN), the traffic to be switched is determined by the total amount of originating and terminating traffic within their catchment areas of dependent exchanges, as well as the levels of transit traffic to be handled by that exchange.

Box 6.6 gives an illustration of the dimensioning of a fixed network local exchange, which has subscriber lines as well as a mandatory route to its parent trunk exchange and several auxiliary routes to neighbouring local exchanges.

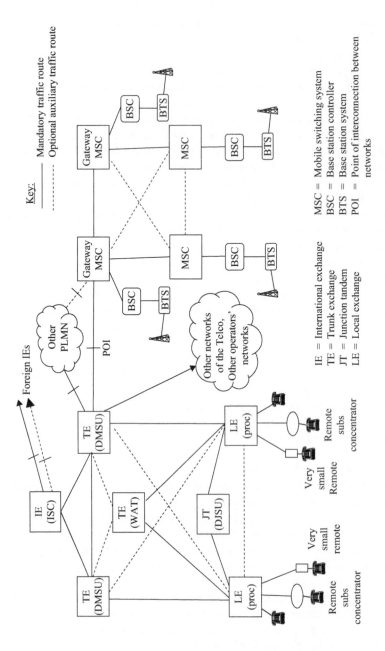

Figure 6.14 The set of traffic routes in a typical fixed and mobile network

Box 6.6   Example of planning a local exchange

The dimensioning for the local exchange covers the following segments (Figure 6.15 refers):

(1)   Number of subs line cards is dimensioned on the basis of one per working subs line, plus an appropriate allowance for contingency and maintenance spares. The line card quantities can be applied on an aggregate basis of all subscriber lines on remote and co-located sub-scriber concentrators, since the cards can be easily deployed from a central pool on a plug-in basis when required.

(2)   Total originating and terminating traffic to be switched by each concentrator is determined using the average both-way traffic rate per subscriber and the number of subscriber lines. In addition any transit traffic carried between other exchanges should be added to create the grand total of switched capacity (in Erlangs) required at the end of design period [12].

(3)   The number of required 2 Mbit/s ports on the switch-block is deter-mined from summing the 2 Mbit/s (30ch PCM) streams from the co-sited and remote concentrators, plus the 2 Mbit/s streams for the routes (mandatory and auxiliary) to other exchanges, together with the 2 Mbit/s streams carrying all the Signalling System SS7 traffic from the switchblock to the SS7 terminal.

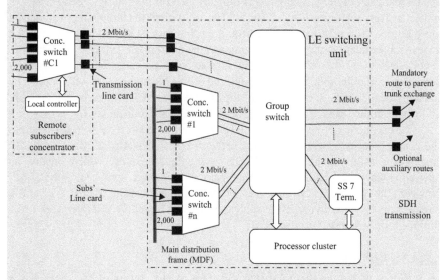

Figure 6.15   Example of dimensioning a local exchange

6.5.5   Data networks planning

The planning of data networking by a Telco is taking on an increasingly important role as the proportion of data to voice traffic continues to rise rapidly, driven by increasing use of data and non-voice services by (fixed and mobile) subscribers. However, the Telcos are also undertaking network transformations (e.g. the shift to NGN and 4G) in which voice services are handled as VOIP – i.e. data – rather than being circuit switched in the MSCs and PSTN exchanges. Even the leased line ('Private Circuit') set of services – originally provided on a permanent circuit basis – have experienced significant run downs, as subscribers are shifted to ethernet and other packet-based alternatives. This profound shift of traffic from the predictable manageable quality controlled circuit domain to the unpredictable, and more difficult to manage packet domain, creates a new set of challenges for the planner. Many of the approaches used for planning circuit-based services and networks have to be replaced by rules of thumb based on trial and experience.

There is a wide range of data and non-voice networking undertaken by a Telco, all of which necessitate an appropriate degree of network planning, as summarised below.

6.5.5.1   Carrier's backbone network

All Telcos have established a backbone data network across their territory. This is usually based on an extensive ATM (asynchronous transfer mode) network which provides a solid Layer 2 (of the OSI model) platform of virtual paths between ATM switches, interconnected by high capacity streams on the SDH (synchronous digital hierarchy) – based core transmission network [13]. The planning objective is to ensure that each of the virtual paths is dimensioned for the future forecast levels of traffic, at least for years 1 and 2 ahead. The key points for the planner to manage are that the ATM platform needs to route the Internet traffic from the broadband access subscribers via the local exchanges to the appropriate ISP point of presence (POP). However, the ATM platform is also used to carry business customer services – e.g. point to point ATM paths and VPNs (virtual private networks). The planning challenge is to cope with all the diverse uses of ATM across the network and ensure future capacity for years 1 and 2 is equitably allocated.

The backbone network has an extensive set of high capacity IP routers across the country, accessed via the Layer 2 network of ATM platform. Many of the IP routers are MPLS (multi-protocol label switching) enabled, which creates an alternative layer 2 platform of virtual paths – i.e. label switched paths (LSP) between the routers [13]. The use of MPLS introduces a degree of traffic engineering to the network of IP routes, as well as enabling VPN services to be provided for business customers [14].

6.5.5.2   ISP role

Most Telcos undertake the role of an ISP. This means that, in effect, the Telco is providing (and operating) a part of the Internet. The planning challenge is that sufficient capacity is required from all of the broadband access nodes (local exchanges), not only from residential subscribers, but also from business customers who will have high-speed leased line links from office locations to the Telco's ISP node. One of the key variables for broadband access is the level of contention at the point where subscriber access broadband lines terminate (aggregation point, as described in Chapter 5).

Another important factor for the planners is the provision of the high-capacity links from the Telco's ISP node to the interconnection with other ISPs, using a range of peering arrangements. There are a range of ISPs to whom appropriate peering connection is required – transit ISPs, access ISPs, and application ISPs [14].

6.5.5.3   Broadband access

The planning of broadband access networks, which are primarily used for Internet-based services, was covered in sections 6.5.2 and 6.5.3.

6.5.5.4   Data service offerings

Apart from broadband access, Telcos offer a range of other data and non-voice services. These include services offered to business customers who run their own data networks within their offices, factories, campus sites, etc. – examples being MPLS, ATM, ethernet and VPN. The planning of coverage and capacity of such services is based first on the forecasts from the relevant product managers, and second on the basis that the Telco will be using parts of these data platforms for its own communication purposes – e.g. the ATM platform serves both external customers and provides a platform within the Telco's network for many of the other data services such as broadband access to ISPs. The planning challenge is then to aggregate the various capacity and coverage demands, as well as to ensure that adequate capacity is provided on the backhaul and at contention nodes to ensure appropriate quality of service is achieved. The planning design periods tend to be relatively long for these aggregated data services to ensure continuity of service, recognising that many of the customers are businesses and large corporates who will tend to favour long-term contracts for service with the Telco.

6.5.6   *Planning the Telco's various OTT functional networks*

The various functional nodes, such as the Telco's data centres housing the range of servers and data stores, the cloud computing servers, etc., are all planned on the basis of forecast capacity for the component usages. While the relevant product managers will probably provide the outline capacity forecasts, these will tend to be vaguer than many of the telecommunications services due to the greater levels of volubility and unpredictability of such services. Other functional nodes, such as IPTV, with its portfolio of services – pay TV (e.g. Sport), video on demand, games, shopping channels, etc., will also be dimensioned on the basis of product manager forecasts. The design periods will be large enough to ensure a wide margin to cope with the peaky nature of this traffic.

A particularly important feature of planning for TV nodes is the time-of-day (and possibly seasonal) nature of the customer demand, which creates some challenges in the dimensioning of the nodes and the planned capacities of the transmission links to the serving nodes in the network. The design, dimensioning and planning of such networks depends on achieving an optimum architecture in terms of how best to distribute video content, which is changeable on demand, to a wide population of customers on the Telco's network. The principle is that the network video streaming costs are optimised by the use of strategically placed local storage nodes (video caching) to hold the more popular video content so that undue multiple paths from source to all the current users are avoided. More information on this vast subject may be found in Reference 15.

6.5.7   Core transmission network planning

As explained in Chapter 5, section 5.4.5, a Telco's Core Transmission Network acts as a common facility to carry the traffic between all the network nodes. The planning task, therefore, is primarily one of assembling the total set of traffic requirements between all the voice switching (PSTN, Mobile, VPN, etc.) units, all the data network units (IP routers, ATM switches, Ethernet switches, etc.), data centres, leased-line cross-connects and multiplexors, together with any interconnect traffic. All the different switching, routeing, cross-connects and multiplexing units are housed in exchange buildings around the country, with a core network transmission unit serving each building (which is usually located on the ground floor close to the cable entry points).

The traffic requirements are quantified in terms of the basic transmission building blocks – i.e. 2 Mbit/s, 34 Mbit/s or 144 Mbit/s in the United Kingdom, Europe and generally internationally; 1.5 Mbit/s, 6 Mbit/s, 45 Mbit/s in North America. Ideally, the traffic figures cover the existing (year 0) with forecasts covering the next 5 years, although it is usually years 1 and 2 that are used for the short-term planning of the Core Transmission Network. The result of this activity is a traffic matrix covering each of the planning years, showing actual and forecast traffic between all of the units across the country.

The next step is to map the traffic demand matrix – which identifies the traffic routes across the country – onto the cable and microwave radio link network routes (known as 'engineering routes') across the country. Since there are more transmission network nodes than there are exchange buildings – (in the United Kingdom by a ratio of about 3:2 [7]) – the traffic routes will often be carried over two or more engineering routes in tandem. At each transmission node the incoming tributary channels are extracted from the cable or radio systems and inserted into outgoing cable or radio systems, allowing the optimum loading of the transmission network, as described in Chapter 5.

Once all the traffic has been mapped to the engineering routes in the core transmission network for years 1 and 2, those routes that will require additional capacity are identified. In many cases, the extra capacity can be easily provided through additional slide-in cards at the SDH multiplexors and termination equipment. However, the planning process usually includes an examination of whether the need for extra capacity can be avoided (or at least postponed for the new future) by the re-routeing of the traffic on to other less full engineering routes, even if this involves some circuitous routeings.

The output from the planning process is the allocation of traffic routes to engineering routeings for the next 2 years, identification of transmission systems that need extra capacity, and any major construction work to create new or enhanced transmission capability.

Typically, the Core Transmission Network planning criteria may be summarised as follows [16, 17]:

1. *Meeting capacity for the forecast period*: Covering the total capacity for the design period for all services carried between the network nodes.
2. *Provision of sufficient spare capacity*: To cover potential growth above forecast, a maintenance allowance and ability to accept re-routed traffic.
3. *Achieving acceptable network utilisation factor (NUF)*: That is, the ratio of working to total capacity meets the target for the company (as described in Chapter 5).
4. *Meeting regulatory requirements*: Covering interconnection to other networks. Also, non-discrimination criteria, such as service transparency and equivalence in dealing with other operators' traffic.
5. *Provision of target network resilience*: In terms of how the traffic is spread across routes – using diversity rules as well as sparing and link protection [12].
6. *Meeting maintainability requirements*: covering network loading, use of new equipment, monitoring capability, routeing flexibility and levels of diversity.

The planning rules are developed to meet the above criteria (1–6) such that the overall capex is minimised. Alternatively, the rules may be designed to minimise the total opex – which is less common and more difficult for the company to quantify. Box 6.7 gives a description of the steps involved in planning the Core Transmission Network.

Box 6.7   The planning of the core transmission network

Figure 6.16 illustrates the planning process, and provides a summary of the points above. It shows a portion of the network with four exchange buildings, A to D, each having a PSTN telephone exchange, a mobile exchange (MSC), several data nodes (IP, ATM, etc.) and a leased-line digital cross-connect unit. The traffic requirements for the units in Exchange A are indicated, and are allocated to the traffic routes between A and B, A and C, and D. The figure also shows how the traffic between A and D is mapped on to two engineering links A to H and H to D, with the transmission node H (not in an exchange building) acting as the 'flexibility point' allowing the optimal packing of the transmission systems terminating there.

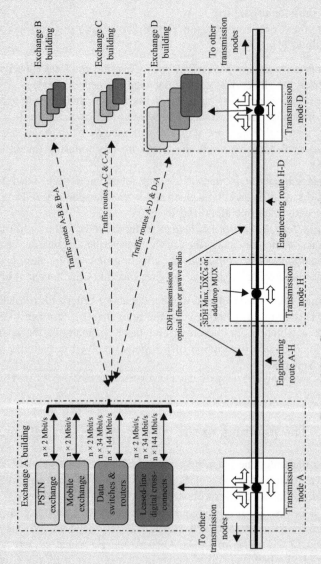

Figure 6.16　Core transmission network planning

6.5.8   Computer-based planning systems

The planning process for each of the network sectors described earlier is, of course, supported by a set of computer-based planning systems. There is a range of software tools specifically tailored for dimensioning copper networks, evaluating the cell site radio transmission properties, optical fibre transmission systems, etc. Importantly, these individual systems are usually grouped into suites of planning and dimensioning tools used in the planning offices. Of course, also needed are the various databases itemising the line plant records, network inventory and configurations, traffic and circuit records, etc., which feed into planning tools. In some cases the full extent of the planning process for each of the network sectors is realised through an integrated suite of computer-based systems. This concept is illustrated in the description of operational support systems in Chapter 9, Figure 9.18.

6.6   Network planning: transformation and conversion

6.6.1   Introduction

Given that the ICT industry is developing at a fast rate, there is a continuing need for Telcos to introduce equipment upgrades (software and hardware) and new technology in order to meet service needs and reduce costs. However, for Telcos this raises two major challenges: first, the upgrades and new technology need to be introduced without breaks in customer network service – unlike much of the ICT industry – and second, the sheer scale of the telecommunications networks coverage and capacity means that to totally replace old equipment by the new technology will take a significant length of time.

The network conversion periods are set by the rate that the new equipment can be produced by the suppliers, as well as the practical limits on how quickly the equipment can be installed. For example, when BT undertook the major programme of replacing its analogue PSTN by digital TDM switching systems it faced the task of converting some 6,500 local exchanges and over 400 trunk, junction tandem and international exchanges. It took several years to build up the conversion rate to the maximum of some 4M subscriber lines per year – that's over 15,000 lines changed over each working day! The total conversion period for replacing all the old exchanges and transferring all 20M subscriber lines to the new network, with almost no breaks in service was some 12 years [18]. In addition to the physical change out of the switching and transmission equipment, the use of the new TDM digital technology warranted a radical change to the network traffic routing structure, as well as the number of trunk exchanges, due to the different ratio of switching to transmission costs (as explained in Box 5.2).

Other examples of network transformation include:

- The introduction of IPv6 and gradual replacement of IPv4 routers in the Internet and private IP networks.
- Introduction of fixed NGN and the replacement of the (digital TDM) PSTN.
- Conversion of LTE version of 4G to the fully IP version, using VOLTE for voice calls.

In this section, we look at the different strategies for network transformation to implement new technologies and network structures. Clearly, given the level of complexity and the scale of the task, the network transformation needs to be controlled by a well-structured programme using project management discipline, as described in the appendix.

6.6.2   The objectives of a network-conversion strategy

Telcos need to follow a carefully devised plan for the implementation of new technology into an existing network, which in the case of a large scale change-out needs to be incorporated into a comprehensive network conversion strategy. Although the length of conversion periods will vary between programmes, they will tend to be at least 2 years and may be as much as 10–12 years for the major change outs. The objectives of network conversion strategies in general are briefly described below. It should be noted that some of the objectives may in practice be mutually counteracting, and so a balanced judgement may be needed in formulating the optimum strategy.

(a) *To make best use of available equipment from the manufacturers.* As mentioned earlier, the quantities of new equipment that can be supplied during the conversion period may be a limiting factor, particularly in the early stages of the build-up, and it is important that the available equipment is deployed in the most effective way to gain maximum advantage. Thus, it would make sense to choose locations for the first wave of equipment where the target customer segment can use the enhanced facilities or services offered by the new network – so gaining early revenues.

(b) *To keep the costs within the capital and current account expenditure within the programme budget.* In addition, the objective is to create an acceptable CapEx profile – namely, one which smoothly rises to a moderately level plateau before gently dropping at the end of the conversion period. In some situations, the rate of installation of new network equipment may be limited by the need to adhere to an acceptable capex profile rather than any constraints on the equipment availability.

(c) *To minimise the amount of inter-working between new and old networks.* Throughout the conversion period there will be a need to provide interconnection between the old and the new networks, in order to keep service continuity. This means that interworking equipment will be needed to convert between the old and new protocols (e.g. analogue-to-digital/TDM, IPv4-to-IPV6; TDM/PCM-to-IP/PCM, and SS7-to-SIP). However, this interworking equipment not only introduces some performance degradation, but it is also ephemeral since it is not required once the network conversion is complete. In some cases the minimisation of the cost and performance impairments introduced by the interworking equipment may be the determinant of the chosen conversion strategy.

(d) *To maximise the availability of new services to the appropriate market segment of customers.* As indicated in (a) above, it is important to consider the service coverage spread of the new network as it is rolled out. If the business case for the programme depended on a focus on particular customer segments then the most demanding driver for the network conversion strategy could be its rapid deployment in specific geographical areas.

(e) *To maintain the levels of network resilience.* Unfortunately as one network is gradually withdrawn and the new one installed there will inevitably be areas where lack of network continuity will give rise to vulnerability to overloads and breakdowns. With this phenomenon occurring across the network the overall levels of resilience will degrade. Remedial action may be necessary in the case of a long conversion period, or the particular vulnerability of important parts of the old or new network – alternatively the conversion strategy may be directed at maintaining overall network resilience.

(f) *To maintain the levels of network performance and QoS.* As described in (c) above the presence of interworking equipment inevitably introduces impairments to the network performance. Therefore, a key objective is often to ensure that the QoS degradation is minimised by carefully routeing the traffic over the old and new networks so that the amount of interworking equipment encountered is kept acceptably low. This may mean that the rollout of the new network, and the withdrawal of the old, is conditioned to facilitate the traffic routeing. For example, calls starting on the new network do not swap between new and old networks more than once during an end-to-end connection.

6.6.3   Interworking and cut-over

As mentioned in (c) above interconnection via interworking equipment is required between the old and the new network during the conversion period in order to maintain full connectivity. This situation is illustrated in Figure 6.17, which

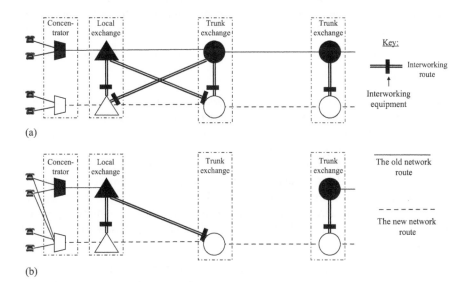

Figure 6.17   *An example of the interworking between old and new networks:*
            (a) interworking routes during the conversion period and
            (b) interworking routes just before cut-over of subscribers

shows the possible points of interworking when a PSTN is progressively replaced by a new network – (a) during the conversion period and (b) just before cut-over of the subscribers to the new network. Interconnection of the old and new network will occur through interworking routes within exchange buildings where both switching units of both networks are present. Normally that would be at trunk exchange and main local exchange sites. Sufficient meshing of the interconnect routes is provided to enable the routeing of calls between the networks to adhere to the rules, thus maintaining QoS, as described in section 6.6.2. Figure 6.17(a) shows that the new network units are installed adjacent to the old-network unit, with traffic carried by both – known as 'overlaying'. When the new network concentrator unit overlays the old unit, the existing subscribers could either remain terminated on the old unit, and new subscribers joined to the new unit (known as 'growth-only overlay'), or more usually, a batch of subscribers are re-terminated on the new unit as soon as it is installed (known as 'replacement overlay').

As the conversion progresses the existing switching units are replaced by their co-sited new units, as shown in Figure 6.17(b), with the subsequent removal of the interworking equipment now made redundant. The final stage of conversion is achieved by transferring the subscribers off the old concentrator unit to the new. The transfer of subscribers is made with just a momentary break in service by swapping between the parallel provision of lines at the MDF (main distribution frame). Once this is done the old concentrator equipment can be withdrawn. Eventually all the old switching units and their traffic routes will have been withdrawn and the new network fully established. It should be noted that the new network structure and routeing rules will be different to the old, and so there may not be a one-for-one replacement of each switching unit or traffic route. This means that when the PSTN is replaced by an NGN, the number of exchange buildings required may be reduced.

6.6.4   The range of network conversion strategies

The strategies for achieving a national network transformation – installing the new and withdrawing the old network – fall into essentially two camps: focused regional and national dispersed. These strategies differ in the way that the transition is organised, although they do, of course, start from the same position (just the old existing network) and finish with just the new network (old network being fully replaced). However, they do have differing characteristics in terms of meeting the objectives (a) to (f), above. Therefore, the choice of which type of strategy to follow depends on the circumstances and which objectives are the most important. In practice, there is also a third group of strategies, which are based on a hybrid of the two main strategy groups:

Focused regional: In this strategy the installation of new equipment is focused into conversion cells comprising a small section of the country. Each year more cells are converted, gradually covering the country (or target area if not the full country) until full conversion of the network is achieved. The effect is similar to completing a jigsaw puzzle. By its nature, the focused regional strategy concentrates the available new equipment onto specific areas, usually in the form of total replacement rather than overlaying the existing units – resulting in good capacity in selected areas, but poor overall coverage in the early days. Generally, the interworking equipment is restricted to the periphery of the conversion cells. Therefore, by careful choosing of cell areas the interworking costs and quality degradation for the network transformation can be minimised with this approach.

National dispersed: By contrast, with this strategy the installation of new equipment is confined to overlay units spread across the country (or target area) alongside the existing old-network units; the latter are frozen and growth in traffic is taken by the new network. Progressively, the new equipment is deployed to all locations and the overlay units then totally replace the old units. The main characteristic of this strategy is the thin spreading of the new network readily across the country, giving early widespread presence in the conversion period, but with restricted capacity – i.e. good coverage but poor capacity in the early days. Because of the widespread nature of the deployment of new equipment the need for interworking can be higher than for the focused strategy, depending on the degree of overlay at each network site and the extent of the traffic dispersion – with the consequential effect on interworking costs and QoS degradation.

6.6.4.1 Hybrid conversion strategy

This category covers a range of strategies which use a combination of the two basic strategies, and has a corresponding mix of coverage and capacity characteristics.

Box 6.8 shows how a simple example of how a target area of network would be transformed to a new network following the focused regional and then the national dispersed strategies. Although the example shows a fixed network situation, the approach is equally applicable to mobile networks, data networks, cable TV networks, etc.

Box 6.8 Conversion of a target area following either of the two strategies

The example target area is shown in Figure 6.18. It comprises two cities (A & E) and three rural areas (B, C, & D), each served by a trunk exchange with the traffic routes between them, as shown in Figure 6.18. The triangle shapes indicate main local exchanges and the figures in brackets show the number of dependent concentrators (remote and co-sited).

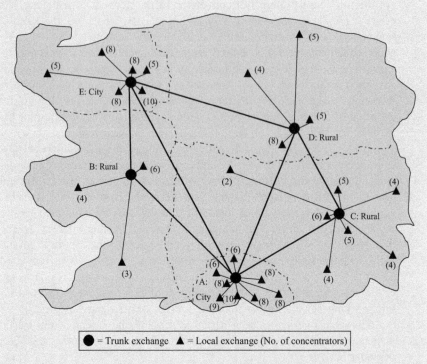

Figure 6.18   The example target area

Focused regional

The application of the focused regional strategy to the target area is illustrated spread over a conversion period split into four phases ($\phi1$ to $\phi4$). The new switching units and transmission links, shown as open and dotted lines, are assumed to totally replace the old equipment. As is typical with this strategy, the conversion begins with the replacement of the two city areas and the main transmission links between them in $\phi1$ and $\phi2$, followed in phase in $\phi3$ by the rural areas. By $\phi4$ the target area is total transformed.

National dispersed

The application of the national dispersed strategy to the target area is spread over five phases, with the starting point as shown in Figure 6.18(a) and finishing with $\phi5$ Figure 6.18(i) (same as Figure 6.18(d)). The roll out starts with installing overlay units at the major centres with links between them, and rapidly extends the new network to a thin overlay across the area. The old units are frozen and subscriber and traffic growth is taken by the new network. Once there is full coverage the capacity of the overlay is increased so as to replace the old units, with the final all-new network in $\phi5$.

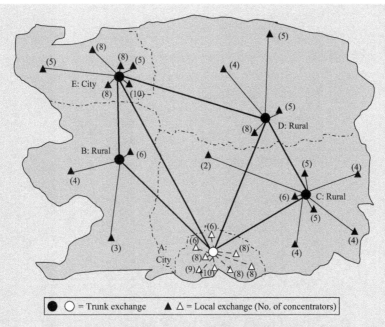

Figure 6.18(a)   Focused regional φ1

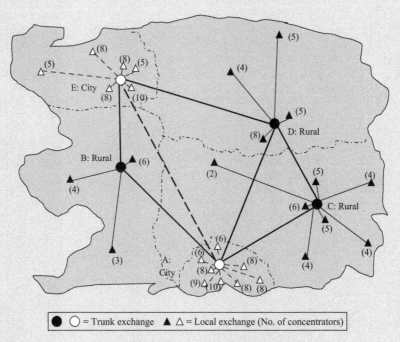

Figure 6.18(b)   Focused regional φ2

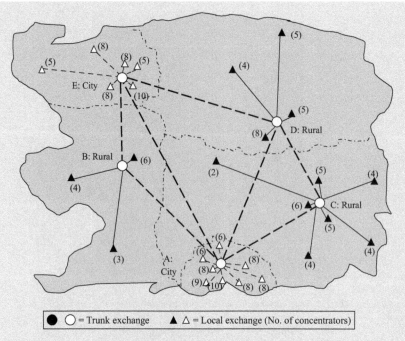

Figure 6.18(c)   Focused regional ϕ3

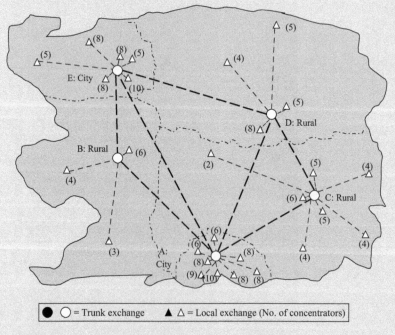

Figure 6.18(d)   Focused regional ϕ4, the complete new network

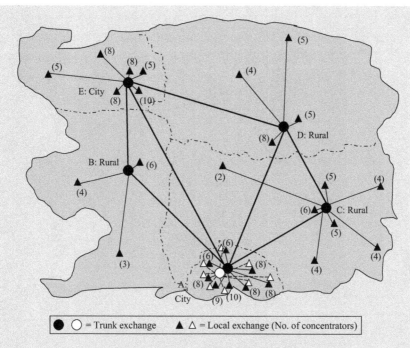

Figure 6.18(e)   National dispersed φ1

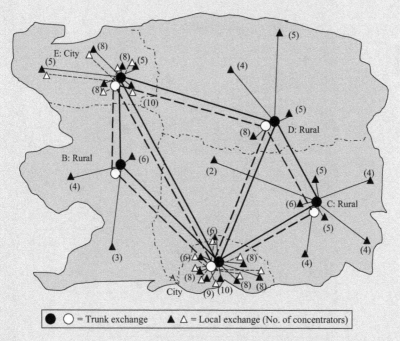

Figure 6.18(f)   National dispersed φ2

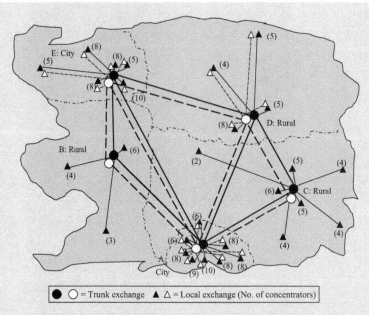

Figure 6.18(g)   National dispersed φ3

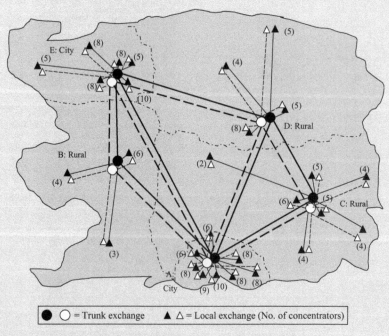

Figure 6.18(h)   National dispersed φ4

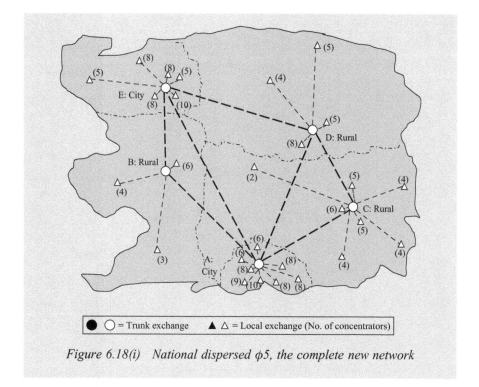

Figure 6.18(i)   National dispersed φ5, the complete new network

6.7   Building the CapEx programme

Now, we will consider how the capital requirements for each of the programmes, as quantified by the network planners, are incorporated into the company's annual CapEx build process. The first factor to appreciate is that of the capital *intermax* phenomenon. This occurs when there is a change of technologies in the network, as illustrated in Figure 6.19. The capital spent on the new network equipment purchased increases from zero to the steady state value for the duration of the conversion period, after which the amount spent remains roughly level to meet the annual growth for the network. In the same period the amount spent on the old network technology should in theory drop to zero immediately, but, in practice some continuing expense on the old network will be required in order to keep service continuity in the areas where the new equipment has not yet been deployed – as we saw in section 6.6. Thus, even though the capital cost of the new equipment is less than that of the old, the total capital requirements during the transition period may well increase from the previous levels before dropping to the new lower capex spend once the new network is fully deployed. This intermax may be an appreciable amount that has to be accommodated by the company's capital budget, and its management is an important factor for the Telco to consider when embarking on a major network-transformation programme.

Unlike the annual OpEx budget for a Telco, which is primarily driven by the size of the work force, the annual CapEx requirement is driven by the planned infra-structure built (i.e., network, functional and data centres). Thus, in theory, the CapEx budget required is the total of all the programmes planned for each of the next five

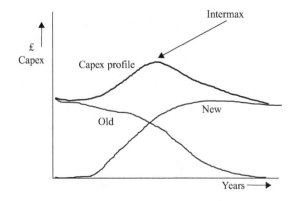

Figure 6.19   Capital profile intermax

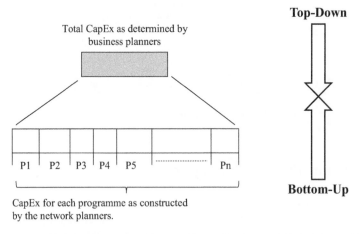

Figure 6.20   Top-down to bottom-up reconciliation

years. Aggregating all the programme CapEx requirements would constitute the, so-called, 'bottom-up' build. However, the capital expenditure each year is one of the major measures of a company's financial performance since it contributes to the depreciation line of the P&L, and presents the need for actual cash spending – often with the commitment for continued expenditure with equipment suppliers during a contract lifetime. Furthermore, the company may not be able to fund the CapEx from reserves and so will need to borrow or issue bonds. Therefore, the maximum allowable level of capital spent for the company will have been determined during the business planning process (see Figure 6.1 and Chapter 4). Invariably, this 'top-down' view of the CapEx budget will be less than the bottom-up figure, as shown in Figure 6.20.

The reconciling of the bottom-up and top-down views of the capital requirements leads to highly energetic discussions within a Telco as part of the annual budget-build process. The planners would argue that they have derived their required total amount spent through legitimate application of the company's planning rules to meet the company's service needs. While the business planners would take the view that the company cannot afford such expenditure – despite the fact that the future revenue projections may depend on the network and systems enhancements forming the new programmes. Normally, the bottom-up view can be massaged down by slipping the start dates of some of the programmes, delaying the rate of build-up of others, possibly seeking new commercial arrangements with the equipment suppliers, and in some cases freezing or deleting some programmes. There may also be the opportunity for some expansion of the top-down figure, subject to senior management approval. Finally, a reconciled annual CapEx budget for the company is agreed.

The CapEx budget for a Telco normally covers up to 5 years (beginning with next year), covering all expenditure on the company's physical assets – network, support systems, functional nodes (IPTV nodes, data centres, etc.). Box 6.9 illustrates the sort of mix of items in such a budget for a hypothetical Telco, showing how the expenditure on the different types of equipment changes year by year as the network is modernised. The total budget size will amount to several millions or even billions of pounds, depending on the size of the Telco and the extent of new infrastructure build.

Box 6.9   Example of capital budget for a Telco

The mix of assets purchased by the CapEx budget over a 5-year period is shown in Figure 6.21 at years 1, 3 and 5 for a hypothetical Telco. Although the quantities shown for each category are illustrative only, they do show the changing mix over the 5 years as the network becomes modernised using NGN equipment, which after an intermax at year 3 results in a lower overall CapEx budget from year 5 onward.

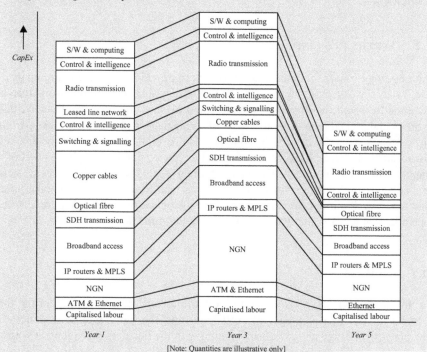

[Note: Quantities are illustrative only]

Figure 6.21   Capital budget over 5 years for a hypothetical Telco

The required CapEx budget also depends on the contractual arrangements for payments to the equipment suppliers, e.g. whether a proportioned amount is paid when contract is let, and how payment is made throughout installation, or deferred until after completion. Such payments can form a significant proportion of the CapEx budgets in early years and restrict the availability of CapEx for new projects.

In addition to the various categories of physical assets purchased, the CapEx budget may include the intangible item known as *capitalised labour*. Typically, Telcos capitalise the current account spent on the workforce undertaking the detailed planning (dimensioning, positioning, etc.) and installation of the equipment – on the basis that the assets cannot be realised unless such work is undertaken. Such costs are added to the relevant asset expenditure and depreciated accordingly.

Finally, the size of the CapEx budget is usually a highly publicised figure, since it gives an important indication of the company's fortunes and its confidence in the future. Most CEOs and company chairmen make a virtue out of the 'investment' in the future that the company is making. The industry analysts also examine the publicised CapEx expenditure, as well as the more detailed figures in the company's report and accounts for the previous financial year, in order to ascertain the network strategy being followed and what this might mean for its future business prospects. Hence the importance for a Telco's network planning activity throughout the company to be of high quality!

We cover the financial aspects of capital planning in greater detail in section 4.4.3.

6.8   Works programme

An important role for the planning process is to provide one of the main inputs to the Telco's work programme. This annual programme covers the work that is required to install the planned equipment and plant in the network. Usually, this work is undertaken by a dedicated team of technicians, and thus constitutes part of the current account budget for the company. However, in many cases some or all of the installation and build work is undertaken by the equipment supplier as part of a supply and install contract. The latter is often appropriate in the cases of new technology being deployed or for those Telcos who have a small technician force and tend to rely on outsourcing for much of the operational work. We look at the operations in general, and the role of network build in Chapter 9.

6.9   Summary

In this chapter we have examined the role of network strategy in supporting the company's business strategy, as well as noting the close interdependence of the network strategy and network planning with the business management within the Telco. In particular the need to gain a coherent story for senior management was examined since this enables the necessary business cases for capital expenditure to be authorised. We considered the specific planning activities for expanding coverage and capacity of the network elements to meet the needs of the various services supported, as well as the introduction of new technology. Finally, the assembly of the capital and current account budget requirements to meet the planned expansions, upgrades and introduction of new services (products) was discussed.

References

1. Valdar, A.R., Newman, D., Wood, R., & Greenop, D. *A Vision of the Future Network*, British Telecommunication Engineering, Vol. 11, Part 3, October 1992, pp. 142–152.
2. Valdar, A. *Understanding Telecommunications Networks*. IET Telecommunications Series 52, Stevenage, 2006, Chapter 11.
3. Grundy, T. *Harnessing Strategic behaviour – Why Personality and Politics drive Company Strategy*, Financial Times Pitman Publishing, London, 1998, Chapters 3–5.
4. Valdar, A. *Understanding Telecommunications Networks*. IET Telecommunications Series 52, Stevenage, 2006, Chapter 6.
5. Ward, K.E. 'Network Planning', Chapter 20 of *Telecommunications Networks*, 2nd edition, IEE Telecommunications Series 36, 1997, edited by J.E. Flood.
6. Valdar, A. *Understanding Telecommunications Networks*. IET Telecommunications Series 52, Stevenage, 2006, Chapter 2.
7. Valdar, A. *Understanding Telecommunications Networks*. IET Telecommunications Series 52, Stevenage, 2006, Chapter 5.
8. Valdar, A. *Understanding Telecommunications Networks*. IET Telecommunications Series 52, Stevenage, 2006, Chapter 4.
9. Schiller, J. *Mobile Communications*, Addison-Wesley (Pearson Education Ltd), 2000, Chapter 4.
10. Lee, W.C.Y. *Mobile Communications Design Fundamentals*, 2nd edition, John Wiley & Sons Inc., 1993, Chapter 2.
11. Temaneh, N.C. 'Estimation of a Probability of Interference in a Cellular Network using Monte Carlo Technique', Satnac Proceedings, 2009, www.home.ntecom/satnac/prodeedings/2009/papers.
12. Flood, J.E. *Telecommunications Switching, Traffic and Networks*, Pearson Education Ltd., 2001, Chapter 4.
13. Valdar, A. *Understanding Telecommunications Networks*. IET Telecommunications Series 52, Stevenage, 2006, Chapter 8.
14. Spraggs, S. 'Traffic Engineering', Chapter 15 of *Carrier-Scale IP Networks: Designing & Operating Internet Networks*, BT Exact Communications Technology Series 1; IEE, 2001, edited by Peter Willis.

15. Simpson, W. *IPTV, Internet Video, H264, P2P, Web TV and Streaming: A Complete Guide to Understanding the Technology*, 2nd edition, Focal Press, Oxford, 2013, Chapter 1.
16. Muir, A.W. 'Transmission Planning', Chapter 16 of *Transmission Systems*. IEE Telecommunications Series 27, 1995, edited by J.E. Flood & P. Cochrane.
17. Manning, T. *Microwave Radio Transmission Design Guide*, 2nd edition, Artech House, Boston, 2009, Chapter 2.
18. Valdar, A. 'Circuit switching Evolution to 2012', *The Journal of the Institute of Telecommunications Professionals*, Vol. 6, Part 4, 2012, pp. 27–31.

Chapter 7

Customers and marketing

7.1   Introduction

As was said in Chapter 1, there is no business without customers. And customers buy products and services to fulfil their needs or to create benefit. Marketing is the process for identifying sufficient customer need, creating products and services to satisfy that need profitably and communicating to the customer that the need can be fulfilled by purchasing products from your business. To be successful in business it is therefore essential to understand marketing and to be able to apply its key concepts and processes.

Marketing ensures that the customer is at the centre of the company's purpose. It can therefore be said that marketing is an orientation rather than purely a set of functions. Nonetheless, to understand marketing fully we will explore some approaches that help to build clear marketing objectives and support successful market planning. We will then look at the four key aspects of the marketing mix – product, price, place and promotion – and apply them to ICT companies. Finally we will explore the fact that market power is shifting rapidly towards technology-enabled, savvy consumers and that commercial success in future will depend on building unshakeable trust through total customer empathy.

7.1.1   What is a market?

It is clear that a key aspect of any profitable company is its revenue; indeed, success is often measured by revenue growth. In turn, revenue is the product of three aspects of customer management:

- winning customers
- retaining customers
- maximising average revenue per user (ARPU)

Therefore, understanding markets, customer needs, aspirations and service requirements is a vital aspect of any company's strategy; in short – companies need a strong marketing input to the business strategy, which in turn will drive the creation of the marketing plan. Indeed, we can go further and say that successful marketing and an effective marketing plan comes with a clear and focused understanding of the target customer group (customer segment in the jargon) and how best to serve them.

In addition, Telcos are in the business of developing, managing and selling products and services to customers (see Chapter 8 for a complete explanation of product management). One of the best definitions of a market is *a place where products meet customers and customers make choices.*

This idea can be represented in a notional matrix where customer segments intercept with products (see Figure 1.10). Most companies will have a clear idea of the market segments that they wish to address. Successful companies will build a strategy around a clear view of the characteristics of their ideal target customer. This may be relatively broad, e.g. Marks and Spencer's customer base – or quite narrow, e.g. Louis Vuitton's target market, but the key to success will be clear focus. This focus will include a detailed understanding of customer needs and attitudes, their demographics, propensity to buy and the likely changes that might occur to these factors. Most companies will address more than one segment or have sub-segments within their target customer base.

Unfortunately, ICT companies are not strong advocates of successful customer segmentation, as they tend to focus on usage rather than on customer needs. For example, a Telco might segment customers into business customers and consumers, and then for business customers into; large, medium and small businesses; or into government departments, wholesale customers, global corporates, national corporates and SMEs (small to medium enterprises). Alternatively some ICT companies segment by product type such as datacentres, network services, IT services and voice services. In the consumer market it is common to find segmentation based on the amount spent or complexity of product mix. Measures such as these tell us something about the customers' current usage but nothing about their future needs. They are therefore rather unsophisticated aids in targeting successful marketing action or driving future business. We will discuss market segmentation in far greater detail in section 7.4.

In fact, many ICT companies are far less concerned with specific target customer segments and are far more focused on what customers buy i.e. products. There are two reasons for this. One is that in ICT – as you might expect – value is often achieved through innovation and technology. A second reason is that many ICT companies are based on a single, multi-purpose network, whether it is a telecommunications network (e.g. BT, Vodafone) or an IP server-based network (e.g. an ISP or Sky). In these circumstances, a winning strategy will be to maximise utilisation by developing a wide range of products to operate across the shared investment, thereby spreading costs and providing a one-stop-shop to customers. This is covered in more detail in Chapter 8.

7.1.2   Company orientation

We might therefore say that ICT companies are far more likely to be product-focused than customer-focused. In fact, there are broadly four forms of orientation that might apply to any company:

> *Product-focused (as ICT companies tend to be).* In this case the marketing plan will ensure that there is a lot of investment and attention in R&D, product

design and features; and marketing campaigns revolve around the excellence of the product. Thus, the profit comes from technical prowess.

Production-focused (such as might apply to chip manufacturers or mobile handset providers). Here the focus is on investment in volume, up-to-date production capacity and on process efficiency and quality. Volumes will be high and unit costs low. Total quality management, LEAN production and 6-Sigma process control will be important. Marketing will focus on cost and quality of the product. Profit from such companies arises from efficiency.

Sales-orientation (such as any major retailer; Dell; Carphone Warehouse). Here the focus is on the buyer/seller interface whether that's online, in a shop or in the form of telesales. Selling and promotion techniques are important and volume and *cost-to-serve* are essential metrics. Profit comes from continual growth in the top-line and tight control of costs.

Customer-focus (Amazon often claims the title of the World's Most Customer-Centric Company). Here the whole focus is on identifying and satisfying customer needs, building a strong reputation and creating trusting relationships. There will be great attention on customer satisfaction and *average revenue per user* (ARPU) and profit arises through satisfied customers and repeat business.

While there will be ICT companies represented in each group, the strongest orientation will tend towards *product-focused*. The most successful ICT companies combine these approaches, trying to be product focused initially but also, as competition evolves and many products become commoditised, introducing a sales and customer-focused orientation.

7.1.3   What is marketing?

Marketing is probably the most misunderstood business discipline, so before we go any further it's worth considering exactly what marketing is, and what it is not. Marketing is a business approach that puts the customer at the centre of everything that the company does. It is not the preserve of the marketing department alone, though certainly most companies will employ a marketing director and a group of marketing professionals. On the contrary, in a market-orientated company everything from R&D, to customer service, to the legal and finance functions will start with the customer. Certainly, they will be supported in their endeavours by the marketing department, but their mindset, the data they rely on to do their job and their targets will be strongly influenced by increasing customer satisfaction. This view of marketing can be shown diagrammatically (Figure 7.1) and was brilliantly captured by David Packard, the co-founder of Hewlett Packard, when he said:

"Marketing is too important to be left to the marketing department."

It was, perhaps, less well expressed when, reputedly, the finance director (FD) of a large Telco bumped into the director of Sales and Marketing. Concerned about poor revenue results, the worried FD asked 'how's marketing and sales'.

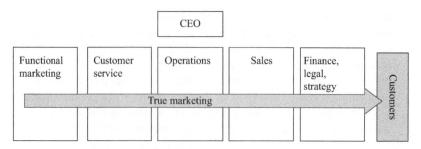

Figure 7.1   Marketing as a cross-functional activity

'Well, sales are right down' replied his colleague, adding 'if things don't pick up we will have to try marketing'.

Exaggerated, perhaps, but the story demonstrates how some companies view marketing, not as *the way we do things round here*, but as an add-on. We need to be clear about the difference. Marketing is about designing and presenting the products that customers might want, while the sales function is about persuading customers to buy those products.

Despite being a mind-set, marketing is also a professional discipline which encompasses a series of specific functions. These include some or all of the following:

- The process of understanding customer needs through market research.
- Holding customer data and using it to respond intelligently (often achieved through a customer relationship management (CRM) system – as described later and in chapter 9).
- Market strategy, planning and target setting.
- Market segmentation.
- Managing the marketing mix (products, price, place or channel and promotion/ campaigns).
- Measuring and addressing customer satisfaction.
- Brand and reputational management.

We will address each of these in the subsequent sections of this chapter.
In conclusion, we might say that:

> Marketing is getting the right product in front of the right people, in the right place, at the right time, at the right price, with the right promotion – and making a profit.

7.2   Marketing strategy and planning

In Chapter 3 we discussed approaches to strategy development. Clearly, any marketing strategy has to start with the company strategy, and can rely on much of the same analysis; e.g., industry and market analysis. If the company has completed its corporate strategy the market planning group will benefit from a clear vision, the agreed company strategic objectives and a planned route to achieve them. The company then needs to develop a separate (but related) market plan to take this thinking further. It is useful at this stage to return to Cliff Bowman's first question: 'Where should we try to compete – which markets and segments of markets?' (see section 3.6.3).

The market planning process there needs to:

- complete a situation analysis
- identify key marketing objectives and set high-level targets and milestones
- allocate resources and monitor results

7.2.1   *Situation analysis*

It is essential to understand where we are before we develop a route forward. Sections of this will have been done as part of the corporate strategy process e.g. economic and industry analysis; Porters 5 Force analysis of the market; and a comprehensive view of the company's competences. We covered each of these in Chapter 3. Nonetheless, it is essential at this stage to review the output and to ensure that the market aspects are all well covered. In addition, a key element of the situation analysis is a marketing audit. The key aspects of a marketing audit are shown in Figure 7.2 but would certainly include both external and internal aspects of the market; the strength, or otherwise, of the product portfolio; sales and profitability; pricing policy; promotion and distribution arrangements. Each aspect of the market audit needs to be supported with current and relevant data.

7.2.2   *Key marketing objectives and high-level targets and milestones*

At this stage of analysis there is a key decision to be made. Put simply, the question is 'Do we do more of the same, or something very different?' A useful way to approach this question is to use Ansoff's matrix (see Figure 7.3). Professor Ivor

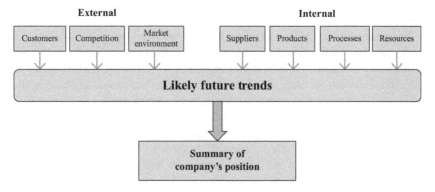

Figure 7.2   Key aspects of a marketing audit

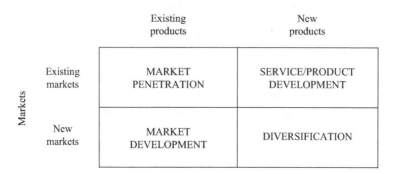

Figure 7.3   Ansoff's matrix

Ansoff, a distinguished mathematician with a successful career in both commerce and academia, developed the matrix in the 1960s while working for leading American companies, such a s IBM, Gulf, Philips and GE.

The model addresses both the current and future product options and existing and potential new markets. We will look at product portfolios in detail in Chapter 8, but it is vital in completing a successful Ansoff matrix to understand how successfully the current portfolio meets customer needs. In assessing the market analysis we should consider potential new markets as both geographical (such as new regions, countries or even continents) and customer-based markets such as consumers, business, public service organisations, etc.

The top-left quadrant of the matrix addresses the existing product set in existing markets. We should ask how profitable the current products are; what market share they do enjoy; what potential for growth is there; if the products providing high perceived use value. Answers to these questions might lead to the judgement that the existing portfolio can be successfully developed to add profit and value to the company in its existing market. *Market penetration* is a relatively low-risk strategy but does require the company to deploy good R&D, innovation and product management. This strategy may prompt a response from competitors.

The lower left quadrant addresses the opportunities arising from taking existing products into new markets e.g. an overseas market; a new market segment; a new channel such as online; etc. We should ask how well those markets are already served; how well our existing product-set would meet the customer needs in these markets; the economic conditions prevailing in the new market etc. *Market development* is a higher-risk strategy, though existing company capabilities in product development, etc., will provide significant benefit. Nonetheless, if we were to settle on this strategy we might consider working with local partners to reduce risk.

The top-right quadrant addresses completely new product-sets in the company's existing markets. Usually, companies will consider product groups that either enhance their existing offering or employ some of the same resources.

For example, a mobile company might add a music streaming product to add to their existing calls and text products, providing a more complete portfolio for their users and hopefully increasing average revenue per user (ARPU). Alternatively, they might add a security product that uses the same cellular infrastructure, thereby sharing costs and spreading risk. Nonetheless, *product development* remains a medium-risk strategy; promoting new products inevitably calls for a different marketing approach and will attract the interest of more-established competitors.

Finally, the lower right quadrant proposes developing new products for a new market. This is *diversification* and highly risky. For this reason it is seldom deployed except by the specialist investment vehicles. The risk profile can be reduced by making acquisitions, buying in expertise or developing a completely different brand.

Completing an Ansoff matrix will help companies identify the correct market strategy and objectives. It will also help to identify the capabilities required to develop the business in the desired direction. In addition, good data analysis of existing and potential new markets, hopefully available from the situation analysis (see previous section), will lead to proposed targets and milestones. Two examples of devising a suitable marketing strategy are given in Box 7.1.

7.2.3   Resources and results

Having determined the desired marketing strategy, it is necessary to identify the actions required to achieve the stated strategic objectives. In Cliff Bowman's terms we are seeking to answer the question: 'How should we change to be successful?' If we are intending to enter new international markets there is a likelihood that overseas offices, partnerships, new recruitment of local staff, perhaps with some key home managers relocated overseas, are all required. Even if the new market is within the same geography, e.g. a youth market, it is likely that new staff with different experiences and outlook will be required. Similarly new product-sets will require different knowledge, skills and outlook. For example, developing a range of e-commerce products would seem, superficially, to be the same as managing telephone calls or Internet information services, but they would certainly require a different approach in terms of pricing, business models, sales channels, etc.

In the 1990s it was evident that slow, traditional Telcos were trying to compete with Silicon Valley start-ups using their existing product managers and marketeers. Steeped in the values of long investment cycles; industry-standard levels of reliability; and universal service, these highly successful companies were never going to invent products, such as Google Search, YouTube or ITunes. Nor were they ever going to tolerate the ninety-nine failed ideas that are necessary to find the one new dot.com blockbuster.

In these circumstances partnerships or mergers and acquisitions (M&A) must be on the agenda. Even here problems remain, largely to do with the conflicting company cultures. It is very challenging for a large, successful Telco to work with a tiny start-up, founded by a couple of visionary engineering graduates with a great idea. Time and time again in the dot.com boom Telcos failed to consolidate or replicate the

Box 7.1   BT and Vodafone use Ansoff to find very different strategies

The late 1990s and early 2000s was a difficult time for Telcos. The world recession of 1997, triggered by an Asian debt crisis, brought a swift end to the dreams of globalisation and deregulation of the ICT industry. The dot.com bubble, developed based on a wildly overly optimistic view of immediate commercial value of the Internet and ecommerce, was well and truly burst. Telco's share prices tumbled and economic growth went into reverse. Clearly, Telcos needed a new strategy to survive. BT, the UK's incumbent operator, and Vodafone, the UKs biggest mobile operator, each undertook a major strategic review and, although they faced very similar problems, emerged with very different strategies.

BT, faced with huge debts and stretched management resources, decided on a *product development* strategy (top-right quadrant in Ansoff's terms). They focussed on their home market in the United Kingdom and rapidly sold their fledgling overseas businesses, e.g. in Japan, Spain, Holland, Korea and Malaysia. Although some of these overseas adventures were beginning to break-even and to show future potential, they were viewed as absorbing far too much resource and too carry far too much risk for the new global economic environment. At the same time, BT began to develop the new product-sets that would maximise the use of their network and skills and add customer value e.g. Internet and multimedia services, managed networks, broadband and, eventually, IPTV.

At the same time, and faced with many of the same strategic imperatives, Vodafone decided to re-double their *market development* strategy. In the same period (1997–2002) they purchased Aircell in Ireland, Airtouch in the United States, Mannesmann in Germany plus a number of holdings in Japanese networks, including, ironically, BT's share in Japan Telecom. Where they couldn't successfully acquire a company they chose to partner; by 2010 they had operations in 21 countries and partners in further 40 countries. They did little in terms of new product development, other than to rationalise product-sets wherever possible, purely to reduce operational costs.

Despite very different strategies, BT and Vodafone would both be judged to be successful, proving that there are no right or wrong answers in developing a strategy.

innovation that was coming out of West Coast America, despite allocating significant resources and making approach after approach to the new kids on the block.

In fact, the problem was not in resource allocation but in identifying the correct marketing objectives. What these companies were trying to do was product development (in Ansoff's terms: top-right quadrant). They were aiming to add a layer of new products to their existing consumer offering. They felt that their huge customer base was of value to these new product start-ups. In fact, what the successful companies eventually discovered was that they should have been pursuing a market development strategy, whereby they created a wholesale market to offer connectivity and bandwidth products to the new, emerging dot.com companies. In this way they didn't have to pick winners and losers in this notorious fragile market, nor provide innovation in the new world of e-commerce. All they had to do to be successful was to sell products that they already understood to the huge companies that emerged from some of the fledgling start-ups, companies such as Apple, Google and Microsoft.

7.2.4   The marketing plan

The culmination of the planning process is the Marketing Plan, a written document summarising the current marketing position, where the company wants to be in the future, and the actions that will get it there. As we have seen, this starts with the corporate business plan (see section 4.4) and is based on the clear strategic objectives arising from the company strategy (Chapter 3). The three stages of the planning process described at the start of this section should, ideally, lead to a plan that:

- Is based on clear, shared marketing objectives.
- Is both quantitative and qualitative in nature.
- Provides marketing action plans – expressed in terms of the marketing mix or 4 P's (see section 7.5).

It will ensure that the company's activities are based on likely developments in the market, are focused on customer needs and are co-ordinated. Crucially it will not be a summation of the sales targets for the year, based on an extrapolation of the past. Nor will it simply seek additional marketing resources without identifying clear customer and revenue benefit. Indeed, it should match required marketing and sales resources against proven and desirable market opportunities.

Finally, a clear, well-written marketing plan that is demonstrably based on the corporate strategy and company objectives will reduce conflict and, together with other functional plans, align the company employees around a single, well communicated plan [1].

7.3   Identifying value through market research

In Chapter 1 we identified commoditisation as a growing threat to telecommunication revenues and profit. Commoditisation is the phenomenon whereby products (or services) are selected by customers based on price alone, with no differentiation of features, service, brand or quality. In some cases, customers' expectation of what they receive for a given price is constantly rising (e.g. the speed of broadband). The practical aspects of commoditisation are falling prices and free offerings such as text messages, bundled services etc. It is the marketing department's job to fight this trend. To do so they need to identify what it is that customer's value and what they are willing to pay for. Most importantly it is marketing's job to identify customers' needs that, either,

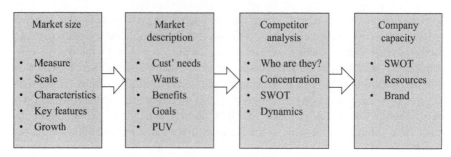

Figure 7.4   Market research

haven't been identified by competitors, or which aren't being addressed. In this way they aim to unlock some hidden value and command a relatively higher price.

In considering strategy we explored the concept of perceived use value (PUV) in the context of competing products. This is a legitimate tool in the marketing of established products but may need to be enhanced when we are considering either new product or market development. While PUV is independent of price we should remember that generally consumers see value as benefits minus price. Benefits may include intangible elements such as style, or a psychological feel-good factor, as well as the functionality of the product, i.e. does it get the job done? Similarly, price will not simply include the upfront cost of the product, but also the switching costs from an existing supplier; any costs of disruption or a perception of poor customer service [2].

The search to identify what it is that customers' value starts with Market Research, a vital tool in any marketing department's armoury. Broadly there are four stages, as shown in Figure 7.4.

However, Stages 1, 3 and 4 have been already been covered in developing the company strategy (see Chapter 3). Indeed, much of this information/analysis will be standing data in any company's routine management information, though clearly in must be refreshed regularly, perhaps quarterly.

Stage 2 *Describing the Market*, is what should interest us here. In seeking to identify customer needs, market research holds a number of significant advantages. First, it is a systematic, scientific approach that – properly designed and interpreted – can provide proven answers and authentic customer insight. It will provide reliable evidence on which to base informed management decisions and it reduces the risk of launching an unprofitable or unsaleable product. In designing market research there are two broad classes of research methods to consider: qualitative research and quantitative research.

Qualitative research aims to evaluate the opinions and motivation of consumers. The primary research methods in use are:

- Face to face interviews (both structures and unstructured).
- Focus groups – normally managed by a professional facilitator, they might be groups of 12 or so people from a cross-section of the target audience. They

may be asked to consider a proposed new service or even to experience a prototype product. The discussion will be free-flowing but nudged forward with appropriate questions. Normally, the name of the company commissioning the research is not mentioned, to avoid pre-conceived ideas.

- Consumer observation.
- Regular journals kept by consumers of their experience and use of products and services.
- Ethnographic research – a research approach whereby researchers actually 'live' with the willing consumer to observe how they live their life and interface with the products in question. For example, a retail supermarket undertook ethnographic research with consumers who had specific allergies. They discovered that the consumers kept a specific shelf in the fridge for all the allergy-free food. From this idea the retailer rearranged its shop layout to put all the allergy-free foods in one section – thereby significantly increasing their share of that lucrative market.

Qualitative research has great value. For example, it provides rich customer insight into consumer behaviour, their use of products and into their perceptions. For these reasons it is extremely valuable in product prototyping and testing; in eliciting views about promotional material; and in determining brand perceptions etc. Nonetheless, the weakness of qualitative research is that it is likely to be based on small sample sizes and therefore not able to provide statistically valid answers. In addition, it normally requires subjective evaluation by the researcher.

For these reasons, companies will normally couple qualitative research with some quantitative research.

Quantitative research aims to use systematic, empirical investigation to identify statistically valid conclusions about consumer preferences. The main research methods are:

- Surveys, whether that is by phone, post, Internet, or face to face. It is vital that any questionnaire be structured, with perhaps tick boxes, numerical ranking of features, preferences, etc., and with the opportunity to provide verbatim comments. Most surveys will take the opportunity to collect demographic details of the respondents, in the hope of making some correlation between, e.g. age and preference.
- Customer usage data. The extensive use of point of sales data, comprehensive customer relationship management (CRM) systems and sophisticated data-mining techniques enables companies to gather quantitative research data directly from the actual buying behaviour of their customers. In particular CRM systems emphasises the value of correlating various pieces of information about particular customer's usage, behaviours and demographics as a valuable source of intelligence for the marketeers. This information can help with product definition and positioning, market segmentation, pricing, etc. We explore an example of this in Box 7.3 where we describe the phenomenal success of Tesco Clubcard.
- Structured retail and dealer feedback.

- Complaints and returns.
- Independent review sites such as TripAdvisor. Perhaps the most important recent development in qualitative research is the growth in data available from search engines, independent review sites and comparison websites (see section 7.8.5).

The aim of any quantitative research is to involve a large sample size, selected at random. This should ensure statistically valid answers that can reliably be used to predict the behaviours and preferences of the relevant population. As no interpretation is required of the statistical elements of the survey, the results are free from bias (apart from possible bias built into the population of respondents). Quantitative research is often used to test a hypothesis such as 'people are happy to view films on their mobile phone'. Finally, as the results are statistically valid and often ranked or scaled, they can be applied to pre-agreed models to set optimum prices, advertising expense, etc.

For these reasons, the main uses of quantitative research are to test pricing proposals; identify the best sales channel for a particular promotion; set levels of customer satisfaction; and determine relative preference compared to competitors etc.

Secondary research: Both qualitative research and quantitative research are normally commissioned by single company, specifically for their use and with a particular purpose in mind. For this reason it is sometimes called primary research and the results are treated with great commercial confidentiality. There is, however, a great deal of other information about consumers, markets and products available in the public domain and companies should always triangulate their conclusions from as many data sources as possible.

Sources of important secondary research information are:

- News feeds.
- Consumer magazines and websites.
- Analysts' briefings (often available from corporate web-sites).
- Specialist industry researches.
- Company reports.
- Regulators, an especially rich source of market and consumer satisfaction information for ICT companies.

Analytical tools: Information and data gathered from both primary and secondary research is of little value unless it is systematically analysed to help define a set of features and price ranges for future products. A wide range of business tools and analytical models are available and are described throughout this book e.g. PEST, Porter's Five Forces, SWOT, Ansoff's Matrix and Boston Consulting Group (BSG) Matrix.

7.4   Market segmentation

There are many definitions of market segmentation. Kotler and Armstrong define market segmentation as 'dividing a market into distinct groups of buyers who have distinct needs, characteristics, or behaviour and who might require separate products or marketing mixes' [3]. Focusing marketing efforts on specific market segments can have great commercial value and is vital in positioning a company and its products in the marketplace. First, the process of segmentation is strategically important, in that it enables companies to create increased value for customers by understanding their needs more clearly than the competition. Indeed, the segmentation process should always be concerned with customer needs (i.e. what customers want, not just what you can provide) and is critical for refining every aspect of the relationship between the company and its customers. In particular, channel strategy, customer service, billing arrangements, promotion and price should all be critically different depending on the segment chosen. (For instance, certain customer segments will pay a substantial premium for broadband delivered over dedicated infrastructure as it provides guaranteed quality of service and resilience.) Because segmentation is such a fundamental aspect of a company's approach to the market it is vital that the results of segmentation are comprehensively communicated throughout the company and that they inform every marketing decision.

A particular form of segmentation will be unique to the company in question. However, broadly, we hope to divide the market up into groups whose members are as homogenous as possible; ideally these groups will be as distinct from other groups as is possible. There is no single standard approach to the process of segmentation and no single *right* answer. However, it is important that market segmentation is based on real customer needs or motivations and is supported by research, evidence and hard data.

Indeed, it is essential to base segmentation on real and up-to-date data. Some of the data typically used in the process might be:

- lifestyle surveys
- demographics
- lifetime value analysis
- customer psychographic studies
- customer buying behaviour
- customer profitability modelling
- analysis of existing customer data

All have pros and cons, although a key point is that any review should encompass the whole market, and not just a company's existing customers. Indeed, Telcos often compound this problem by not only focusing solely on their own customers, but also by focusing exclusively on their customers' spend with them as providers! Similarly, an analysis that relies solely on demographics or profitability crucially overlooks customer motivations and the benefits that customers seek from products and services.

While there is no standard process, the steps shown in Figure 7.5 will achieve a workable and commercially valuable segmentation.

Having identified some possible segments it is worth testing them against the following criteria. If you can put a tick against each item the likelihood is that you have a viable set of targetable customer segments. They should be:

- measurable
- distinctive
- substantial
- accessible
- actionable

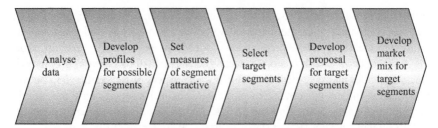

Figure 7.5   Steps in segmentation (source: Prof Moira Clark, Henley Business School)

Having checked the validity of the segments it is essential that the company prepares in every possible way to address the agreed customer segments and their needs e.g. they can be used to define the sales structure; determine the service approach; and inform the market plan. In particular, the agreed segmentation should drive all elements of the marketing mix (product, price, place and promotion) as discussed in section 7.5.

Boxes 7.2 and 7.3 describe some real-life examples of market segmentation.

Box 7.2 Market segmentation: a telecommunications example

A scientific approach to market segmentation becomes far easier and more rewarding as Telcos collect more data on the behaviours of their customers. One mobile service provider operating service across Western Europe developed a segmentation based on the behaviours and usage of their customers. They identified the following five key segments:

> *Quick talkers* – People who made voice calls, but kept each call short in duration
> *short message service (SMS) users, but will talk* – people who mainly use SMS, but who also make voice calls.
> *Valuable roamers* – People who travel and use their phone abroad.
> *SMS Users, with high top-ups* – people who text and go beyond their contract allowance
> *Outbound voice sociable* – People who initiate lots of calls and like to chat.

Having identified the key segments it proved relatively easy to place each customer and potential customer into the appropriate segment by asking seven simple questions:

1. average quarterly bill size
2. average recharge value
3. preference for scratch-card recharge

4. voice usage vs SMS usage
5. inbound/outbound balance
6. average call duration
7. size of calling circle

This first-cut segmentation was cross-checked with a classification of the overall value of the customer based on their profit contribution per month. For example *SMS Users, but will talk* made-up 43% of the high value customers, whilst the *outbound voice sociable* segment provided a further 25%.

Armed with this analysis it proved relatively simple to devise appropriate marketing action plans that maximised attractiveness to each segment and focusing action and resources on those segments with the greatest profit potential. Table 7.1 shows some marketing plans matched to segment [4].

Table 7.1 Marketing plans matched to segments

Segment	Potential action to increase profit
Quick talkers	Encourage to make longer calls and expand their calling circle; short marketing messages
SMS users, but will talk	Communicate through text messages; encourage them to make voice calls; provide a deal on multimedia messaging service (MMS)
Valuable roamers	Link marketing promotions to travel benefits; expand calling circle and encourage social calls.
SMS users with high top-ups	Communicate through text; provide more contract benefits; encourage voice calls
Outbound voice sociable	Communicate through outbound calling; provide benefits to a wider calling circle; offer on-net benefits to the whole calling circle

If we look at some specific cases applying currently in the ICT industry we can see a great variety of approaches and results.

Traditional network providers (e.g. France Telecoms) often use existing customer data to segment their market, either by physical orders (e.g. single-line users; multi-line users; etc.) or bill-size (e.g. light users, heavy users, etc.). There are advantages to this approach. It is easy to apply, to measure, and to use. However, it is not needs-based or based on customer motivation and it may overlook valuable segments that are not currently represented in the existing customer base.

Mobile operators (e.g. EE) try to use lifestyle analysis to identify key segment (e.g. single professional people, young families, etc.). This approach is often needs-based, and is relatively easy to identify. However, the data may prove expensive to collect and some segments might not be substantial enough to be identified separately.

Social networks (e.g. Facebook, Linkedin, Twitter, Tencent) seem to use demographics, linked to needs; specifically they segment tightly on age groups and

Box 7.3   Tesco and the segment of one

The market segments that companies target can be broad e.g. all consumers, or they can be very specific, e.g. professionals in their 30s, living alone. Clearly, the first is easy to address through mass marketing and blanket promotion, but the response rate is likely to be very low. The second group, on the other hand, is far harder to identify and to reach, but if you can address them directly, they are likely to be much more responsive to your offer. Many companies see the progression of market segmentation as going from the macro to the micro. But what if you could separately identify each and every one of your customers and to understand their specific needs; and what if you could do it relatively cheaply.

This was the objective when Tesco – the UK grocery retailer – introduced their revolutionary Clubcard in 1995. Initially introduced as a discount scheme to encourage loyalty, it quickly became clear that the data being collected on the purchasing habits of individual customers was massively valuable. In the words of Sir Terry Leahy, the architect of the scheme and later the highly successful CEO of Tesco worldwide:

> 'At its heart lies a very simple idea: people join (the Clubcard scheme) and receive 1% discount as a reward. In return we (Tesco) gather vital data about the products the customers buy.'

Despite knowing the individual names and demographic details of their customers Tesco chose to anonymise the data to avoid data protection concerns and to maintain the integrity and reputation of the scheme. They also faced significant technical challenges in both collecting and analysing all that data. Nonetheless they persisted and quickly grew the scheme to include 10 million members in the United Kingdom alone.

Initially the data was used to group customers by lifestyle e.g. students, affluent parents, etc., and to create summaries of likely purchases. Overtime this analysis gave way to a classification based specifically on what people purchased, e.g. vegetarians, people on a budget, devotees of organic and healthy eating etc. This approach proved even more successful.

So how has Tesco used this information to grow its business? Well, in a number of ways. For example, they can tailor offers to individual customer groups or segments; they can identify customers that may be defecting to a competitor and understand the reasons; they can use local customer preferences to design and stock individual stores; and they can suggest related purchases to a customer, based on what they already buy and what other with the same needs bought previously.

Very specifically, every quarter Tesco mail each and every Clubcard customer with rewards, targeted coupons and product information. Each mail is unique to the individual customer, creating millions of variants. With that level of micro-segmentation it isn't surprising that response rates are 30 times greater than the industry average and that Tesco progressed from number three in the market in 1995 to number one in 2011 [5].

Note: Since then Tesco have faced massive price competition and their fortunes have declined severely. Time will tell if a forensic customer-focused strategy can help them recover or if they choose to reposition themselves.

specific customer needs/obsessions (bands for kids; social life for 20-somethings, work for 30+). This approach is based on customer needs and provides added-value to advertisers. It also ensures that users self-select the social network that is best for them to join. The worry with this approach is that it is easily replicated and may not identify enough added customer-value to increase revenues or provide sufficient differentiation from competitors.

Customer equipment suppliers (e.g. Sony, Dell, Apple and RIM). It might be tempting to think that these companies are *product-focused* rather than *customer-focused*. Certainly, they attempt to make standard products in high volume and sell them to as wide a market as possible (e.g. the iPad). On the other hand a company such as Apple is very determined that the whole package (product, customer service, online presence) supports a certain lifestyle. We might conclude then that psycho-graphic segmentation plays a part. In addition some form of market categorisation may be applied such as the notion that people are more or less inclined to adopt technology, e.g. innovators; early adopters; early majority; technophobes, etc.

A final thought on segmentation. Customer needs and expectation change and they change especially rapidly in the ICT world. It is therefore vital to continually keep segmentation under review and to be ready to adjust or change or add new segments.

7.5   The marketing mix (The 4 P's)

The *marketing mix* refers to the actions taken by a company to promote its brand, products and services. Professor Moira Clark of Henley Business School often likens the concept to the idea that a marketeer is required to make a recipe from a limited number of ingredients, possibly in varying quantities depending on the cake being baked! What is certain is that Edmund Jerome McCarthy, Professor of Marketing at Michigan University, used the term *marketing mix* to christen a marketing model that's simplicity and practicality has kept it in the forefront of marketing practice since the 1960s [6].

The four elements of the marketing mix are Product, Place, Price and Promotion (collectively described at the 4 P's). We will consider each in turn.

7.5.1   Product

As we have discussed previously, ICT companies tend to be product-orientated. This is a natural feature of the network and technology bias of their industry.

This means that the development and management of their product portfolio is both important and well developed. Key features of this element of the marketing mix will include the management of the product portfolio and especially, the number and fit of product-lines, the range of products in each product-line and the positioning of those products.

Many ICT companies see original research and development as a key to creating sustainable competitive advantage. Others will adopt a *fast-follower* approach. All new product developments will be concerned with design, quality, features, operating manuals and style. In addition, the less tangible aspects of the product, such as the service level, help desks, resilience, time to repair and guarantees all need to be specified.

However, simply launching new products isn't enough. Active, whole-life management of products is essential and calls for an understanding of product life-cycles. Once the position in the life-cycle has been assessed, a variety of different management actions should ensure maximum product profit.

Given the importance of product management, it is often seen as a separate discipline and even a separate department of the company. Chapter 8 provides a more complete understanding of all the elements of product management.

7.5.2   Place

This element of the marketing mix is often misunderstood as meaning the geographic market being addresses. Certainly geographic coverage must be addressed but a clearer explanation of Place might be *channel to market* (but channel doesn't begin in P!). This *channel* is two-way and forms a conduit for many things. It is where customers discover information about products, prices, availability; it may be the conduit for the sales process (see section 7.5); it may be the way that they place an order and track delivery and it may also be the route for payment. Finally, increasingly the channel is also the distribution route, whereby products such as digital music or books are actually provided.

As part of the marketing mix it is essential to determine the best combination of channels to market, whether that is a sales force (either face to face or via a call-centre), retail outlets on the high street or purely online. Each of these channels has advantages and disadvantages, and so it is essential to match the preferred channel with both the types of product and the target market segmentation.

For example, maintaining a specialist sales force is expensive, but probably vital if the products are complex, bespoke and high value (Box 3.6 describes a situation when Nortel got this wrong). An out-bound call centre will certainly be cheaper, touching far more potential customers, but perhaps only able to sell simple products which need little explanation.

Retail outlets, i.e. shops or retail concessions within bigger shops, are certainly expensive to open and maintain, especially in prime locations. Traditionally, Telcos have wanted to have a high street presence but found it hard to justify. For example, in the 1990s BT opened a range of expensively fitted-out shops with the purpose of selling and providing customer service, take bill payments, etc. Unfortunately, after a number of years making losses it was discovered that the prime use of the shop was for customers to take advantage of promotional discounted prices, or to negotiate preferential payment terms when they were unable to pay their bill. It became essential to find cheaper ways to serve these needs without the expense of high street shops.

Conversely, shops are a key element of the mobile marketing mix. With the advent of the smartphone the size, weight, touch and feel of the actual physical phone are a key element of the buying decision, not only of the phone but also of the network. (In fact, the price of even the most expensive phone will be subsidised or spread across the network contract.) Interestingly, these shops are almost entirely showrooms, in that, once the customer has selected the phone they want, everything else done in shop could be done by the customer online at home. The shop assistant will access the exact same Internet sales homepage.

Finally, online channels are clearly the main route for many ICT products. This is partly because they provide efficiency for the company (i.e. they are cheaper, can handle both volume and complexity of transactions and can be kept current – a growing concern for all industries). However, there are also many advantages for the customer. They are available 24/7, fit many peoples' shopping preferences, provide both advice and the opportunity to purchase, as well as post-order tracking. This neatly reinforces the final point about Place. In considering the right channel to market, we should not only consider the product and the company's distribution needs, but also the preferences and likely lifestyle of the specific target segment [7].

However, the above description of separate channels is rather misleading in today's world. Companies need a multi-channel strategy, whereby all the channels recognise the same customer, with the same attributes, and offer the same proposition, in whatever form the customer chooses. This requires good co-ordination, excellent system support and tight channel management. A more proactive stage of channel integration might be deemed an omni-channel strategy. In this case consideration is given to mounting each element of a company's marketing strategy across a range of channels (online store, telesales, apps store, social media, physical billboards, etc.) and ensuring that the various channels are co-ordinated and interact with one another. For example, cinema adverts might include a call to action involving online order-fulfilment or a YouTube tutorial.

7.5.3 Price

The third element of the marketing mix is Price. This requires marketer to consider the standard aspects of pricing, such as list prices, discounts, promotions, etc. Indeed, the starting point would normally be the current price position in regard to competitors and the price of all the other relevant products in our portfolio.

However, for ICT, a vital aspect of the marketing mix is also pricing *structure*. Indeed, real competitive advantage can be gained by creating well-developed, innovative pricing strategies. For example, when dial-up Internet services were launched it soon became clear that traditional, per minute, pricing was not going to attract business or create customer satisfaction. A move to subscription (or, in the vernacular, *all-you-can-eat*) pricing was required. Similarly, IPTV services often compete with satellite and terrestrial TV, not on the content that they carry, but on the pricing packages that they offer. These decisions are far from easy. They often undermine existing revenue streams and reduce profit in the short term, with the promise of more business in the future.

Pricing is a vital element in any ICT company's competitive armoury and there is a great deal of both theory and practical judgement involved in being successful. We cover it in much greater detail in section 8.2.3.

7.5.4   Promotion

Promotion encompasses all of the methods of communication that a company may use to promote its offerings to those who may wish to purchase. It includes promotion of not only the company's products and services but also the company's image, brand and reputation. For this reason the company's public relations (PR) strategy should be covered here.

A company's brand represents everything about the products, services and reputation of the organisation in the consumer's eyes. It goes far beyond the common misconception that it is just the company's logo. Rather, it should be seen as the personality of the company and is a strong, often unspoken, influence on buying decisions. A brand represents an emotional tie between both the customers and employees of the company and the company itself. It is based on a collective truth (the vision and values of the company) and should not only define the company but also shape its future action.

Incidentally, when we discussed strategy development and communication in Chapter 3 we referred to the importance of a strong company vision. Such a vision is vital to aligning the company employees in delivery of services and products (see section 3.7.2). It is clear that the company vision should be strongly linked to the marketing plan and especially, it should inform the brand positioning and even the language that it expresses. Specifically, the company vision might form a marketing strap-line, neatly combining the internal and external essence of the company.

In financial terms, brand would be categorised as an intangible asset; nonetheless, it is hugely valuable. While Coca-Cola remains the world's leading brand ICT doesn't fall far behind. Brands such as Facebook, Google and even Vodafone have massive recognition and strong pulling power in user's decisions. Brand management is a perfect example of the fact that marketing is the responsibility of everybody in the company. Building a strong brand takes a concerted effort over a long period of time; destroying brand value can happen overnight with the launch of a single weak product, a catastrophic failure of service, or even a reputational scandal.

Once again, it is vital to match the form of promotion (advertising campaigns, promotional offers, merchandising) with the type of product-line in question and the market segment being targeted. For example, there is little point advertising a high-end product targeted at senior business managers on TV; it is likely that your target customer has little time to watch commercial TV. Similarly, price promotions are unlikely to be effective in a market segment that is price-insensitive, such as the luxury goods industry.

Orange, the mobile phone company, got this right when they launched in 1994. They identified a very clear, tightly specified segment and then ensured every element of the marketing mix was correctly focused on that segment. The results of Orange's strategy were highly impressive, with high growth and strong margins (see Box 7.4).

Box 7.4   Birth of a brand – Orange

When Orange was founded in the United Kingdom in April 1994 they were the newest and smallest mobile operator in a highly competitive market. They, therefore, defined their target segment very clearly as young, professional, creative types who would see a mobile phone as a life-style choice, not just a utility to make phone calls. Orange presented their products with bright, quirky packaging, eye-catching promotional material and advertising that would appeal to the younger end of the market. For example, the TV and cinema adverts provided little information about the product or its price, but concentrated on presenting a lively, connected, exciting life. Perhaps this new approach to tackling a specific market segment is best summarised in Orange's strap-line:

The Future's Bright, The Future's Orange

This strap-line has proved so strong that it is now firmly embedded in the consumers' psyche and has become part of the language –see Horse and Hounds article.

How successful was this holistic use of all the marketing mix? Not only did Orange grow faster than the other mobile operators, they consistently had the lowest churn-rate and highest of margins in the UK market throughout the first 5 years of the mobile industry. As a part of France Telecommunications Orange remains one of the strongest worldwide ICT brands.

Crucially, promotion is about communication and communication is a two-way process. Bombarding consumers with inappropriate and ill-judged communications (junk mail, unsolicited promotional calls, text messages, pop-up advertising) that doesn't provide an opportunity for the recipient to interact may elicit a negative reaction. The aim is to devise the right form of communication that fulfils the consumers' need for information, at the right moment in their decision making. It should provide an opportunity to complete the deal once the decision is made.

Another key aspect of communication is that it will only be effective if the source is trusted and seen to be independent. For that reason, consumers are far more likely to listen to personal recommendation, social networks, comparison sites and their own web searches than to official promotional material. In section 7.8 we discuss the importance of ICT and its ability to support empowered consumer in their buying decisions.

7.5.5   Three more P's

Since Professor McCarthy first developed the 4 P's there have been numerous attempts to expand the *cake ingredients*. Possibly the most useful – aimed at marketing services rather than products – is the addition of People, Process and Physical Evidence.

Clearly, service-based industries require not only skilled *people* but people with a customer-orientated frame of mind. These may be recruited, trained or simply inspired by a company that truly embraces the values that the leaders champion. However, it is only achieved when a company that truly aspires to provide great customer service matches its people's attitudes, frame of mind and approach, to the customer needs that they aim to fulfil. In section 3.7.2 we discussed the example of Pret-a-manger but every restaurant, hospital and travel company will know the value of promoting attitude as well as competence.

However, those same companies will design very tightly specified processes to deliver high-volume, consistent service each and every time. While customer service may appear to be individually crafted and focused on *me*, the truth is that there is a lot of high-volume repetition involved. Delivery against that *process* (the sixth P) and consistency of output will lead to increased customer satisfaction.

Finally, the seventh P is sometimes called *physical evidence*. This recognises the fact that service, when it is well done, is often invisible to the consumer receiving and paying (either directly or indirectly) for the service. The consumer, therefore, needs to be given a physical sign that they have received the service, such as a swing card in the car that's just been valeted; or a hygiene wrap around the toilet in the hotel bathroom. Apple shops are an excellent example of physical environment (décor, etc.) providing evidence of the lifestyle choice that comes with buying Apple products. If we consider another ICT example, perhaps broadband providers should include a pop-up online gauge telling you, in real time, what download speed you are achieving!

7.6   Sales

7.6.1   Sales planning

In Chapter 4 we looked at business planning and, in particular, the planning hierarchy. Any financial plan requires a clear view of forecast revenues for the coming year (and probably the following year as well). Ideally, this is based on a sales plan that will detail sales volumes by product, customers and market segments. It looks separately at acquiring new business (i.e. winning new customers), achieving growth from the existing customer base and churn rates among the current base of customers. It includes assumptions about prices, discounts and sales promotions. It should also detail available sales resources, sales teams, websites, promotional material, etc. In many cases it includes a geographic analysis to help accurate allocation of resources and to target promotional activities.

There are a number of purposes for such a plan. As we have seen, it forms a vital part of the financial plans of the business and is central to the budget-setting process. It also is a key input to target-setting and the remuneration of the sales teams; finally, it forms a vital link between customer intelligence and the whole business proposition.

7.6.2   The relationship between marketing and sales

Most ICT companies spend far more on their salesforce than they do on their marketing department. It is therefore essential that the planning and management of the sales effort is supported by good marketing plans that are based on rigorous, data-rich analysis. This may seem obvious but the temptation is that when a company is concerned about its short-term financial results it mounts sales campaigns to fill the gap, with less attention to profit and the longer-term relationship with their customers than the company strategy might dictate. So the question is how do marketing and sales interact.

As we have seen (section 7.5.2) one of the 4 P's stands for Place, that is to say the route to market. Channel design must take into account efficiency and cost (often measured in terms of *cost-to-serve* per customer or margin created); the complexity and value of the products; the type, preference and sophistication of the customer; and the market segment to be addressed. Channels have a number of functions – to inform, to distribute, and, crucially, to sell. In short, to ensure the right product is available, in the right place, at the right time to enable customers to buy. Sales capability comes in many forms, from entirely automated and web-based systems to relationship-selling involving large teams of account managers focused exclusively on a single, very important account. To be successful ICT companies need to employ a mix of different channels and regularly assess the relative effectiveness of each. The most important aspects of channel design are to set clear demarcation and to avoid competition between channels.

7.6.3   Indirect sales channels

Sales channels can be both *direct* and *indirect.* Indirect sales channels include agents, wholesalers and distributors. The decision as to whether to use a direct sales approach or to go to market via an agent is often based on cost. Employing a sales team is expensive, whereas agents may well be able to provide greater market coverage at lower costs by sharing overheads with a range of other companies' products. There may also be other benefits such as access to new markets, sharing of risk (e.g. inventory and bad-debts) and additional services such as customer training. The potential negative is, of course, a lack of control. Before any decision is made to use intermediaries we need to be clear:

- Will our agent also carry our competitors' products, and how will we fair in that situation?
- How will we manage the tension between direct and indirect channels?
- Will they provide enough sales effort and promotion to achieve our sales plan?
- Does their brand fit with our strategic positioning?
- Are they someone we can trust and work with; are they financially sound?

ICT companies are increasingly using intermediaries either as aggregators or as a franchisee. In the latter case a Telco will license a local business to set up and own, say, the small and medium enterprise (SME) market for a particular geographic area. They may be restricted to selling only the Telco's product set, or may enhance their product range with complementary products. They will be able to buy-in product at discounted price and sell at an agreed price, bearing the risk of bad-debt themselves. This franchise model is proving successful where product margins do not support an expensive, inflexible direct sales team. On the other hand, they are increasingly in competition with, e.g. direct telephone sales operations.

7.6.4   Salesforce management, reward and motivation

There are a wide range of sales force configurations that might be appropriate, depending on the type of market and customer being addressed, the complexity or standardisation of the product set, and the affordable costs. We will explore the characteristics, management, motivation and reward of each in turn.

Telephone sales. This approach is normally focused on mass consumer or small and medium enterprise (SME) markets where the products are relatively simple and easily understood, e.g. broadband, CPE. However, product margins are thin and therefore the cost per sale is a major focus. For this reason the relatively young sales staff typically work in large call centres, with close management (backed by specific key performance indicators, such as call duration, sales conversion ratios, etc.), and receive minimal training. Wages are normally relatively low, and include an element based on sales; hours may not be guaranteed. The more successful operations employ sophisticated customer relationship management systems (CRM) that provide customer data, product suggestions based on specific customer needs, dialogue guides and order-completion – all on screen.

Traditional outbound telephone sales (where an agent rings unexpectedly) is losing credibility and is being replaced by inbound telephone selling. In this situation the customer is prompted by a promotion to ring the company with an enquiry about a certain product or service and the sales dialogue progresses from that initial contact. Alternatively, the customer might ring for a totally unrelated reason such as a billing enquiry or customer service issue and be engaged in a sales dialogue. Inbound sales teams are proving remarkably successful with a conversion rate some five times higher than outbound.

Generic sales field force. These sales people typically focus on medium and large businesses, meeting decision makers face to face and discussing the customer's specific needs. They are often organised into industry sectors and, ideally, have a good understanding of their specialist sector, e.g. retail or manufacturing. They also have a good knowledge of the products and propositions that have been developed with their sector in mind. They often have some limited flexibility to agree product modifications, negotiate on price and to introduce third-party products to complete a customer solution.

The management of generic sales teams is highly focused on setting monthly or quarterly sales targets with perhaps more emphasis on revenue than margin or profit. Sales bonuses and rewards may follow the same approach. For that reason there is a danger that generic sales or sector-focused teams may ultimately sell the wrong thing to the wrong customer, at the wrong time. The solution to this problem is to develop clear sales plans for each customer or group of customers and to overlay the traditional revenue-based sales targets with an element of product-mix, costs and margin/profit. This last target requires an ICT company to understand its product profitability profile in some detail, which is far from straightforward, as we explore in section 8.2.2.

Major account management. As the name implies, these teams are focused on a single (or a few), highly valued customer. In terms of Telcos the major accounts are likely to be other telecommunications providers, multinational companies, media content providers, major banks, government departments, etc. The sales teams need to be highly trained, spend considerable time getting to know the target company and its business strategy, and aim to build close, personal relationships with the key decision-makers. Armed with this insight the account manager seeks to put together comprehensive 'solutions' to the company's business problems. These 'solution' sales consist of a wide range of products, possibly including some non-standard items, brought together in a comprehensive system to address, not simply price or functionality concerns, but the whole-business needs of the target account.

Customers may invite 'bids' or 'invitations to tender' for major sales contracts and it will be the account team that uses their expertise and in-depth knowledge of the customer's business to respond to this invitation and, hopefully, win the business.

This emphasis on relationship and 'solution' sales places the major account manager in a very different relationship with their own company. First, they require strong technical support in the form of systems engineers to design, sell and implement the solution. Often this technical expertise is directly associated with the specific major account team and will be rewarded in comparable ways. Second, the major account manager aims to place himself or herself between the customer and the company. They, therefore, become the focus of a wide range of issues, ranging from customer service failures to high-level corporate-reputation concerns. In an ideal world they will have the authority and status (the *clout*) to get these issues resolved, whether that be through managing a dedicated customer service call centre or orchestrating a meeting between senior managers from the customer with their company CEO. Too often Telcos say that they 'put the customer first', but then do not support the major account managers when they are battling to resolve problems for their customer.

Successful account managers are cut from a different cloth than generic sales people. They emphasise relationships, understand a single business in detail and have a much greater concern for the long-term development and profit from the relationship. Typically, a high proportion of their personal remuneration is in the form of a discretionary bonus based on a wide range of measures, certainly including revenue and profit but also including customer satisfaction, future systems planning style and behaviour [8].

7.7   Customer service and satisfaction

Clearly customer service is a key aspect of the marketing function and central to a company's offering. For example, the level of service offered may be a key marketing differentiator. Recognising this, companies might offer tiered service levels (gold, silver and bronze for example) and price accordingly.

It is vital to ensure that the quality of customer service provided is appropriate to the product and to the market segment that is being targeted. Businesses will systemically survey customers to ascertain their level of satisfaction and their attitudes towards every aspect of the service they are receiving. It is essential to act on the insight provided by such surveys and customer feedback. For example, there is a very clear link between customer service and brand reputation.

These subjects are fully covered in Chapter 9.

7.8   The empowered consumer

There can be no question that the Internet has changed how business is done and, in particular, how consumers make buying decisions. In this section we look at some examples, not only of how consumers purchase ICT services, but also how they use ICT to ensure the very best deals in other spheres. In short, how ICT has created empowered consumers.

7.8.1   Disintermediation and the death of the middleman

One great advantage of the internet has been the opportunity to empower the consumer and to cut out the middleman. If we think of information, news, music or fiction, it is now possible for the consumer to go directly to the source. Personal blogs enable people to read the opinions of the people who interest them without buying newspapers, magazines or attending meetings. The move from what John Naughton, Professor of Public Understanding of Technology, the Open University, calls *push production* of media content to *pull demand* [9] ensures that a far wider range of articles and blogs, and therefore a wider range of opinions have become available. Similarly, in music, Napster (and other peer-to-peer sites) enabled the sharing of music without the involvement (and profit) of the music studios, retailers, etc. This quickly led to the fundamental shift that saw new music-makers putting their material directly on the Internet without signing up to a producer or studio. (We explore this in greater detail in section 7.8.3.) A lot of news now comes from direct feeds from eye witnesses using camera phones, twitter posts, etc., without the need for journalists, press photographers or newspaper editors. Indeed, the boot is now on the other foot with mainstream journalists regularly trawling the Internet for stories, news pictures and film. Print newspapers and TV news channels struggle to remain financially viable in the face of so much online do-it-yourself news, comment and opinion.

For the consumer, this move towards consumer production of media and *disintermediation* (cutting out the middleman) has many advantages. In particular, it enables events to be viewed without the filter of an editor or political spin, and gives wider choice. However, ironically, the sheer volume of material has led to the development of aggregator sites to pull together the most useful and interesting pieces. Spotify, iTunes and other music sites offer aggregation services effectively selecting a playlist for you, based on your choice of music (just like an old style radio station!). Online newspapers, such as *The Huffington Post* and *The Onion*, aggregate blogs, news feeds, political opinion and online gossip into a single source – the first destination each morning for many, just like their old daily newspaper [10, 11]. Initially, aggregator sites were free, paid for through advertising, but it may be just a matter of time before they become subscription sites (as many traditional newspapers now are).

In terms of e-commerce: high street travel agents, insurance brokers, recruitment agents, check-in desks have all been disintermediated as consumers, empowered by technology and information, have done their own research online and filled in their own application form, placed orders or whatever. Even retail banks and credit card companies will be disintermediate as digital money (e.g. value stored on your mobile phone) replaces coins, notes and credit card transactions. And this trend will continue into public services with passport offices, doctors' surgeries, benefits' offices, even professors, being cut out of the space between the consumer and the content that they value. Of course, many of these middlemen have simply changed their model and gone online themselves; the good ones have reinvented themselves and are doing very well indeed, such as recruitment agencies and travel companies – or they add value in selecting and presenting information.

At the same time as we see the death of the middleman we are seeing a brand new Internet phenomenon – the comparison website (otherwise known as a middleman!). In this case, it's an entirely automated service aimed at providing a review of all possible matching services (e.g. car insurance for a professional couple living in London suburbs with clean licences, and driving a 1.6 Ford

Mondeo Zetex). Broadly, they compare prices and provide the cheapest ones, although they may also help with particular features of the service, terms and conditions, etc.

And as ever, with every action there is a reaction. Some of the strongest brands now exclude themselves from comparison sites, hoping to use their name and reputation to attract customers direct.

Perhaps the greatest example of disintermediation is eBay, the online auction site that has effectively disintermediated retailers altogether. Essentially, the service consists of consumer selling direct to consumer, but its radical new business model relies heavily on trust, a key feature of e-commerce that is going to grow hugely in importance in the future (see section 7.8.4).

7.8.2   The empowered consumer dictates the right channel

Has online retailing made shops obsolete? If you believe a lot of the technical press you would be inclined to say yes. Some believe that the birth of Amazon was the death of high street shops (see Box 7.5 to understand the phenomenon that is Amazon).

However, if you listen to the empowered consumer you might arrive at a different answer. For example *Clicks and Bricks* (Cs & Bs) is an alternative retail model that is proving very successful. Combining physical shops with an excellent online service gives customers the opportunity to see, touch and even try-on goods, and then order them online for convenience, delivery and better deals. This is a highly successful model in the fashion world. In other sectors, e.g. white goods and cars, consumers may visit a well-stocked department store/car showroom to check out the full range and then go online to buy at the best price (not necessarily with the same department stores or car showroom).

Alternatively, in the Cs & Bs model, customers will research online, checking a vast array of products and technical details before confirming their final choice in store/dealership. Ironically, cars also fall into this category (70% of purchasers will research online before buying) demonstrating that personal choice is highly influential in deciding which model an empowered consumer might follow. No longer do we all buy certain goods in one particular way – we choose the channel that suits our disposition, free time, lifestyle, etc.

Groceries fall into the Cs & Bs model for different reasons. Initially, consumers were nervous of buying food online because quality requires a visual check of freshness; after all, brand loyalty is an important aspect in the consumers' mind, ensuring the right quality across the whole shop. However, noticing how heavy their shopping had become and that the weekly list included many of the same items every week, consumers found they liked the convenience of delivery. So, they might divide their shopping into a major heavyweight shop, which they do online and have delivered, and quick visits to a conveniently positioned (high street, railway station) fresh food store. So powerful is this trend that major retailers have stopped the trend for out-of-town megastores and replaced them with well-located convenience stores, such as Tesco Express, Sainsbury Locals, etc.

Box 7.5   Amazon – building an online brand

Worldwide, people are now buying both digital and physical books online. Despite the huge growth in sales of e-Readers, such as the Kindle, there remain a large number of physical books still being delivered by the 100-year old mail services. Christmas 2009 saw Amazon delivering more e-books than paper versions in the United States; and Christmas 2010 saw blanket advertising of the Amazon Kindle e-Reader in the United Kingdom. This proved the starting gun for the United Kingdom to fully embrace e-Books. In May 2011 Amazon announced that they were regularly selling more eBooks than physical books. Christmas 2011 saw the Kindle as the biggest selling product across the whole of the EU.

Perhaps E-book sales is tolling the death-knell for physical bookshops, as music downloads have done for Music shops.

But Amazon's business now depends more on consumer goods than books. Whatever it is that they sell, Amazon is a massively trusted and well regarded brand that in 10 short years has gone from a freakish online book seller to a worldwide phenomenon. At the beginning of e-commerce, many people (the authors especially!) were sure that successful online brands would be built by existing trusted brands going digital. Amazon proves otherwise. Existing booksellers (Borders, Waterstones, etc.) have struggled to keep up; their response (e.g. to turn themselves into coffee shops, 2 for 1 offers on the best-sellers, to add a website to their range of outlets, etc.) was slow and only partly successful. Responses from consumer-durable companies are likely to be equally poor.

So how did Amazon build this position? In a single word: execution. In particular, their catalogue could be built to include virtually every book ever published. Compare this with even the biggest bookshop that can carry no more than a few thousand titles. They quickly built a reputation for excellence in what they do. Their website worked, their catalogue of books was comprehensive, deliveries arrived when they were expected; above all they gave no doubt that the privacy and the security of your credit card was 100% safe. Part of this was built on the back of very robust management and information systems, but it was also achieved through web design that gave the consumer confidence. Finally, it is important to note that Amazon succeeded where many other *dot.com* companies did not, because their business was supported by a large infrastructure of efficient warehouses, well-trained staff and excellent inventory systems. The big lesson from the history of Amazon is that successful marketing must encompass the excellence of the back office processes, as well as the more-visible front office activity (e.g. Amazon's website).

Finally, in yet another form of evolution, we see previously exclusively online traders appearing on the high street. There can be no more iconic online operation than eBay, but they now find great value in having delivery points on the high street rather than relying entirely on the mail service. Elsewhere we see *click and collect* services developing for groceries, white goods, etc.

Online shopping has created a period of great flux and uncertainty for the high street, but it may be that, in time, retailers will largely reinvent themselves as collection points for goods purchased online.

Again, consumer choice and convenience has called into being a new business model.

7.8.3   The empowered consumer reconfigures an entire industry

Perhaps the most radical shake-up of any industry has come with music. Four separate factors came together in the 2000's to produce a radical shift in the industry:

- the explosion of consumer-produced content, in particular the use of YouTube and similar sites to reach an online audience and to self-publish material
- music, created as a series of digital files, was easily downloaded, even on relatively slow broadband
- a technology company (Apple) ready to stake its future on MP3 technology
- the buying power of technically savvy young adults

Success was apparently guaranteed, although it very nearly wasn't. The slowness and reluctance of the record companies to make their content available to iTunes and other digital catalogues led to problems of piracy and illegal file-sharing which, in turn, threatened the reputation of the online music world. Illegal file-sharing continues to dog the record industry around the world, cutting profit and impinging on copyright. In some domains the concerns have led to proposed legislation that, if enacted, would criminalise downloaders and risk their broadband service. What has proved to be a very profitable venture for online retailers (such as Apple) has damaged record companies and put them at loggerheads with their customers.

Questions remain. In particular, a debate continues to rage about how music production will be funded in a digital world, although there is no doubt that, for some musicians, publishing online is a cheaper option than going to a major music company. Indeed many acts have grown successfully through self-publishing online.

What is clear is that an orderly market, with people paying a fair price for digital downloads, will be a more effective approach than piracy. While some think the record industry is losing out, others recognise that music as a whole may gain from downloading. For example, single track digital downloads are three times higher than their physical counterpart. While this increase doesn't fully compensate for the lost revenue from albums it may point the way for future success. Also, research from Forrester [12] seems to suggest that downloaders actually spend more on music than is avoided by their illegal downloading. Part of this is that illegal downloaders are interested in music, partly that they skew their expense away from CDs towards live music events; in addition, they may be downloading to *try before you buy.*

It is clear that consumers, empowered by technology, have changed the face of a major sector of the entertainment industry. Developments in the music industry between 1995 and 2015 certainly provide vital insights into how movies, literature, sport, education and many other content-based industries will go in the future.

7.8.4   The importance of trust

As Web 2 applications, such as EBay, start to dominate ecommerce, consumers will value new aspects of the online relationship. Trust will be a key component. We saw earlier that privacy and security were vital in the early days of ecommerce to build customer confidence and encourage the use of these new and innovative services. This goes much further now. Sites will not only need to look inviting and be easy to navigate but will include other features aimed at building trust. For example, a wide range of accurate and detailed information about products, presented in an easily understood way, with intuitive search and filter facilities will be vital. Excellence execution (delivery, customer service, help-lines) will also be important in building ongoing trust, as is an absence of in-your-face advertising. Empowered consumers want to feel a sense that what's important is *My Needs, not Your Business.*

There is no more contentious brand on the Internet than Google. This is not surprising as the ubiquitous search engines hold a huge amount of highly sensitive data about all of us. Consumer trust is highly important to them and they nurture it through a very carefully thought-out strategy and a wide range of safeguards. Despite some bumps along the way they have continued to retain the trust of the users and to be the default search engine for the world (Box 7.6).

Going forward, therefore it is clear that companies (and especially marketers) need to continue to respond to the challenges and opportunities arising from the newly empowered consumer. More and more we see e-commerce companies embracing customer advocacy.

7.8.5   Customer advocacy

E-commerce has developed hugely since the Internet first enabled it in the early part of the 21st century. Nonetheless, radical new features are already influencing future services; one is peer recommendation. EBay succeeds because strangers are willing to trust other strangers, basing their confidence on feedback ratings posted by people who are entirely unknown to them. This is perhaps surprising, but it works and is a hugely powerful tool in commerce. Many independent websites and comparison sites now offer customer reviews (when did you last book into a hotel without checking what others had to say?). Shouldn't we then also expect any company offering a product or service to quote customer reviews themselves, the good and the ugly, perhaps verified by an independent third party?

Given the level of communication now progressed through social networking, customer advocacy becomes an essential tool of success. With a large and growing group of people recording their experiences and posting them on social network sites, this is an obvious source of consumer information; customers advocating your product in their tweets, etc., will be essential to success. (Incidentally, it is said that

Box 7.6   Trust: Can anyone challenge Google?

We have identified the importance of privacy and trust in building customer confidence. An interesting example in this area is Google. Having started as a typical West Coast technology start-up (the founders – Larry Page and Sergey Brin – were PhD students at Stanford University as recently as 1998), their business is totally dependent on the trust of their users. Users need to trust that the search engine will find all the relevant material; that searches will not be influenced by sold advertising; and that the vast quantity of personal data held will be used only for good, not evil. (The company motto is, extraordinarily, 'Don't Be Evil'). Google don't make any pretence that they aren't in business to make money. They are a massive corporation that earns billions of dollars from paid-for advertising. What is more, that advertising is personalised to a very great extent based on the clicks of individual users. Armed with a detailed knowledge of your searches, an intelligent observer could easily determine your location, your lifestyle, your health, your family situation, etc. Despite all this, users continue to trust Google and to associate it with the strongest of brand reputations.

How does it achieve this? Part of the answer has to be the sheer value offered to the user. A free service coupled with the excellence of their product has ensured that the Google experience is valued as the most valuable tool on the Internet by astonishing 90% of internet users.

But the other part of the answer is the philosophy of the company. Paid-for links and sponsored results are clearly identified; information collection is transparent; and the quality of customer service is very high. There are no 'in-your-face' banners or other advertising. And when the Chinese Government wanted to suppress search results, Google eventually pulled out of China (having made an early mistake of staying for too long).

So the future challenge for Google will be to try to retain our trust by staying true to their start-up values (fairness, openness, excellence and customer focus) in the face of rising concerns about privacy and big business manipulation. And the future challenge for every other online enterprise will be to emulate their success.

if a celebrity tweeter recommends a product or service the relevant server immediately collapses, unable to process the mass of new traffic! – that's a very real sign of customer advocacy working.) Others have chosen to formalise this and to build online businesses by harnessing individual's recommendations and reviews to advocate and earn commissions. Probably the most successful is TripAdvisor [13].

However, while companies will certainly seek to benefit from social network endorsement, the contrary is also true. It used to be said that if you enjoyed poor service you would, on average, tell 10 more people. Today that can be 10 million or more. Consumer terrorism/activism is an established tactic to counter poor service.

Anti-McDonalds sites are well known, but similar sites apply to IBM, British Airways, Microsoft, Starbucks etc. And you don't even have to set up a website. Individuals can use the power of digital communication to inform millions of their complaints by posting tweets, blogging or clever use of YouTube. A particular favourite example is a disgruntled musician's complaint to an airline about his broken guitar. This song and video posted on YouTube not only damaged United Airlines reputation, it proved a launch-pad for Dave Carroll's musical career [14]. Successful companies (and public services) of the future will find ways to harness the power of social networks and use viral marketing to their benefit, rather than fall victim to it.

7.8.6 Customer empathy

It is one thing for the customer to be your advocate but, as we have seen, the power is in their hands. It is therefore incumbent on cutting-edge companies to think about how they can turn things on their head and become advocates of the customer themselves. As we have seen attractive, safe, secure access, with easy-to-use online facilities will help to build the trust required to persuade customers to do business. But how do companies really show they are on the customer's side? A few innovative ideas are beginning to take shape, such as online guides to making the best buying decision; sites that compare a company's offering with rival products; and independent reviews and comments from previous customers' – good and bad.

Indeed some of the best examples come from the ICT industry. For example:

- Far greater emphasis on really understanding individual customer needs. As an example a particular fibre-based Telco selling high-speed digital connectivity and dark fibre has banned voice answering machines and telemarketing, relying solely on a personal commitment to speak directly to customers about their requirements. Every customer has a named salesperson, a named service coordinator and ready access to senior managers.
- Online portals are now routinely provided to major clients, allowing them to make network changes or order online, giving them full control over their services.

Finally, it's clear from research that personality and cognitive style play a major role in determining how we respond to information presented in digital form. Some of us like graphs; some prefer raw data; some are persuaded by statistics, others by images or customer testimonials. By analysing how individuals navigate around a site we can determine their likely preference and re-format the site to suit their style. It's likely that presenting material in a format that suits the viewer builds further trust and leads to a greater propensity to buy. Site morphing, as it is called, is still at a very early stage of technical application and field results are unclear, but it may be key to doing business online in future [15].

7.8.7 The empowered consumer: conclusion

We continue to see further, far-reaching developments in e-commerce that will put the consumer even more in the driving seat. Customer-centric sites, of the type described here, go well beyond the idea of excellent customer service, relationship marketing or

THEN   NOW

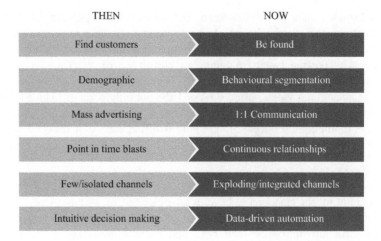

Figure 7.6   The six new rules of marketing [16]

the 4 P's. This has led some people to suggest that we need to completely rethink the rules of marketing, one example of which is shown in Figure 7.6. Time will tell how things do develop, but the new marketing order will certainly have total customer advocacy at its centre. When the company has the customers' interests entirely at heart – and demonstrates this completely through their communication and actions – they build trust and loyalty that will ultimately lead to success in the market place.

7.9   Summary

We began this chapter by saying that there is no business without customers and that marketing is an orientation rather than purely a set of functions. Nonetheless, we have explored some approaches that help to build clear marketing objectives and support successful market planning. We then looked at the four key aspects of the marketing mix – product, price, place and promotion – and applied them to ICT companies. Finally, we recognised that market power is shifting rapidly towards technology-enabled, savvy consumers and that commercial success in future will depend on new marketing rules. More than anything else success will rely upon building unshakeable trust through total customer empathy.

References

[1]   Mc Donald & Chrisopher, M. *Marketing: A Complete Guide*, Palgrave MacMillan, 2003, Module 2, Chapter 6.
[2]   Kotler, P. & Armstrong, G. *Principles of Marketing*, Pearson/Prentice Hall, 2008, Chapter 1
[3]   Mc Donald & Chrisopher, M. *Marketing: A Complete Guide*, Palgrave MacMillan, 2003, Module 1, Chapter 3.

[4] Bayer, J. 'Customer Segmentation in the Telecommunications Industry', *Journal of Database Marketing & Customer Strategy Management*, 2010, Palgrave MacMillan, http://www.palgrave-journals.com/dbm/journal/v17/n3/full/dbm201021a.html

[5] Leahy, T. *Management in 10 Words*, Random Books Business Books, 2012, Chapter 2.

[6] McCarthy, E.J. *Basic Marketing: A Managerial Approach*, R.D. Irwin, Homewood, Ill., 1981.

[7] Mc Donald & Chrisopher, M. *Marketing: A Complete Guide*, Palgrave MacMillan, 2003, Module 3, Chapter 16.

[8] Mc Donald & Chrisopher, M. *Marketing: A Complete Guide*, Palgrave MacMillan, 2003, Module 2, Chapter 12 and Module 3, Chapter 15.

[9] Naughton, J. 'Our Changing Media Ecosystem'. From a collection of essays prepared for the UK Office of Communications. London, 2006, pp. 41–50.

[10] The Huffington Post, online news aggregator: http://www.huffingtonpost.co.uk/

[11] The Onion, http://www.theonion.com/

[12] Mulligan, M. & Wiramihardja, L. *Music Industry Meltdown: Recasting the Mold*, Forrester, January 2010, https://www.forrester.com/Music+Industry+Meltdown+Recasting+The+Mold/fulltext/-/E-res56147

[13] TripAdvisor, online trip planning and booking, http://www.tripadvisor.co.uk/

[14] Carroll, D. *United Breaks Guitars*, Dave Carroll Music, July 2009, http://www.youtube.com/watch?v=5YGc4zOqozo&NR=1

[15] Urban, G.L. 'Online Trust and Customer Power: The Emergence of Customer advocacy'. From 'ICT Futures', edited by Paul warren, John Davies, David Brown, Chapter 4.

[16] Dholakia, S. 'The New Rules of Marketing. Marketo's Secret Sauce Case Study', http://www.slideshare.net/marketo/the-new-rules-of-marketing-sanjay-dholakia

Chapter 8

Product management

8.1 Introduction

The role of product management is today widespread throughout companies in all industry sectors around the World. There is always a manager, team, or whole department of people dedicated to the day-to-day and longer-term steering of the delivery of a particular product, the monitoring of its performance, and marshalling of the company's resources to ensure its well-being. Put succinctly, the role is to champion the product throughout its life. In this chapter we examine the various facets of product management, followed by the associated role of managing a portfolio of products within a company – all in the specific context of the tele-communications industry. We conclude the chapter by exploring the important contribution made by innovation and R&D to product management.

Before going any further into the subject, it is useful to remind ourselves of what is said in Chapter 1 concerning what we mean by the term 'product'. Strictly speaking, it is generally accepted in all industries that a product is any tangible item sold by the company – e.g. mobile handset or smart phone, and home WiFi router. Whereas, a 'service' is a non-tangible item sold by the company – e.g. broadband access, the ISP role, and telephony calls. However, despite this useful distinction the two terms are frequency used interchangeably in practice across all industries. Even for a telecommunications operator, whose business is primarily concerned with delivery of services, the term 'product management' is invariably used in preference to the more cumbersome 'product and service management'. In this book we will refer to the management of both products and services as 'product management'.

The broad scope of product management is clearly demonstrated by the following outline of the activities involved.

(a) *Whole-life management of the product*: covering the initial development of concept, its specification and design, its launch, on-going management, mid-life enhancements, and eventually the withdrawal of the product from the company's offerings – a period ranging from several months to several decades, as appropriate. To successfully manage the whole life of a product the following functions would be deployed.

(b) *Product costing*: covering the determination of direct costs incurred by the product as well as the apportioned indirect costs that must be included in any product P&L reckoning, and the internal management reporting within the company.

(c) *Product pricing*: in which the position of the product in relation to its competitors is determined, and also its relative positioning within the company's portfolio of products.

(d) *Distribution plan*: What sales channels (telesales, retail outlets, specialist sales teams, etc.) are to be used, and what incentives and promotions will be required to support them, as discussed in more detail in Chapter 7 (Sections 7.5.2 and 7.6).

(e) *Product launch management*: providing the oversight of all the activities in the company required for a smooth trouble-free launch of the new product, covering the sales and post-sales activity, on-going operations, and the geographical spread of the rollout.

(f) *Demand forecasting*: in which an estimate of the number of products that will be sold in the future is made (covering several years), and hence the expected revenue that can be included in the company's budget. This is a difficult enough task for established products, but can prove to be extremely challenging when dealing with new products, particularly in the pre-launch stage when developing the justification for creating the product.

(g) *Management reporting*: this normally covers the assembly of achieved sales volumes and revenues, together with any customer issues relating to quality of service, or the identification of the need for feature upgrades to the product – all presented on a regular basis to senior management within the company.

(h) *Developing and presenting product business cases*: this covers the big challenge of creating a business case for the initial launch of the product, based on cost and revenue estimates, an analysis of the market, and a convincing case for the company's Board to commit sufficient resources. Business cases will also be needed for subsequent feature upgrades of the product, or the investment in further rollout of network capacity to support a bigger geographical spread or increases in volumes.

(i) *Tracking R&D and bringing innovation* to the product as appropriate to keep it competitive.

We will now consider how this wide range of activities are applied to products in a Telco.

8.2   The product management roles

8.2.1   *Product life cycle*

The classical life cycle of any product is shown in Figure 8.1, in which the sales volumes are plotted against time. Although the sales volumes cover the time from the product launch to when it is no longer sustainable and is withdrawn, the life cycle model usually also includes the period before the product is launched when there is the main design work and expenditure on network and support-system infrastructure. Interestingly, the sales profile of all types of products and services follows the basic shape of the standard life cycle, despite the wide range of lives and the variation in the industries concerned. This makes it a useful tool to systematise the product-management activity.

The life cycle falls into five phases: pre-launch, introduction, growth, maturity, and decline. We now examine each of these in turn, noting the various product management activities and the business objectives involved.

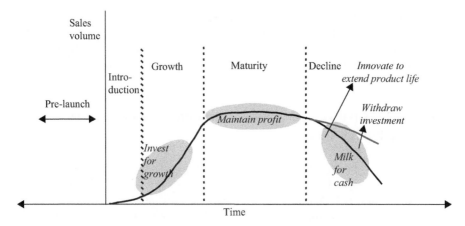

Figure 8.1 The product life cycle model

8.2.1.1 Pre-launch Phase

Since the pre-launch phase encompasses all the work within the company to define the new product, assess the costs of implementation and the potential revenues, and then seek authorisation for expenditure, it can cover many months or even a few years depending on the novelty and complexity of the product. The phase begins by identi-fying the need for the product and the market opportunity, as discussed in Chapter 7. It is at this point that the product management team start the definition of the product, based on the market opportunity and the current offerings by the competition.

There will always be choices about which set of features should be included – e.g. speed, need for new customer terminals, and network–user interfaces. In many cases a trade-off may need to be made between the inclusion of certain features and the implementation costs; the market activeness may be at the expense of undue devel-opment delay or extra cost. Often, where a range of requirements are identified, the best approach is to define a set of products, perhaps with different levels of service which might be aimed at different segments of the target market. It is the responsibility of the product manager (PM) to be clear about definition of the new product(s), so that the implementation costs and timescales can be assessed within the company.

Invariably, the choice for the Telco is whether to extend the capabilities of the existing platforms (i.e., network and support systems), e.g. increase of speed, or to deploy some new equipment. Either way, there will need to be capital expenditure on the network and/or the support systems before the product is launched. It is worth noting that many products are provided just by software upgrades to the support systems and do not involve any network equipment – e.g. new billing packages.

The costs and timescale involved in this network or software upgrade activity will need to be fully assessed and included in the product-launch business case. As mentioned earlier, this activity can be lengthy, particularly if the Telco has to seek proposals from the equipment manufacturers, as described in Box 8.1. Once sufficient information has been gathered a feasibility study is made to ensure that there is a practical and affordable network design to support the new product. Often, there are optional ways of implementing the new network product (service) and these need to be considered by the PM. If the feasibility study indicates some insurmountable problems with certain product features the PM will need to modify the product definition. Invariably, this raises a challenge for the PM because the original marketing opportunity, as defined, may not be fully realised by the modified product.

Box 8.1   Involvement of the equipment manufacturers in the pre-launch phase

Telcos are highly dependent on the equipment manufacturers for building the network platforms. Because of this importance, many Telcos have a close, and often strategic, relationship with particular manufactures over a long period to ensure continuity of supply and after-sales support, as well as obtaining early sight of the manufacturer's developments in equipment and new technologies. In many cases these manufacturers would be directly invited to supply the additional features on the existing equipment already installed in the network. However, Telcos are normally required by law to give all suitable manufacturers the opportunity to bid for the new network provision, and the following process applies:

Request for Information (RFI) is issued by the Telco to all eligible manufacturers, seeking an indication of what equipment they could offer against an outline description of the new network features required. Responses to the RFI enable for the Telco to conduct feasibility studies and can form an input to the initial business case for the product(s).

Invitation to Tender (ITT) is issued by the Telco to a short list of suitable manufacturers based on an assessment of the RFI responses. This comprises a technical specification of the required network equipment, and a set of technical and commercial questions (prices, terms, timescales, after-sales support, etc.) concerning the manufacturer's ability to meet the specification.

Tender is prepared by each invited manufacturer addressing the specification and questions. This document is usually voluminous and represents a great deal of preparation by the manufacturer's bid preparation team over several months.

Tender evaluation and adjudication is undertaken by a team within the Telco, including the PM, in which the manufacturers' ability to meet the technical and commercial requirements are assessed against the previously stated criteria. A final decision on which manufacturer should be chosen to supply the equipment is usually endorsed by senior management, maybe even at board level.

Contract awarded to the chosen manufacturer for delivery and installation of equipment. This might include the need for development and modification to the equipment in order to meet the Telco's requirements.

In parallel with the technical feasibility work, the PM will start to assemble the elements of a business case to seek authority within the company for the launch of the new product, and allocation of the financial and manpower resources involved in launching and running the product. It should be noted that much of the information, particularly on costs and customer take-up, may be speculative at this stage. Thus, many companies follow a two-stage approach: first, an initial outline business case is made to seek approval for continuing the technical and commercial work on defining the product and its implementation; second, the definitive business case is produced later, once firmer cost and timescale information is available to the PM. We look at the subject of business cases in more detail in Chapter 4.

The PM will depend upon a number of company resources to develop and deliver a business case. These could include professionals from such disciplines as network and platform design, procurement, sales and marketing, commercial, legal and regulatory, and finance. For significant projects a PM should consider establishing a product team of nominated representatives from the appropriate disciplines. The team's activities will be at their most intense in the period up to launch, but ideally should also extend to in-life product management.

Once the business case has been authorised, the process of product launch begins within the company. This important activity is covered in section 8.2.4.

8.2.1.2   Introduction phase

The Introduction Phase of its life cycle begins as soon as the product is launched. Obviously, this is an anxious time for the PM since there will be concerns about whether the product works in practice as planned and that there is good customer take up. There will undoubtedly be many teething problems with the product and with its operational support – especially the order taking and maintenance processes (see Chapter 9) which the PM will need to address. For this reason it is important for the company to consider prior to launch what metrics and reports are required to monitor product performance and what capabilities need to be in place to support sales, installation and maintenance staff dealing with the new product or service (see also section 8.4.2).

It is interesting to consider how the product design criteria during the pre-launch and introduction phase differ from that in the later phases. The first big difference is the overall objective – normally, there is a big push within the company for the product to 'get to market' as soon as possible. Indeed, there may be an opportunity window in the market, before the competition is expected to enter, which the company is targeting. This means that the initial implementation of the product may be based on proprietary, possibly prototype, rather than established technology meeting international standards. There may also be a regulatory holiday for this product, in that there is not a requirement for the Telco to provide interconnection to other operators while the product is in its nascent state, thus simplifying the initial technical design of the product. On the other hand in some jurisdictions an agreed interconnection regime, or new wholesale service may be a pre-requisite for launch of a new retail service. The other big difference in the design criteria is the primary aim being to achieve an early presence of the product in the market, which may mean that the design is suitable for small volumes only and is not cost optimised. Finally, the support systems available for the product in this phase may be rudimentary, with improvised arrangements – this is acceptable, given the low sales volumes to allow an early launch date – rather than incur the delay for developing new operational support systems (OSS, see Chapter 9). An example of improvised OSS in the introduction phase of a product is given in Box 8.2.

Box 8.2 Example of improvised OSS during product introduction phase

MCI was the first operator in the United States to compete in the long-distance market with ATT. The company rapidly became known for innovative service, but it was still very much the underdog in competing with the ATT giant, and it was desperate to gain a critical mass of customers. Around 1990 MCI launched a creative tariff package called 'Friends and Family (F&F)' – an attractive loyalty programme. This product enabled the MCI customers to call a maximum of 20 nominated MCI customers (friends and family) at a high discount (30%) on the price.

Quite rightly, MCI considered this a bold move and the F&F product was developed in utmost secrecy and with great urgency, since there was a fear that ATT might come up with something similar for their customers, as a spoiler. The F&F product was therefore launched before the necessary modifications to their billing engines could be developed – a bold move! This meant that for the first few months of service the complex billing arrangements for F&F customers were processed by hand at the MCI offices. Clearly, this approach was only acceptable while the volumes of customer take-up of the new product was low (i.e. in the Introduction Phase). Within a few months the computerised OSS was introduced and the F&F product moved into the growth phase to become a widespread great success [1].

8.2.1.3   Growth phase

Following a successful introduction of the product, there will be a big drive within the company to expand its geographical coverage and to increase the capacity of the network to allow the growth in the sales volumes. In this growth phase, the main objective for the company is to establish the product in the market. Therefore, the PM may need to produce further business cases for the necessary investment in network capacity and possibly some changes in technical design to cope with the larger capacity. There may also be a need to invest in the development or purchase of new OSS to support the product sales and maintenance of increasing customer base. Hopefully, this requirement will not be seen as new and will have been flagged-up at the initial business case stage.

At some stage during the growth phase the national regulator may require the product to be opened up to interconnection with other operators. This may require some modifications to the technical design of the implementation of the product. Again, this need should have been planned for – it may be unacceptable to discover interconnection is not feasible or technically complex and costly after the launch of a retail service.

8.2.1.4   Maturity phase

The growth phase of a product may last many years – e.g. rolling out the geographical coverage of a new generation of mobile network – during which the sales volume will continue to rise. However, all products eventually reach a point of market saturation where the annual sales volumes cease to rise and the life cycle curve is essentially flat. The company's objective during this maturity phase now becomes one of maintaining a successful product in service and maximising its profitability. This means keeping as high a price as possible in the prevailing market conditions, and reducing as much product cost as possible through some or all of the following:

- Improvements to the technical design of the product implementation, maybe involving network equipment upgrades.
- Improvements to the various operational processes being followed.
- Changes to the OSS.
- Exploiting the fact that the level of customer usage of this product is now well understood to optimise the product design.
- Reducing advertising costs, since the product is now well known.
- Resolving identified issues relating to sales and installation costs, or to causes of failure.

8.2.1.5   Decline phase

The sales of all products eventually start to decline, as a result of changes in the market in terms of competition, substitute products, and new technology. In this decline phase the product is approaching obsolescence. Normally, a company will try to prolong this period as much as possible in order to continue receiving revenue (at a declining rate) from what is now a mature, cost-optimised product. In some circumstances, the company may add selected enhancements to the product – and so incur some new capital and possibly current account spend – in order to reduce the decline in sales. Even the modest reduction in the decline can prove to be worthwhile, given the profitability of the product and the high level of sales at the start of the phase. The objective for the company, of course, is to maximise the area under the sales–revenue curve.

Eventually, the product becomes unsustainable, with little or no new sales and a rapidly increasing rate of customer withdrawal. At this point the product moves into negative profitability. The product then becomes a candidate for withdrawal from the company's portfolio. In some industries a company can withdraw their products whenever they choose, though there is always a dilemma as to when to migrate existing customers to a new and unfamiliar product, because they may choose to migrate to a competitor's product instead. However, for a Telco there are also regulatory constraints on how and when products can be withdrawn, in order to protect customers who rely on the continuing use of the Telco's services to run their businesses. Invariably, Telcos are obliged to give advance notice (typically 18 months) of product withdrawal. In addition, the Telco may wish not to upset an important vestige of their long-term customers, even if it means running their obsolescent products beyond the economic withdrawal date.

8.2.2   Product costing

The top management level of a company will have a clear view of the total level of revenues, operating costs, and capital employed, and therefore the profitability of the business each year. In order to get a view of the profitability of its product portfolio, these revenues and costs need to be associated with each product line. It is usually relatively straightforward to identify product revenues from the company's revenue systems. However, attributing operating and capital costs to each product is more challenging, particularly for telecommunications businesses, where a large proportion of the cost base relates to a capital-intensive network shared by a large number of different products. In this section we consider the various costs that a PM in a Telco needs to track and how they might be managed.

The PM needs to know the costs incurred by the products for two main reasons. The first is so that they can construct business cases seeking authority to launch the product, or subsequent enhancements or capacity and coverage expansions. The focus here will primarily be on cash flows and therefore on the cash costs, both current costs and capital costs, associated with the business case. The second is, of course, so that the product's price can be set, and the profit margins and P&L determined and managed. A PM has a responsibility not only to know his product's costs but to manage them down to improve profit. For many companies, and especially in the telecommunications industry, there is a further important reason for knowing the products' costs – that of reporting to the national regulator. As described in Chapter 2, the regulator may require quite detailed product cost information, especially from the incumbent operator(s) to ensure fair competition for retail products (i.e. to the end-user) or wholesale products (i.e. to other operators). Also, the regulator needs such information to monitor, or even set, inter-connection charges (i.e. the conveyance of traffic originating or terminating in another operator's network).

It is important at this stage for us to consider the different types of costs that feature in the P&L for a product. These fall into the two broad categories of direct costs (incurred because the product exists) and indirect costs which the company has to bear irrespective of whether the product exists, an appropriate proportion of which has to be included in the products P&L. A range of costs that feature in a Telco's network product P&L include:

(a) *Direct network costs*: These cover the maintenance and depreciation of the network equipment and plant used by the product. Some of these will be used exclusively by the product, others will be shared with other products and apportioned appropriately.
(b) *Indirect network costs*: These are the costs involved in running the network (e.g. power, transport, planning, building security, network IT) which are usually shared across all products.

(c) *Non-network direct costs*: This category covers those non-network costs attributable to the product and often under the PM's control (e.g. marketing and sales, stores, billing, some R&D).

(d) *Business overheads*: These are the indirect costs incurred in running the company (e.g. the various corporate management (HQ) departments, general R&D, public relations, estate management, and central IT support). This category may also include the costs associated with redundancy payments incurred by the down-sizing of the company, as well as other structural changes to the organisation.

Figure 8.2 provides a simple pictorial summary of the way that direct and indirect costs are associated with the various products (services) run by a Telco. For example, we can see that Product A incurs its apportioned share of the Telco's indirect costs (as do all the products); it also shares some direct costs with Product B; and fully carries its own direct costs. An estimate of Product A's profitability can be made by associating the sum of the direct and indirect costs with its revenue. We can also see in the bottom line of Figure 8.2 that the total set of direct and indirect costs can usefully be grouped into organisational or business units within the Telco – e.g. wholesale product division, consumer product division and business product division – to enable management board-level tracking of profitability by product and by business unit.

While the top-down approach described earlier gives a systematic way of reporting on the profitability of a portfolio of well-established high-volume products, it is not well suited to new, low-volume or fast expanding services. A further disadvantage of this approach is that it often gives only limited visibility of how a

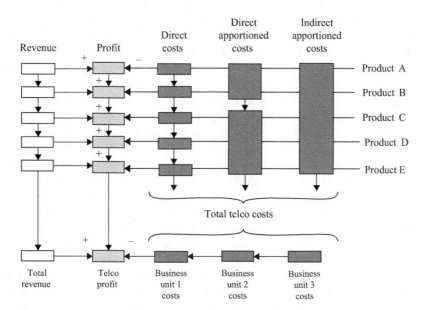

Figure 8.2   Allocation of costs to products within a Telco [Reproduced with permission from K.E. Ward]

product's costs are made up. An alternative approach, which is useful for costing all products but particularly for new, small scale or fast growing products, is to focus on the unit costs of using network components. Unit costs are calculated for each network component, e.g. the costs per unit of traffic or capacity carried, or per line (depending on the cost driver for each component). The product cost is then derived from the sum of the volumes multiplied by the unit costs of all the network components used, as described later in this section. For call products this approach will give a unit cost per call-minute; for private circuits, a unit cost per circuit of given capacity and length; for line services, e.g. line rental, a unit cost per line. In effect, this technique gives an incremental cost for the product – i.e. the cost per added Erlang, circuit or kilobyte. The unit costs when considering the products P&L comprise the operating costs for the network elements, including the depreciation on their capital cost. Network costs are examined in greater detail in Chapter 5.

We can now consider a little more closely how the direct costs related to network usage are allocated to a product. All network products offered by a Telco need to carry the direct costs of the network usage in their accounts, and the cost allocation should be as closely related to actual usage as possible. However, precise tracking of the network usage for each product can be difficult, and could incur unduly high administration costs over the course of each year. Therefore, a simplified and practical approach is required which gives adequate recognition of the costs that should be incurred. Most Telcos adopt the Route Factor method of cost allocation, in which the proportions of the various network elements (e.g. exchanges and transmission links) used by each product over a year are monitored. Take for example the voice-call product of a Telco, with a two-level network as shown in Figure 8.3. Over the year some 20% of calls terminate on its local exchange ('own-exchange call'), 50% go to adjacent local exchanges (LE) and 20% go via its parent trunk exchange (TE) to distance LEs, and 10% go via the parent TE and one distant TE to terminate on distant LEs. Thus, the direct costs of using all the traffic-sensitive network elements involved – local exchange, routes to adjacent LEs, route between LE and parent trunk exchange, etc. – can be apportioned to this product.

Figure 8.4 shows a simple form of routeing usage matrix, with all the network elements across the top and all the network products down the side. Along each product row the appropriate proportions – known as the route (or sometimes routeing) factor – are stated at the cross point with the appropriate network elements. The unit cost (e.g. cost per line or cost per Mbyte, etc.) for each element is its total annual cost divided by the sum of all the capacity (number of lines, Mbytes) of the products using that element. Therefore, the cost apportioned at the cross points along each product row (see Figure 8.4) equals RF x PV x EC – where RF is the route factor, PV is the product volume and EC is the network element unit cost for that cross point. An example of the use of route factors for a broadband access to ISPs product is shown in Box 8.3.

Finally, it is important to note that network operators will often use the route factor method to determine the appropriate costs – and hence prices – for interconnect traffic, as shown in Figure 8.5. In the case of an incumbent operator the national regulator will take close interest in the costs claimed and may seek quite detailed examination of the Telco's derivation of the direct costs. Interconnect product prices are usually set to cover the overall unit costs per call minute – i.e., the unit operating

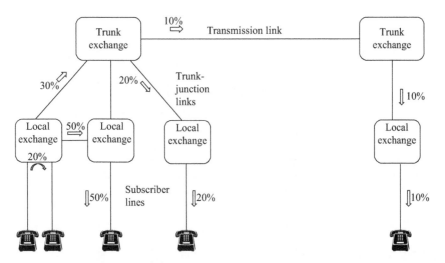

Figure 8.3   Basic PSTN network diagram

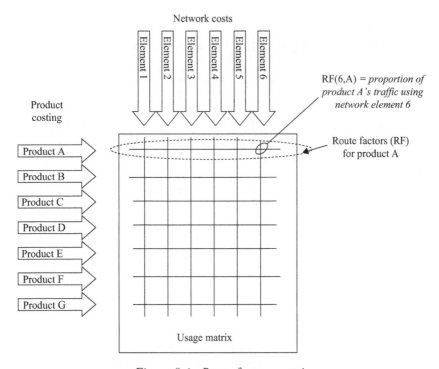

Figure 8.4   Route factors matrix

costs (including depreciation) plus a unit return per capital (unit capital employed multiplied by a regulated cost of capital). The way that an operator's traffic is routed over the incumbent's network may therefore be contentious, since it directly affects the interconnection charges incurred (see section 2.2.4).

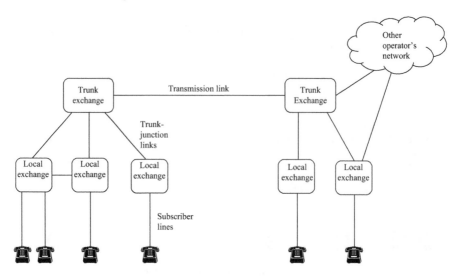

Figure 8.5   Interconnection to other operator's network

Box 8.3   An example of the use of route factors

In the example shown in Figure 8.6, the Telco is providing a fixed line (VDSL) broadband access service to three ISPs (i.e. for Internet access) across the country. (Although in practice subscribers have the choice of a large number of ISPs – about 700 in the United Kingdom.) Clearly all broadband traffic uses the subscribers' line (cost element a) and the termination at the local exchange (cost element b). Forty per cent of the traffic is routed from the LE directly to ISP-1, incurring cost element g, leaving 60% of traffic to use the route to the trunk node incurring cost elements c and d, and so on.

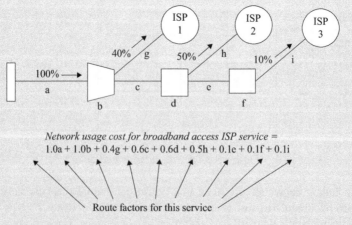

Network usage cost for broadband access ISP service =
$$1.0a + 1.0b + 0.4g + 0.6c + 0.6d + 0.5h + 0.1e + 0.1f + 0.1i$$

Route factors for this service

Figure 8.6   Example of the use of route factors

8.2.3   *Product pricing*

Pricing, which is a major part of the marketing mix (see section 7.5), generally presents one of the biggest challenges facing the ICT product manager. Successful price setting requires a good understanding of the marketplace and an appreciation of the value customers are likely to place on the product. This section looks at the factors that need to be considered when setting the price for an ICT product.

The classic economic theory of supply and demand states that the equilibrium price for a product is reached when the supply of those goods exactly meets the customer demand, as shown in Figure 8.7. This theory assumes that the customer demand is inversely proportional to the price – i.e. the higher the price the fewer items sold – and that the supply will increase in line with the price [2].

Although this theory is limited to commodity market situations, in which the customers make a simple choice based on price level, it does give a useful start to understanding the more complex markets like the ICT and telecommunications world. One way of capturing the customer reactions to price where there are several factors influencing buying choice is through evaluating the price elasticity of demand, often called just 'price elasticity'. This quantity, e, is defined as the proportional change in product demand following a given price change – thus: $e = \Delta$ demand (%)/Δ price (%). The demand for a product in a defined marketplace is deemed to be 'elastic' if the volumes of demand change with price. Normally, a value of $e > 1.0$ is deemed elastic, while e between 0 and 1.0 is deemed small or inelastic. This is an important property for the PMs to know. For example, if a shop needs to make more money from its range of men's socks it has the option of either raising or dropping the price. If it drops the price by 10% and the sales increase by only 6% it will lose 4% of its revenue (indicating an e of less than 1.0); if on the other hand the product demand is elastic, with an e of 1.5 a reduction of 10% in price would lead to an increase in demand of 15%, giving an extra 5% of revenue. Conversely, trying to increase revenue by raising the price of socks by 10% in the case of e of 1.5 will lead to a drop in sales of 15% causing a loss of 5% of revenue!

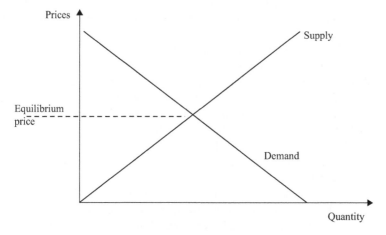

Figure 8.7   Equilibrium price

Let's now identify the main considerations for pricing of products in the ICT, and particularly the telecommunications market. It is useful to separate those considerations that are external to the company from those that are internal.

8.2.3.1   External considerations

(a)   *The market conditions.* Of course, the foremost consideration in price setting is understanding the market, as described in Chapter 7. This means assessing customer needs, likely demand for products, as well as the evaluation of the current and potential competition from both similar products and substitutes. However preparing a new product for launching into an, as yet, untried market can be difficult. An example of this was Apple trying to assess the market for their novel i-Pad prior to its launch. However, even in this case Apple would have identified the target customer segment, which would have given them a reasonable idea of the type of people likely to buy the i-Pad initially, and the value they would put on a new genre of high-tech device.

As mentioned earlier, it is important to understand the price elasticity of demand for the target market since this will be a key factor in the PM's decision on the type of pricing and the level. In the situation where there is expected to be a spread of elasticities across the customer segments there may be an argument for defining a range of product variants, each addressing a specific segment. For example, business customers may value a short-time-to-repair guarantee more than a low price for their broadband line, and be willing to pay extra for this – so a higher priced business-customer version of the product could be established in addition to the price-sensitive consumer product.

(b)   *Regulation.* The telecommunications network services in most countries operate under a regulatory regime which attempts to control the prices of any operator which it judges to have 'significant market power' (usually the incumbent Telco) to ensure a competitive market is established, as described in Chapter 2. The regulation addresses both the retail market – i.e. the price of services to consumer and businesses customers – as well as the important wholesale market.

For the retail market, the regulator may require some shifting in the relative and absolute prices of a collection of the relevant products (e.g. calls and subscribers' lines) – usually resulting in the price capping of a so-called basket of products. This allows the prices of the products within the basket to be varied as long as the aggregate price fits the target level – often a reduction in relation to inflation (e.g. 'RPI-3%' per annum). This pricing control regime may last several years until the desired rebalancing of prices within the market is achieved. (See Box 2.3 for a description of the history of price cap regulation.)

For the wholesale market the price control covers the interconnection between the incumbent and other operators for call completion, and the set of wholesale products, e.g. local loop unbundling and leased lines to provide links for the other network operators and virtual service providers. Normally, the regulator will ensure that the incumbent operator, when having 'market power' (see Chapter 2), sets prices on their wholesale products on the basis of cost plus reasonable return.

It is important to note that the regulation of the telecommunications market only relates to market sectors where the incumbent has market power and this will change over time as the market becomes progressively more competitive, meaning that the number of products which are under regulatory price control should generally reduce. By the same token, the regulatory control of pricing will follow the changes in the market structure as new players enter and new services and customer needs develop. In determining an appropriate business objective it is important also to be aware of relevant legal constraints. Competition law is complex, especially for suppliers with market power. Most firms will have legal expertise to advise on what is or is not permissible in specific jurisdictions.

(c)   *Macroeconomic environment.* It should not be forgotten that the product will be offered to customers who are living and doing business in an economic environment. Therefore, the current and future states of the national economy will be a factor in people's buying decisions, and this needs to be taken into account in price setting. The economy will also influence the general state of the telecommunications and ICT markets, and it may influence the likelihood of new competitors entering the market and even the launching of new technology substitutes.

8.2.3.2   Internal considerations

(a)   *Product costs.* It is axiomatic that the costs of the product need to be understood when setting the price. However, as described in section 8.2.2 this is not necessarily straight forward. The foremost consideration is how the costs vary with the key parameters, e.g. distance, speed or throughput, and quality level. The cost usually varies according to the geo-type (i.e. urban, suburban and rural) of the area served. Thus, the cost of a broadband line in rural areas will be several times greater than for a similar product within a city centre. Even if there is some differentiation in price to reflect this variation, the need for a simple pricing structure or a common price across the country will involve using average costs, so leading to some degree of cross-subsidising across the customer base.

(b)   *Company policy.* Companies will have their own policy with regard to the prices of their products. This is partially set by the nature of the brand – market leader, close follower, innovator, 'safe pair of hands', etc. – and partly set by the current business strategy being followed. The policy will influence the structure of the prices, the nature of their transparency and ease of understanding for the customers, as well as their level being at the top, middle or bottom of the price ranges in the market.

(c)   *Life cycle position.* The position within the product life cycle, as described in section 8.2.1, will be a major factor in setting the price. In general, the PM will consider changing the prices as the product moves through the various stages of the life cycle. So, e.g., the prices during the decline stage will be directed at milking the product's revenue for as long as possible before withdrawal, in contrast to the careful positioning objective when launching the product into, what might be, an untested market. We will be considering the various pricing strategies latter in this section.

(d) *Positioning within the company's portfolio.* As soon as a company introduces a new product into its portfolio the question of its relative price positioning in respect of the other products becomes important. Thus, in the case of a Telco offering leased lines at various rates (for both the retail and wholesale market), the questions arise about how should the prices be set, what differentiation should there be? The questions become even more complicated when the various features of each product in the range are also considered. Also to be considered is whether the product will be offered as part of a bundle of services.

There are two reasons for a company to be careful about the positioning within the portfolio: first, there needs to be a credible story for the customers, and second the introduction of a new cheaper product which might be a substitute for existing products will cause their customers to migrate. This is a classic case of shooting one's self in the foot! Understandably, one of the first questions addressed by senior management within a Telco when authority for pricing a new product is being sought is what level of customer migration between its products will be caused? Clearly, existing customers moving to the company's lower-priced product will reduce the revenue for the old product, which will need to be offset by greater sales volumes in the new product.

(e) *History.* The current set of prices in the market will already set customers' expectations about relative positioning of products. Similarly, the pricing history for a company will create part of its established brand. The product manager, therefore, may be constrained to some extent by the historical pricing policy within the company and the established order within the market.

Having outlined the set of factors above that need to be considered by the PM we can now think about setting a product price. The fundamental principle of pricing is that it is a device to help achieve a business objective for the product, product line or company. Therefore, the first step is to establish the objective for the product price, e.g. breaking into an established market and so competing with several other companies, or launching a new product into an untried market and so setting a new price level which is uncontested. Other pricing objectives might be to comply with the regulator's price controls while minimising the revenue loss for the company, or gaining rapid customer take up of a new generation of mobile networks (e.g. 4G and 5G).

Having determined the objective, the appropriate policies and strategy for pricing the product can be set. Let's consider first the policy choices, referring to Figure 8.8.

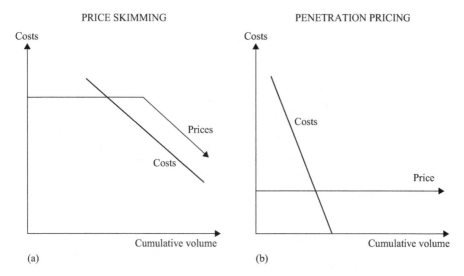

Figure 8.8   Skimming and penetration pricing policies [7]

In the case where the product is new and unique, it is likely that the pricing demand will be inelastic, and this gives the company an opportunity to exploit the desire of certain customer segments to be one of the first to own such products. This is particularly true in the case of the ICT high-tech market (e.g. tablets and smart phones). Entering the market at a high price (significantly above cost) will enable the company to cream off the revenue from this segment, as well as establishing a quality, brand and price expectation for the market – this is known as the 'Skimming policy' [2]. When appropriate the prices can then be progressively reduced to include the other customer segments to increase volume of sales and establish market dominance. The timing of this reduction is important because keeping the product price too high will enable new entrants to enter the market below this price but still high enough to justify the expense of their product launch (as in the UK Tablet market).

The alternative policy is known as 'Penetration' [2] where the pricing aims to be sufficiently attractive to quickly generate sales and hence obtain rapid growth in market share. This policy is appropriate for elastic price demand and where competitors already exist or are likely to enter the market shortly. It has the advantage of enabling the company to exploit the economies of scale cost reductions as soon as possible.

Table 8.1 presents the commonly deployed range of pricing strategies [2, 3]. The table describes the situations where the various strategies are most appropriate. Of course, in some situations the PM has no choice of strategy, e.g., where the regulator has set a cost-plus regime. In such cases the PM may consider different pricing schemes, as described below, in order to improve the financial performance of the product. Having taken the considerations described earlier in to account, the PM will chose the appropriate pricing strategy and policy.

Table 8.1   Pricing strategies

Type of strategy	Explanation
Cost plus	*Price based directly on costs, with an added margin.* This strategy tends to be favoured for prices that are regulated, especially interconnection (conveyance) charges.
Target	*Price based on achieving a target return on investment (ROI).* This is an essentially inward looking pricing strategy set by the internal rate of return on investment set by the Company's management during the financial authorisation process (business case) for the product.
Historical	*Price based on previous prices.* This strategy is deployed where the established market and products have set the customer expectations and there is little room for manoeuvre.
Portfolio	*Price based on position within a range of products.* Telecommunications products are usually in a set of similar products each differing by speed, volume, etc., requiring setting product prices in a sensible relationship.
Competitive	*Price based on target competitors' offerings.* This strategy is favoured by new entrants entering an established market, in which the prices are set to be lower than the main competitor (usually the incumbent operator).
Market-based	*Price based on the perceived value in the market.* This strategy meets the marketers' dream of setting a price in relation to what the market will bear. Apple seemed to have followed this strategy when launching their i-Pad, knowing that the novelty value and their brand would sustain a high price.
Promotional	*Price for a bundle of products, e.g. 'Quad Play'.* This strategy is particularly attractive for companies that want to capitalise on their freedom to bundle several of their products in a way that offers customer advantage, e.g. discounts or synergies across the products.
Selective	*Different prices for the same product.* This strategy is based on establishing sets of prices for a product for different target customer segments. Examples being: • Customer-group pricing – Prices aimed at defined groups of customers (students, seniors, etc.). • Peak pricing – The prices are set to manage the demand at peak times, with discounted levels at other times of day. • Service-level pricing – The differentiation of different service levels foe same product (time to repair, contention ratio, etc.). • Volume contract pricing – The use of volume discounts.

Finally, there is the need to construct the actual pricing scheme, which for telecommunications services are usually based on one or more variables (e.g. data rate, distance, time duration and time of day), rather than a simple fixed price. Examples of some pricing schemes are given below.

Flat rate. This scheme, in which a single price applies independently of the variables, is usually popular with customers because of its simplicity and the predictability of how much they will need to pay. It was first introduced in 2003 to the United Kingdom when access to the Internet (pre-bandband) was by dial-up over the PSTN – known as 'FRIACO' (flat-rate Internet access call origination) [4, 5]. However, the 'all you can eat' for a fixed price approach does suffer from the inevitable disconnect between the actual costs involved and the price. So, with the flat-rate broadband access price scheme the Telcos can encounter unexpected additional costs when customers start to download data for longer periods and at different times of day than predicted, leading to network contention and degraded QoS. On the other side, many studies have shown that customers will often pay more as a predictable, fixed fee than they would for the component elements, e.g. priced by the minute. This seems to be because they value predictability.

Usage-dependent charges. This is the traditional pricing scheme for tele-communications, where the price will vary with either throughput (i.e. volume of data downloaded), duration, time of day or distance (e.g. local, national, international). It has the advantage for the Telco that the prices track the actual cost incurred reasonably well. However, many customers do not like such schemes because the charges incurred are not easily predictable. (A well-known example is the mobile data download bills incurred by unwary travellers abroad!)

Dynamic charging. One way for Telcos to manage the problem of QoS degradation due to customer data downloads exceeding the planned capa-city, is for the price of data usage to be varied according to the instantaneous loading. This might be manifested through customers receiving an alert on phone that data rates are currently discounted as an inducement to get cus-tomers to shift their data downloading to quieter periods. The scheme has been identified as an effective way of Telcos not only managing the ever-increasing data usage of mobile (broadband) customers, but also a method for monetising data usage over mobile and fixed networks [6].

Content charging. In this scheme the prices vary according to the type of content being conveyed or downloaded over the network product, e.g. a first run Hollywood blockbuster; time-shifted classic TV show and recorded music.

Conveyance (interconnected) charging. Network operators charge each other for carrying calls passed to them through interconnection points. The char-ges are usually based on the number of call minutes incurred and the dis-tance travelled by the call within the receiving network. There are also arrangements for charging for the interconnection of data traffic, usually based on the volume conveyed (measured in Megabytes). Charges will normally be calculated using the usage matrix and routeing factors method, described in section 8.2.2 above.

Some examples of the range of pricing schemes developed specifically for mobile data are given in Box 8.4.

Box 8.4   Examples of pricing schemes for mobile data [7]

Shared data plans. This scheme accommodates the trend for individuals to have more than one mobile data device (several smart phones and tablets) as well as several users within a household. The scheme allows the main device owned by the account owner is the 'leader', and all members of the family can become part of the shared group – but the leader has visibility of all users. This has proved to be both popular in the United States and an effective way of encouraging data usage across the shared group, hence requiring a shift to a higher fixed-rate tariff.

Application-specific plans (ASPs). One way to encourage data usage is to give a special price rate (usually flat rate) for specific applications, e.g. social networking and instant messaging).

Zero-rated applications. This is a variant on ASP in which the special rate for the chosen application is zero, i.e. free.

Sponsored data. Instead of the cost of data usage being paid by the user or absorbed by the mobile operator, this pricing scheme relies on a third party to sponsor the cost. Typically sponsors are companies providing adverting platforms.

Fast-lane access. This scheme enables users to gain a temporary speed boost for their data downloads, so in effect guaranteeing the quality of service under all network loading conditions.

Prioritised transit for Content. Here the content provider rather than the user pays for the fast lane access.

8.2.3.3   Gaining authority for price changes

Normally, the PM has the responsibility for producing a business case justifying the prices being proposed and presenting this to either the appropriate governing body within the company, which may be the management board or, in the case of a large Telco, a special-purpose pricing committee. Following the product launch and moving through the product life cycle, the PM will need to keep a watch for changes in the pricing consideration points in pricing– customer reaction, competitor responses, changes in regulation, etc. – and be prepared to make changes to the absolute price, its structure or the scheme being followed, as required.

8.2.4   Product launch

The launching of a new product into the marketplace is a complex task which, by its highly visible nature, is a huge responsibility for the PM. While it is widely understood that successful product launches need to adhere to a comprehensive project management process, with its clearly defined decision points and checks, invariably there will be pressures from within the company, and possibly externally (e.g. government or regulator), for hastening the introduction of a long-awaited or much-needed product. Unfortunately, there are many examples of ICT and tele-communications products being launched prematurely due to such pressures, only to be followed by embarrassing withdrawals from the market for remedial action. There may also be adverse public press coverage and public reaction to badly managed product launches. However, 'time-to-market' can create an advantage for a company in a highly competitive market, so there is a balance needed between undue haste on the one hand and over caution on the other.

The development of new ICT equipment – both hardware and software – will require thorough testing before it can be deployed. Normally, new versions of the equipment will be tested in the manufacturer's laboratories and factories before being installed and 'handed over' to the Telco for operational use. In the case of new software or software upgrades the suppliers undertake in-house (i.e. labora-tory) testing – often known as 'Alpha trials', the objective being to check com-pliance with the agreed specification. However, all ICT products need also to be tested in the real-life conditions of the network or operational buildings and inter-acting with existing equipment – often referred to as the 'Beta trial' stage. Normally, Beta trails will be run in operational conditions usually isolated from the normal users (e.g. the Telco's customers). Telcos may also extend the Beta trail of the equipment into a full testing of the operational procedures for providing service and repairing the new equipment – known as a 'field trial'.

In a nutshell, the launch process should ensure that the product has been fully tested, that the operations (e.g. maintenance, billing and order handling) procedures and documentation are in place, and that the sales channels are appropriately primed. The key stages in this process with their decision gates are shown in Figure 8.9. It all starts with the management authorisation for the product business case. Resources can then be allocated within the company and development started for the various support systems – order handling, maintenance, etc., as required, depending on whether or not the existing support systems are flexible enough to cope with the new product. An important activity at this stage, particularly for telecommunications network products (e.g. broadband access services), is the undertaking of a field trial. The extent of such trials depends on the complexity and novelty of the product and the level of functional enhancements in the network, but they usually involve customers. These may be selected from a panel of willing participants or those within the trial area who are offered some inducement (e.g. free service during the trial period) to test the product and give feedback.

Once the product development phase has been successfully completed (see Figure 8.9), authority to proceed to the pre-launch phase can be given. This phase addresses all the procedures and documentation needed within the company to successfully manage the product on a day-to-day basis. It is only when the PM can demonstrate to senior management within the company that all necessary procedures and documentation are in place should approval for the product launch be given.

However, a word of caution before we leave this subsection on product launch. As with any fast moving marketplaces, like that of ICT, the need for systematic and thorough testing (as described earlier) must be tempered by an appropriate degree

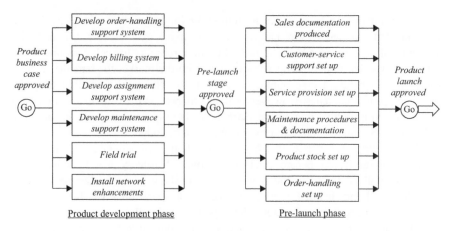

Figure 8.9 Product-launch process

of pragmatism, or else undue caution will mean that any competitive advantage through early entry market may be lost. Box 8.2 gives a perfect example of where the elaborate product launch process was curtailed enabling MCI to hit the market with an innovative product – which proved to be a great success. Some companies make a virtue out of launching immature products (usually software based), knowing that 'tech savvy' customers will tolerate teething problems in order to be in the first wave of users. Getting the balance right between the two approaches is a real challenge!

The appendix gives an introduction to the techniques of project management, which form an important management tool for any company undertaking a product launch.

8.2.5 Product forecasting

It is generally recognised that the most fundamental requirement within a company is the forecast of future sales of its products, since this gives an indication of the revenues and direct costs for some time ahead – usually up to 5 years in the case of telecommunications operators. The PM has the prime responsibility for deter-mining the forecasts of demand and hence revenues for its products. However, despite the importance attached to such quantities, it is a difficult and imprecise task to estimate future sales of a product or the amount of use of a network service. For telecommunications operators the problem is compounded by the need to have the forecast quantities relate to specific parts of the country, so that they will be able to plan sufficient network capacity to carry the resulting traffic from each network node and the links between them (as discussed further in Chapter 6).

A generic diagram showing the stages involved in producing a forecast for an ICT or telecommunications product is shown in Figure 8.10. The main inputs to any forecast are ultimately driven by an assessment of the market (as described in Chapter 7), which takes into account the macro-economic factors, regulation,

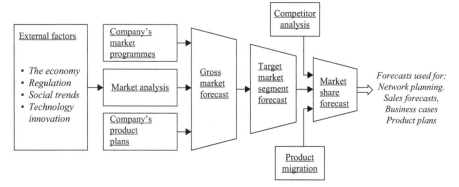

Figure 8.10   Product forecasting

social trends influencing customer behaviours and expectations, as well as expected improvements and innovations in technology. As outlined later there are a number of methods used to derive forecasts, using the market analysis and an estimate of the effect of the company's marketing campaigns and the attributes of the product itself (as described in the product plans). The forecast then needs to be scaled to reflect the specific customer segments being targeted. Finally, the future expected market shares and the effects of any migrations from the company's other products will scale the forecasts for the specific product. It should be emphasised that all forecasts are intrinsically uncertain. This uncertainty is usually reflected in the use of ranges rather than spot figures. While a business case may need a specific forecast, it is best taken as the centre case of a range of forecast outcomes.

There are three main categories of forecasting methods used by all industries, namely: time–series (historical), casual, and judgemental. As we discuss later, the choice of method broadly depends on whether the forecasts relate to a new product in an untested market or whether there is a history of sales and customer usage.

8.2.5.1   Time-series forecasting methods

These methods rely on the availability of historical data on sales to make projections for the future. Figure 8.11 illustrates the principle, whereby the projection is derived from determining the trend indicated by the curve fitting the past performance for the product. At its simplest this projection, or extrapolation, extends the gradient of the last few years' sales, on the basis that a curve or line can be derived that fits the data points reasonably well. As Table 8.2 shows there are a number of techniques for achieving this good fit with historical data.

Having produced a reasonably smoothed curve fit to the historical data there is still the challenge of how to project forward into the future. In most cases a simple straight line projection may be sufficient for the near future – say within a year – but this approach always has the danger that impending changes of direction may be missed. Figure 8.12 illustrates how simple extrapolation can be misleading. In the case of a 'style' paradigm applying, there will be a cyclical shape to the future curve which may extend over several years. When a 'style' gets adopted in the

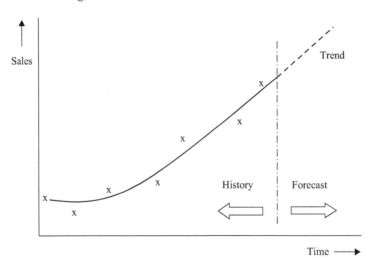

Figure 8.11   Trend forecasting method

Table 8.2   Time-series forecasting methods

Name	Description	Characteristic
Moving average	Sums the values over the last *n* periods and divides by the number of periods (*n*).	Tends to smooth out irregularities in the historical data. However, the smoothing can also mask important changes indicating new trends.
Weighted moving average	This is a variant on above, in which the more recent values are given higher weightings than the earlier results.	Can capture changing trends by emphasising the more recent values. It does require good quality historical data points.
Exponential smoothing	New forecast = last forecast + α times the difference between last forecast and the actual achieved value. α is a smoothing constant between 0 and 1.	The set of data points is progressively smoothed by adding proportions of the error between forecast and actual data points. By adjusting α different weighing can be given to historical data.
Curve fitting	Method of lease squares finds the best fit to a set of data points by finding the line/curve that minimises the sum of the squares of the difference from the data points.	There are many software packages that will perform this curve-fitting function.

marketplace it becomes the 'fashion', and the demand curve goes up significantly; after the peak it usually follows a gradual decline. However, a 'fad' exhibits rapid growth in demand shortly reaching a high peak, followed by a rapid decline. The challenge for the PM is to identify which trajectory is about to occur!

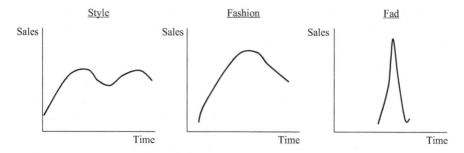

Figure 8.12   Detecting trends: style, fashion, or fad

8.2.5.2   Causal forecasting methods

This set of methods attempts to understand the causes of customer demand and create models that exhibit similar behaviour so that future demand under specified conditions can be forecasted with good accuracy. The models need to contain the variables that significantly influence the demand. One technique for identifying which variables should be included is by calculating the correlation coefficient [8] between the historical demand data and each candidate variable. Correlation coefficients close to 1.0 indicate a dependent variable, while values of less than 0.5 indicate no significant dependency. A model of the demand can then be constructed which leads to an equation of the form:

$$y = a + bx_1 + cx_2 + dx_3 + \cdots$$

where *a*, *b*, *c*, *d*, and so on, are constants and x_1, x_2, x_3, and so on, are the dependent variables. This approach is often referred to as Multiple Regression Analysis [8,9].

Forecasting the demand for each of the players in a market requires more complex techniques, e.g. the Markov method which uses matrix calculations to generate separate forecasts [9].

The causal methods can lead to reasonably accurate forecasts, provided that the right set of dependent variables has been identified and that sufficient good quality historical data is available to calibrate the constant coefficients. Examples of dependent variables are the measure of economic growth (GDP) and the take-up of mobile phones; the number of households with teenager and 20s in residence, influencing the take-up of high-speed broadband. Statistics for such variables, including the range of households, the population age profiles, etc., are readily available from government and other websites.

8.2.5.3   Judgemental forecasting methods

These are qualitative methods that allow subjective assessment to be applied to the forecasting process, which is particularly valuable in cases where there is little or no historical data, or a big shift in customer behaviour or technology is expected in the not too distant future. An obvious example of where judgemental techniques would have been appropriate is back in the early 1990s trying to assess the impact of the arrival of Internet-based services and applications. There are two major judgement techniques, as described below.

(a)   *Delphi method.* This technique, which originated at the RAND Corporation in the Cold War period of the 1950s, is now one of the most widely used judgemental method. It provides a forum for a panel of experts to apply intuitive and knowledgeable opinion on how a market will develop and construct a set of forecasts. The techniques revolve around several iterative sessions of questions and debate of small (5 or so) groups of people leading to a consensus. Successful application of the method requires an appropriate set of questions to be framed by an expert panel for consideration by the groups. Although it can lead to good quality forecasts, it is considered to be time consuming (several weeks) and heavy on manpower [10, 11].

(b)   *Scenario planning method.* An alternative casual method is that of scenario planning, which has its origins in the oil industry, but is now widely used in many industries (including ICT and telecommunications), as well as the military. The idea of there being several possible futures which differ from today's situation the further out in time is illustrated in Figure 8.13. The scenario planning method is ideally suited for tackling the expected major shifts – societal changes, new commercial models, disruptive technology, etc. The technique relies on a panel of experts studying a large set of proposed drivers and possible future events and determining the effect of these on several possible futures (the 'scenarios'), as explained in Box 8.5. We cover the use of scenario planning in determining business strategy in section 3.6.4, and Box 3.5 gives a real-life example of the use of scenario planning [12].

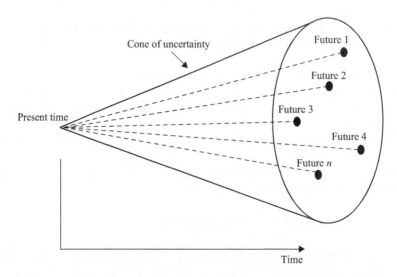

Figure 8.13   The cone of uncertainty and the range of possible futures

Box 8.5   The scenario planning method

The steps of the Scenario Planning process (Figure 8.14) are as follows:

1. Map identified basic trends and driving forces on an importance vs impact graph.
2. Discard the unimportant, to create the stack of key uncertainties (A to Z).
3. Define and describe a wide range of possible futures (scenarios).
4. Reduce the number of scenarios to two or four (avoid having an even number to prevent people going for the mid-range).
5. Allocate the relevant key uncertainties to the scenarios, as shown earlier.
6. Assess the consequences of each scenario on the demand for the product.
7. Identify business and product strategies to cope with each scenario.
8. Identify signposts for each key uncertainty – so that the identity of the actual future scenario becomes apparent as time progresses, and the appropriate strategy can be followed accordingly.

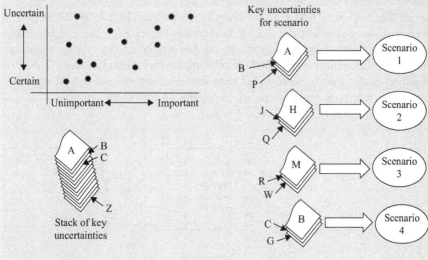

Figure 8.14   The scenario planning process

8.3   Portfolio management

8.3.1   Product mix

Having looked at the management of an individual product, we now consider how a company can gain business advantage by forming a set of several products into a managed portfolio. In fact, the ability for the telecommunications companies to offer a comprehensive set of products to meet customers' needs creates an important competitive advantage. For the Telco's the delivery of a coherent set of products to individual customers, particularly large business customers, provides economies of scale and other cost savings. Supplying a range of services to a customer tends also to increase loyalty and thus reduce churn. Therefore, the challenge for a Telco is to create a portfolio of products that effectively addresses the target market segments on the one hand and incurs the least operational costs on the other hand. Of course, full exploitation of such a portfolio requires appropriate operational marketing support – through advertising and promotional activity – as well as a sympathetic sales force. Often, the close association of the Telco's account managers with the main business customers can not only lead to successful selling of the portfolio of products, but also provides a channel for customer input of ideas for future product development.

The way that a coherent portfolio of products can be directed to target business market segments is illustrated in Figure 8.15. Here, we can see the set of products, A to I, in the Telco's main portfolio. An appropriate mix of these products can be offered to each of the target customer segments, e.g. finance and retail. These segment portfolios may have a dedicated team within the PM's team to assemble the products and create specific segment sales packages, if appropriate. The individual customers within the segments will then take a selection of the products as required. In addition there may be a case for some specialised products to be developed for major business customers, e.g. J, K, L, as shown in Figure 8.15. These, separately or in association with the generic products, can be targeted on the specific major customer (e.g. the military or an off-shore oil company) on a bespoke basis. This form of selling carries a high sales cost but represents considerable value to the customer. It involves not only crafting specific solutions to customers' problems but also

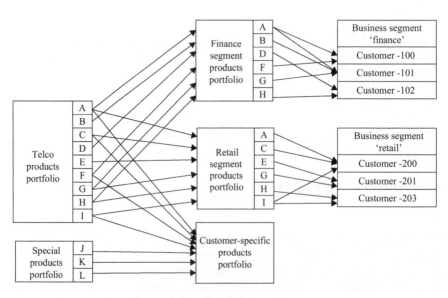

Figure 8.15   Portfolio management

maybe tailoring the products to meet their needs. Pricing of such bespoke packages is often based on competitive tenders.

8.3.2   Managing the product portfolio

Management of the product portfolio is closely associated with the business development process within a company (see Chapter 3). The need for additional product revenues and associated profits is identified and scoped within the business planning and development activity. This process may identify market opportunities, potential partnerships, and perhaps acquisitions by the company which could provide opportunities for new markets and products.

The main objective for a successful business is to have a portfolio of products that collectively maximises the return on the investments over the years. This generally means that companies have a mixed portfolio so that as markets change over time some products will need injections of cash to exploit revenue growth opportunities, while others may be in low-growth situations, require little or no investment but receive good levels of revenue. The famous Boston matrix, developed by the Boston Consulting Group (BCG), provides a ready way of evaluating the mix within the portfolio and determining how to maximise its return [13].

Figure 8.16 shows the BCG matrix, which has one axis representing relative market share and the other axis representing relative market growth, onto which each product in the portfolio is plotted. For convenience each of the quadrants of the matrix forms a category of business strategy for the plotted products, namely: stars, cash cows, dogs and question mark.

> *Stars* are the products that currently have a high market share (greater than 35% of the biggest competitor) with a high growth rate (i.e. greater than 10% above inflation). The suggestion is that a company should focus its investment into its stars in order to exploit the market potential.

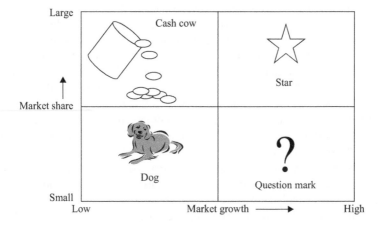

Figure 8.16   BCG matrix

Cash cows are the products that are currently profitable and have a high market share but are within a static or low growth market. The company should not invest further in these products, but should instead 'milk' them for revenue and profit for as long as possible.

Dogs are the opposite of stars – low market share in a low-growth market. The company should not invest in these products, in fact those that are not profitable should be divested or withdrawn from the portfolio.

Question marks are the products that are in a potentially high growing market but they currently have a low market share. The uncertainty associated with these products is because it is not certain whether investment will improve their position.

In practice, most products occupy these four categories during their life cycle. Newly launched products are usually in the 'question mark' category, because they are in a growth market but have not yet gained much presence. So, should more investment be put in at this stage? Ideally, the products gain market share and become 'stars'. As the market saturates, the product gradually moves into the 'cash cow' quadrant. Eventually the product moves into the decline phase of its life cycle and becomes a 'dog'. A successful portfolio will at all times consist of several products in each of these categories, with the PMs and Portfolio managers being aware of the progression of the products through the matrix. Given its dependency on the assumed relationship between investment and improved positioning of a product within the market, the BCG matrix should be used as just one of several analytical tools to evaluate the mix of a portfolio, and hence the investment strategy required to maximise its return for the company.

Figure 8.17 sets out the elements of the continuous process of portfolio management within a company. Usually the vehicle for capturing the current status of each product is a product plan, which records its specification, prices, target market segments, main competitors, as well as the current strategy for development and future plans. Where new products need to be developed and then launched business cases are produced by the PM, based on product costs and revenue forecasts, as described earlier. An important part of managing the portfolio is the investigation of how the bundling of several products could lead to successful increases in overall revenue and market penetration. During our discussion of product pricing in section 8.2.3, an important factor for consideration identified was the relative positioning of a product within a range of similar products within the company's portfolio. The portfolio management responsibility is to create a sensible structure for the products so that the external world (customers, regulators, competitors, etc.) can accept the differential in price between products – e.g. broadband services at speeds 200% apart are priced with 25% differential – while not creating unwanted migration between products. Of course, there may be circumstances where from a portfolio perspective encouraging migration of customers from an old product to a new product may actually be wanted – in which case the pricing of the two products should reflect this. The latter is often the best way of managing a run-down of customers off a product due for withdrawal from the portfolio.

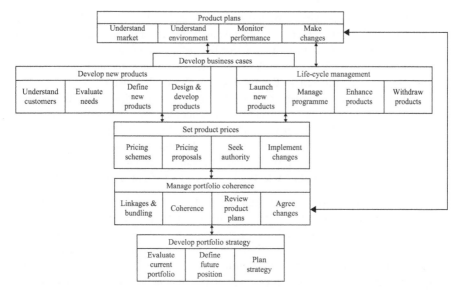

Figure 8.17 Elements of portfolio management [Reproduced with permission from K.E. Ward]

There is a further complication in product price positioning, particularly for telecommunications operators, in that there needs to be a sensible and defensible margin between wholesale and retail prices. This creates a big challenge for Telcos in managing their retail and wholesale portfolios, given the need for coherence within the two separate markets on the one hand, and the inextricable linkages between them on the other, as described in Chapter 2.

8.4 R&D and innovation

8.4.1 The roles and objectives of R&D activity

ICT and telecommunications are considered to be 'high-tech' industries. This means that the products and services are derived from networks, servers and support systems, which rely on recently introduced highly functional technology. Furthermore, the applications used by the customers and their terminals (smart phones, tablets, etc.), well as the over-the-top service providers using the networks, are also 'high-tech'. Therefore, all parties are aware of the current state of technology developments and there is strong pressure on the network and service providers, and terminal device manufacturers, to adopt the latest technology advances.

There are three main reasons why a company might adopt new technology:

1. *To be able to provide new functionality.* In the case of terminal manufacturers, development in screen technology, say, could offer competitive advantage in the highly stressed smart phone market. For a Telco, the new functionality may

lead to wider deployment of high-speed broadband services. Therefore, the functionality improvements would normally be apparent to the customers in terms of features or appearance/feel of the service or device.

2. *To reduce costs.* The cost reductions due to the new technology may not necessarily be reflected in the product prices, so the customers may not be aware of its use. However, cost reduction is a major driver for Telcos and service providers in the adoption of new technology. We look at the economics involved in the adoption of new network technology in Chapter 5.

3. *Improvement to brand image.* Because of the high-tech nature of the market mentioned above, companies wanting to appeal to certain customer segments may demonstrably adopt new technology in order to improve their brand image.

Companies normally marshal their ability to deal with hardware or software technology within R&D units. In fact, one of the important indices used by analysts to ascertain the technology activity of a company is its expense on R&D as a proportion of total turnover. Values of this index, which is often referred to as 'R&D intensity', would normally be around 2–4% for a Telco; while some very high-tech manufacturing companies might have indices as high 25% [14]. However, care is required in interpreting this index because there are various definitions of what constitutes R&D spend by a company. Nonetheless, it is important for the senior management of a high-tech company to decide what R&D intensity is appropriate and affordable, and hence set budgets accordingly. Let's now consider what R&D means in the ICT and telecommunications industries, how this affects their ability to innovate, and its importance to their product and portfolio management activities.

Research is defined as the investigation of (hardware and software) technologies and processes made at the scientific level. Usually this investigation does not have any direct association with the company's products. R expense is therefore considered as entirely an overhead for the company. It is important to note that although a team of research scientists are working within a Telco's laboratory on, say, a new form of semiconductor technology which could be used in certain network equipment, there will also be (probably) larger teams of scientists working on the same technology in the labs of equipment manufacturers, as well as many researchers in several universities around the country. While some important breakthroughs have been achieved in the labs of Telco's in the past [15], this is not normally the objective of their research activity. Instead, Telco's direct their research activity to the following:

• To create a good understanding of relevant technologies and how they might be developed in the future. They will also look at how technologies might be exploited by the Telco – and indeed what competitors might be able to achieve with the technology.

• To become an 'intelligent purchaser', ready for when some years downstream manufacturers will be offering the Telco equipment using the new technology.

• To influence international standards bodies, who will be addressing this new technology in the near future.

Development, on the other hand, is essentially an engineering activity which is usually directly or indirectly associated with one of the company's products or group of products. *D* expense may therefore be apportioned as a direct cost within the company to the network, or to the relevant products. The objectives of development activity within a Telco are as follows:

- To adapt prototype or immature equipment to be able to work within the Telco's network. This will include changes to the hardware and software of the prototypes, as well as the construction of mediation systems to allow interworking to the established interfaces within the Telco's network and support systems.
- To work with the PM's team to specify the products that could exploit the features of the new technology.
- To make ongoing improvements and upgrades to the suite of network management and customer-service support systems, relating to the existing as well as new services run by the Telco. (We look at this aspect further in Chapter 9.)

Generally, development expenditure within a Telco or an equipment manufacturer will be higher than its *R* expense. However, the research centres funded by government, national institutions, or consortium of companies, will be almost 100% focused on research.

One of the tangible outcomes from a company's R&D expense is the creation of a set of patents protecting innovations that could enhance their products. Of course, in many cases the patents may not offer an immediate benefit for the company and so they form a library of potentially useful 'intellectual property' (IP) for later exploitation [16]. Such an IP portfolio represents an intangible asset for the company, which in many high-tech industries, like ICT, can be viewed as significantly valuable – particularly when the company is being evaluated during a merger and acquisition (M&A) process.

8.4.2   The introduction of new technology

In today's technology-aware society some customers, and most managers within ICT and telecommunications companies, get excited by the emergence of new technology. There can be pressure within the companies for deployment of the technology just because it is new. However, history has shown that invariably the introduction of new technology is not straight forward nor rapid. In fact sages within the industry will quote the '7-year rule', which says that it takes this length of time to get a new technology from successful demonstration in the laboratory to a state where it can be deployed in the network. Of course, many people are surprised, and not a little frustrated by the difficulty of incorporating new technologies into robust equipment (with appropriate software) which can be deployed in the network.

Interestingly, the sentiment of the '7-year rule' has been captured by the now famous Gartner Hype Cycle, as shown in Figure 8.18. Gartner, the industry analysts, observed that all new technologies were subject to big changes in their acceptability during the early stages of their life, but that eventually their adoption

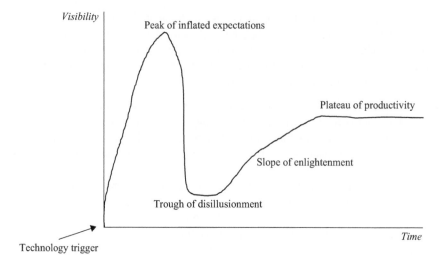

Figure 8.18   The famous Gartner hype cycle

follows a gentle increase. The problem is that when new technologies are first demonstrated there is usually a great deal of excitement, with much hype about what the technology could do – shown as a rapid rise up to a peak of expectations at the start of the curve (Figure 8.18). This period is followed by a rapid drop in expectations, labelled the 'Trough of disillusionment', when neither the imminent arrival nor the technology's attributes seem to be as expected. After some time, the necessary development work enables the potential of the technology to be realised – during the 'slope of enlightenment'. Finally, the technology becomes deployed in quantity and fully accepted. Each year, new versions of the curve are produced in which the current positions of all the topical (hardware and software) technologies are plotted [17]. It is interesting to note that although the shape of the Hype Cycle remains much the same, the time taken for the mass adoption of many new technologies (i.e. to achieve 50% penetration) is gradually reducing.

For the PM there are also other considerations concerning the introduction of new technologies, namely the acceptability to customers. While there may be some customers who will willingly pay a premium price to be one of the first to obtain a new device or technology, as discussed in section 8.2.3, the majority of customers will hold back and wait for the technology to settle in. They may also take some time to accept the new features or complexity involved. Geoffrey Moore produced an analysis of the way that markets adopt new technology [18] in which the buying population is allocated to five segments (Figure 8.19 refers), namely:

- *Innovators*: The first customers to buy the new technology at a high premium price, as described earlier.
- *Early adopters*: These customers follow on from the Innovators, they are later to adopt the new technology, but they are bigger in volume. They may be prepared to pay a small price premium.

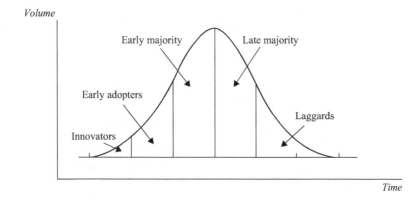

Figure 8.19 Market adoption segments [18]

- *Early majority and late majority*: The bulk of the purchasers fall into these two categories; pricing should reflect the generally accepted value of the technology at this stage.
- *Laggards*: These are the not-tech savvy types of people that are reluctant to accept the general use of the new technology and will take some persuading to discard their old devices and services.

8.4.3 Innovation

It is generally regarded as a necessary attribute for successful ICT players to bring innovation to the product portfolio, as well as the network and servers, operational support systems and business processes. Also, many mangers now recognise that innovation is needed throughout the company, and is not something that is relegated to the R&D community. However, without doubt the R&D resources in a Telco or service company should act as the driver of awareness of what new technologies are on the horizon and how these might affect the markets and customer behaviours, as well as creating ideas for future products. These views need to be put in the context of future business models and the expected paradigm shifts in the way that business is done – e.g. the disintermediation introduced by the Internet (see Chapter 1). ICT companies can achieve these perspectives by developing future scenarios covering the big picture of the new business environment, and visions of where the company might want to be, using techniques similar to the scenario planning described in Box 8.4. Many companies have established small teams of 'futurologists' to help investigate and articulate these perspectives [19].

We should conclude this section by noting that there are essentially two forms of innovation within the ICT industry [20]. These are normally classified as follows:

- *Sustained technology*: which does not create a new market or commercial models, rather it evolves the existing technologies and markets.

- *Disruptive technology*: which is innovation that creates new marketplaces and eventually disrupts the established markets. The Internet is an obvious example of this.

The challenge for any ICT company is to be able to evaluate new technologies so that their impact on the company's business can be assessed. Most PMs will focus on the sustained technology advances, looking for incremental improvements to their products. But there will be some PMs who are seeking the disruptive technologies that will, potentially, create new market opportunities in the future.

As a final thought, we noted earlier that innovation and R&D give rise to intellectual property which is for the PM to turn into company value. That could be through extending a product's life cycle, by broadening its appeal through new features, by enabling higher prices to be realised (known as 'innovation rents'), or by gaining competitive advantage through uniquely available features protected by patent. In turn this means PMs should work closely with their R&D divisions to ensure that developers are informed by an awareness of what customers would value, or what would create competitive advantage for the company.

8.5   Summary

In this chapter we have examined the wide-ranging role of product management, covering the full life cycle of the product. In particular, the important aspects of deriving product costs and the various strategies for pricing were identified. We also considered the difficult, but vital, task of forecasting the customer demand and the associated revenues, noting that different techniques were required depending on the availability of historical sales data. The challenge of managing a successful product launch was highlighted, noting the consequences for a Telco when the launch is mismanaged. Finally, the advantages of managing a set of products as a coherent portfolio were described, together with the importance of understanding how innovation can support this objective.

References

[1]   http:// coveycomsumer.com/mci/mciff.html.
[2]   Mercer, D. 'Marketing', 2nd edition, Blackwell Business, Blackwell Publishing Ltd, Oxford UK, 2001, Chapter 7
[3]   Blois, K. (editor). *The Oxford Textbook of Marketing*, Oxford University Press, 2000, Chapter 10 'Pricing' by Gijsbrechts E & Campo K.
[4]   Altmann, J. & Chu, K. 'How to Charge for Network Services – Flat-Rate Or Usage-Rate Based?' *Computer Networks*, 2001, Vol. 36, 531–591.
[5]   Ofcom. 'Review of the Adjustment Rate for DLE FRIACO'. Consultation document, 2003, stakeholders.Ofcom.org.uk/dlefriaco/
[6]   Rhiad, A.R. 'Dynamic Pricing and 4G Data Monetisation', *The Journal of the Institute of Telecommunications Professionals,* Vol. 8, Part 1, 2014.

[7] Allot Communications. 'App-Centric Operators on the Rise', Allot Mobile-trends charging report, h1, 2014, www.allott.com

[8] Cass, T. *Statistical Methods in Management*, Cassell & Co. Ltd, London, 1973, Chapter 5.

[9] Makridakes, S.G., Wheelwright, S.C. & Hyndman, R.J. *Forecasting: Methods and Applications*, 3rd edition, John Wiley & Sons Inc., New York, 1998.

[10] Skulmoski, H. & Krahn: 'The Delphi Method for Graduate Research', *Journal of Information Technology Education*, 2006, Vol. 6, 1–21.

[11] Evans, V. *Key Strategic Tools*, Pearson Education Ltd. Edinburgh, 2013, Chapter 20.

[12] Chermack, T. *Scenario Planning in Organisations – How to Create, Use, and Assess Scenarios*, Berrett-Koehler Publishers, 2011.

[13] Ten Have, S., Ten Have, W. & Stevens, F. *Key Management Models*, FT Prentice Hall, Pearson Education Ltd., 2003.

[14] Europa: 'Europe 2020 Targets: Research and Development', www.ec.europa.europe2020/pdf/themes/15_research_develpment.

[15] Hayward, W.S. (editor). 'The 5ESS Switching system', *AT&T Technical Journal*, 1985, Vol. 64, No. 6, 1305–1312.

[16] Goldstein, L. *Patent Portfolio: Quality, Creation and Cost*, True Value Press, 2015. Chapter 1.

[17] Gartner. 'Gartner's 2014 Hype Cycle for Emerging Technologies Maps the Journey to Digital Business' August 2014, www.gartner.com/newsroom/

[18] Moore, G.A. *Inside the Tornado*, Harper Business Essentials, Harper Collins Publishers, 2004.

[19] Pearson, I. 'Over the Horizon', *ICT Futures: Delivering Real-time and Secure Services* (edited by Warren, P., Davies, J., Brown, D.), John Wiley & Sons Ltd, 2008, Chapter 7.

[20] Tidd, J., Pavitt, K. & Bessant, J. *Managing Innovation*, John Wiley & Sons, 2001.

Chapter 9

Network and service operations

9.1 Introduction

The telecommunications business is centred on the delivery of service to customers using network infrastructures. Once the networks have been built and the services established, the mission of the operator is to provide the required services at a standard of quality that satisfies its customers and, of course, at a sustainable level of operating cost. This continuous activity is usually referred to as 'Operations'. It is no exaggeration to say that the way that operations are managed will ultimately determine the viability of the operator as a business. On one hand, since all of the operators potentially are able to use the same technology for the network equipment it is the way that the services on the network are run that creates the differentiation – and hence the potential sustainable market advantage for the company. On the other hand, the operational annual costs are a significant component of the company's P&L account and so need to be minimised.

Unfortunately, operations are not viewed as glamorous by managers in the companies, who tend instead to be pre-occupied with the attractions of the latest technology or equipment updates. Not only does this result in new networks and services being launched without adequate operational support – with the consequent embarrassment of retrospective adjustment to services, or even their withdrawal until the necessary operational systems can be included – but in some cases the network company business has collapsed as a result. The Ionica Company in the United kingdom is an example of such an occurrence, as described in Box 9.1. In addition, the development of appropriate operational systems and procedures may often become the key component of the product-launch process (see Chapter 6), and so determine the launch date – much to the annoyance of the marketing and product managers in the company!

In this chapter we look at the world of network and service operations from a business and management perspective and consider the effect they have on the customers' perception, noting the axiom that in a competitive market it is the satisfied customers who stay with you. First, we look at the nature of customer satisfaction and how this relates to the quality of the service provided by the operator. We then consider the nature of operations and their cost drivers for a Telco. Having established what the Telco's are trying to achieve, we examine how the operations are managed,

Box 9.1   The Story of Ionica

The Ionica Company was established in 1992 with the main aim of providing telephony service to residential and small business customers, in competition with the incumbent BT. The unique selling proposition was that they used a new form of wireless system to provide a two-channel local line more quickly and cheaper than the copper local-loop service provided by BT. Calls to other networks were passed over to BT at the interconnect points. The target market was initially rural areas around Cambridge. They then expanded coverage to parts of Yorkshire and the Midlands. The first customers were connected in 1996, and the service attracted customer demand far more extensively and quicker than forecast. At first they created quite a stir among customers and investors, and the company had its IPO in 1997. Despite the expected technical teething troubles with the new technology, the major problems for Ionica were related to the processes around installing new connections. These incurred extensive radio-propagation measurement surveys, and often repeated customer-site visits to reposition the radio antenna. As demand for new connections grew, the cost and delay in provisioning became ever more of a problem. Unfortunately, because of these difficulties Ionica was not able to gain the necessary extra finance for expanding the customer base and the company was declared bankrupt in October 1998. All 62,000 customers were transferred to BT by early 1999 [1].

in terms of dealing with the set of customers and dealing with the equipment and network plant. Underpinning all this activity is the suite of support systems – computer software applications and data bases – which help to manage the complexity, provide levels of automation, and improve the customer experience. Finally, we consider how all the different support systems and databases should be linked and structured (i.e. the architecture), and how the various actions, like providing new broadband lines, are undertaken (i.e. the processes). Although we concentrate on the network operations, we do also consider the running of computing and storage activities in data centres (e.g. providing cloud services).

9.2   Customer satisfaction and quality of service

9.2.1   Customer satisfaction

Customer satisfaction is clearly an important factor in the management of a telecommunications operator's business. In fact, it is usually one of the key parameters that are monitored regularly by the company's main board. Serious customer complaints about service are flagged as a 'Chairman's case', and given high-level visibility and priority within the organisation. Externally, the local and national

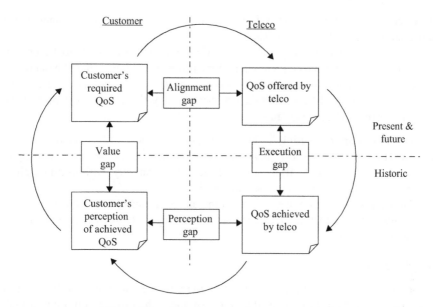

Figure 9.1   Customer satisfaction gap model

press may become involved in publicising bad service and adverse customer reactions, and local and national government may also apply pressure on the operator. In the case of the United Kingdom, Ofcom monitors and publishes annual service performance statistics of all main network operators, as well as intervening in unresolved disputes over poor service where necessary.

For a Telco there are a range of aspects that influence customer satisfaction, many of which are influenced by the way that the network is operated, but there are also many non-network activities which affect the overall perception for the customers. Figure 9.1 offers a helpful model that demonstrates how customer satisfaction is affected by a set of gaps [3]. The top half of the model represents the present and future actions, while the lower half plots historic activity. Starting with the top-left hand quadrant, we can identify the set of quality of service (QoS) requirements that a customer may wish for their application. (We will look more closely at the term QoS, in section 9.2.2.) However, the quality offered by the Telco for their service may be different to that required by the customer. Despite this 'alignment gap' the customer will still order the service. In reality, of course, the actual QoS achieved by a Telco will differ from that offered – and this is known as the 'execution gap'. A Telco will undertake a wide range of measurements of the performance of its network continually in order to establish a view of the QoS being achieved. Inevitably, for a variety of reasons, there will be a difference in the QoS that the customers perceives and that which the Telco measures, i.e. the 'perception gap'. Finally, to complete the picture, the customer satisfaction is largely influenced by the 'value gap' between what the customer originally wanted from the service and their perception of what they are actually receiving.

It is generally recognised that this perception gap is a major determinant in the satisfaction felt by the customers.

From the above gap analysis, we can see that a Telco needs to be careful which network performance parameters it measures in order to gauge the degree of customer satisfaction being achieved, recognising that the customers' view of the service they receive differs from that as measured by the Telco. Examples of the reasons for the differences being:

- Customers are measuring different parameters to those of the Telco.
- Reaction of customer's applications and equipment, e.g. time-outs occurring in the customer's application software during, apparently within-limits, short breaks in the Telco's network transmission.
- Issues with the customer's own equipment or application which they incorrectly attribute to their telecommunications supplier, or interworking issues between the two.

9.2.2   Quality of service

The term 'Quality of Service' (QoS) is an all-embracing expression of the merit of a service in operation as perceived by the customer. The official definition of QoS is given in the ITU-T recommendation E800 as: 'The totality of characteristics of a telecommunications service that bear on its ability to satisfy stated and implied needs of the user of the service' [2]. It is fair to say that this definition is not necessarily restricted to telecommunications, but rather it applies to any ICT services. The challenge for a network operator is to match the QoS achieved on the network to the particular requirements of the customers, noting that the latter will vary depending on the applications the customers are using (e.g. emails will be less demanding than TV streaming). In many cases the acceptability of the QoS achieved by the Telco will also depend on the type of customer – i.e. the customer market segment – since they will value service quality differently against its price.

We can get an idea of the range of factors that affect the QoS as perceived by a customer in the mapping in Figure 9.2. This shows that customer satisfaction is related to the extent that user requirements are met, and that QoS is not the only factor involved – value for money and other factors, like brand awareness, peer-group pressure and fashion, may also be important.

The QoS itself is dependent on many factors, some due directly to the performance of the network, while others are driven by the non-network-related QoS factors of support systems and human factors, e.g. time between ordering and provision of service, time to repair faults, billing accuracy, and courtesy of the help desk staff. As Figure 9.2 shows, the performances of the various parts of the network (or networks) contribute to the QoS actually achieved and hence offered to the various services running. This is shown as a 'mapping' rather than a direct link because the relationship between the many network elements involved and the service provided can be complex, with parts of the network shared by many services and users at any time. There is also a mapping between the achieved QoS of each of these services and that as perceived ultimately by the user. So, the important conclusion is that there is inevitably a mismatch between the actual

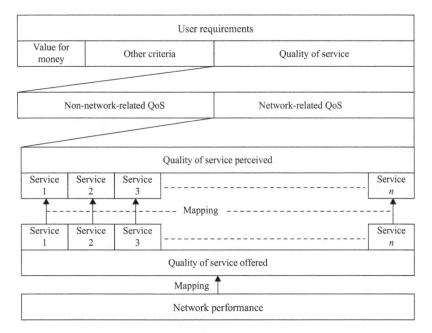

Figure 9.2   Relationship between QoS and network performance

achieved performance of the various elements of network as measured by the Telco and the QoS perceived by the user (customer). Therefore, the challenge for the Telco is what needs to be measured, and how is this done, in order to get the best estimate of the QoS actually being delivered to the user and their perception of it.

Network performance is relatively easy to measure in a quantified way, and there are several well-established international standards for acceptable levels of performance. Many of the parameters are constantly monitored (often automatically) within a typical digital network. In some cases, equipment will automatically be taken out of service when the threshold for a particular parameter is exceeded (e.g. digital line systems will shut down when the digital error rate exceeds 10^{-3}). Most of the performance parameters relate to how well the digital signals (i.e. 1s and 0s) – as individual bits, bytes, cells or packets – are conveyed to the far end in terms of incurred errors in the bits being delivered late or lost, and the extent and nature of the overall delay of the signals. All of these various types of degradation to the communications being carried will be perceived in various ways by the users (see Box 9.2 for examples). Obviously, the Telco aims for the level of network performance to be perceived by the users as very good, but in practice faults in the equipment, ageing of components, breakdowns in the cables, the effects of overloads, stormy weather conditions, etc., will impair performance, perhaps to unacceptable levels for the users. Undoubtedly, the most important of all the performance parameters is that of network availability – i.e. the percentage of time during the year that the network is working, and hence service is available. Table 9.1 gives typical values of network availabilities.

Box 9.2   Examples of network performance parameters

Parameter	Perceived effects on the user
Transmission	
Digital error rate (DER) of transmission systems (i.e. the proportion of bits received in error) [5].	*Audio*: Heard as occasional clicks; becomes noise as DER worsens; *video*: Seen as specks on the screen; increasingly becomes breakup of picture as DER worsens.
Propagation delay of transmission line systems (i.e. the time taken for the signal to travel along the cables or radio links) [5].	*Telephony*: Echo becomes a problem with moderate delays; two-way conversations become more difficult as delay increases above about 250 ms (e.g. with satellite systems).
Circuit switched network	
Grade of service (GOS): Proportion of calls in a circuit-switched exchange (fixed or mobile) meeting congestion (GOS is the main parameter used for exchange equipment dimensioning).	Users will experience increasing chances of no dial tone, or call set-up not being completed as GOS worsens.
Packet-switched network	
ATM cell error ratio (CER) and cell transfer delay (CTD).	Affects the link layer (Layer 2) of a packet network which normally provides managed pathways for IP packets, so will have the detrimental effects associated with delay.
Various measures of IP packet loss (e.g. percentage of packets lost, or not delivered in specified time), or received with errors [6].	Effect depends on application or service, e.g. for VOIP is heard as clicks or noise, for video is seen as specs or picture breakup; for other applications remedial actions in the network or by the application software can ameliorate effects.
End-to-end delay in IP packet transit through a router network.	For voice over IP services/applications will cause distortion and break up, eventually with high and/or variable delays causing mutilation of sound.

Table 9.1   Typical values of network availability

Availability (%)	Unavailability (%)	Outage in mins/year	Comments
99.87	0.13	683.28 (11 h 23 min)	This is the minimum standard for most Telcos
99.99	0.01	52.56	Typical 'High availability' level (also known as 'four 9's')
99.999	0.001	5.26	'Very high availability' (also known as 'five 9's')
99.9999	0.0001	0.52 (32 s)	'Extremely high availability' (also known as 'six 9's')

Many of these parameters are also used by the network operator in the dimensioning of the network – i.e. determining what capacity is needed at various points in the network to ensure that the required QoS for the overall network is achieved with the forecast level of traffic, as described in Chapter 5. More information on the vast subject of network performance can be found in References 3 and 4.

An important consequence of the fact that network performance can be measured quantifiably is that Telcos are able to predict and monitor the key performance parameters for the network. They can then make a business-case evaluation of the chances of the performance levels not being met and so offer their customers some level of guarantee about the QoS that they can expect. In some cases Telcos are prepared to pay customers compensation or a fixed penalty for prolonged absence of service due to breakdown or network failure (e.g. loss of broadband access for more than two days). Telcos take this approach further by creating service-level agreements (SLA) with their business customers, many of whom rely totally on the availability of the Telco's telecommunications, and maybe other ICT services for the operation of their business.

The SLAs are usually prepared jointly by the Telco and the customer, specifying in great detail the required QoS. A typical SLA for a medium-sized business customer with several sites (factories, warehouses, data centres, office buildings, etc.) is a substantial document valid for a number of years, covering all the telecommunications services: leased lines, fixed and mobile telephony, PABX lines, data servers, etc., and the levels of QoS required for existing and future provisions. Invariably, there will be a set of penalty clauses relating to payments due from the Telco for defaulting on the QoS levels achieved (usually assessed once a year). In the case of a multi-national customer, the SLA is even more complex and onerous for the Telco, given the nature of global services and the involvement of other operators on a sub-contract basis. The extent to which an operator meets or even exceeds the levels of service specified in the SLA will be reviewed periodically during the time of the contract. Based on the experience obtained, some customers may even pay a premium to upgrade the SLA. In general, the SLA represents a form of binding agreement between the Telco and the customer, which can help in developing a long-term mutually beneficial relationship.

As indicated in Figure 9.2 there are other factors apart from the network performance that influence the QoS as perceived by customers and hence their level of satisfaction. Examples of non-network-related parameters include:

(i)   *Customer-required-by date (CRD)*: The day when all the services are provided (this may be a simple single broadband line, or a large complex customer network of many lines and sites).

(ii) *Mean time to repair (MTTR)*: The average time taken to repair a fault and restore service to the customer. An alternative measure is the percentage of faults fixed within the specified repair time. However, some customers may be more concerned about the worst case of repair time – i.e. the statistical tail.

(iii) *Reliability of provision or repair activity* – i.e. provisions are free from early life failures (ELF) and repairs are free from 'repeat faults'.

(iv) *Appointments met*: The percentage of appointments made with customers (e.g. for a site visit) that are met.
(v)   *Accuracy of billing*: The correctness and level of detail of the items on a bill, their traceability and the ease of understanding for the customer. In some cases, this may also apply to automatic stops on service when items on a bill exceed a pre-arranged threshold – e.g. for data downloads on a mobile phone.

Many of these non-network related parameters will also be included in SLAs, as described earlier, since they are both quantifiable and important factors in the overall level of QoS for customers. Later in this chapter, we will look at how the Telco's support systems, processes and management of the call centre and field force influence the achievement of the targets for these parameters.

9.2.3.   Quality of experience

The advent of more subjectively challenging services like IPTV (i.e., video content delivered over the Telco's fixed or mobile broadband network) has led to the development of the concept of quality of experience (QoE). In practice, QoE is essentially the same as customer-perceived QoS, but the term emphasises that the quality assessment of the service or application must be from the user's point of view. So, e.g., QoE in relation to IPTV encompasses such factors as the acceptability of the picture and sound quality, the time taken to change channels, the ease of use of the service and the granularity of the image. Figure 9.3 shows the range of human components in the make-up of the QoE for an IPTV service.

A number of methods of predicting the likely customer acceptability of network services have been developed, one of which is the 'e-model' specified in the ITU-T recommendation G107 [8]. This model identifies all the (quantifiable) network-related parameters that will affect the transmission of the service, e.g. sending and receiving loudness ratings, transmission loss, weighted echo-path loss, packet-loss probability, one-way delay and quantisation distortion. An order of merit, known as the 'Rating Factor R', can be evaluated at the design stage of a

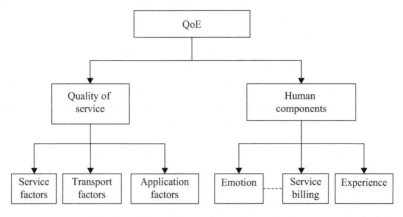

Figure 9.3   Quality of experience dimensions for IPTV services [7]

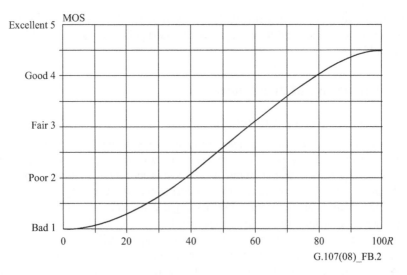

Figure 9.4 Equivalence of the mean opinion score and the rating factor R in the e-model [8]

network by comparing the Telco's parameter values with a specified ideal set. For example, applying the default set of values to the e-model for the transmission of an IPTV service gives an R of 93.2. The normal way of assessing the subjective acceptability of telecommunications service is through the testing with a panel of human users, each giving a score from 1 to 5 for the perceived results of the various parameter settings. The empirical relationship between the values of R and mean opinion score (MOS) ratings is given in Figure 9.4, showing that an R of 93.2 equates to a very good opinion score just below 4.5.

Finally, a Telco's managers can get some idea of the customers' level of satisfaction with specific services by assessing the results of the following methods:

(a) *Use of customer call-back enquiries*: Where the Telco's call centres follow up every site (or home) visit by a telephone, email, text, IM, etc., enquiry about how satisfied are you with the work done by our engineers during the visit?
(b) Customer complaints, especially the number that were escalated through to 'Chairman's cases'.
(c) *Customer bouquets*: where customers send unsolicited thank-you's and compliments to the Telco.
(d) Press coverage.
(e) Data on QoS and customer surveys from the National Regulator.
(f) *SLA monitoring*: Where the performance of the Telco against all its SLAs gives a measure of measure of customer satisfaction.

On a broader front, analysis of many customers' reactions and comments has enabled a customer experience model to be developed. This work, at the Henley

Centre for Customer Management, has identified the factors that influence the customer's experience and how these can be managed for consumer and business markets [9].

9.3　The nature of operations

9.3.1　*The role of operations*

The role of operations for any telecommunications company is to manage the networks and data servers so that the services are delivered at the appropriate quality to its customers, such that the overall cost is acceptable to the company. The challenge increases with the number of different services offered by the Telco, and can become particularly difficult when serving a wide range of customer segments (retail, wholesale and global), since they will have corresponding sets of expectations and needs. As we will be examining in this chapter, delivery of the operations depends upon the full mix of human contact at call centres, the technicians to work on the plant and equipment on the one hand, and computer support systems with their databases on the other. This mix is illustrated in Figure 9.5.

The inter-relationships between the human and computer activities are guided by a set of processes and policies. The important point is that successful delivery of operations depends on each of these elements. Thus, just having some particularly sophisticated computer support system to manage customers' orders, but without well-managed people at call-centres following efficient and sensible procedures, will not deliver good customer satisfaction. Even with the increasing trend for customers to interact with the Telco via a web-based portal for ordering, fault reporting and repair tracking, there will still be a need for customers to be able to speak to a help desk on occasions. The human factor will still be important.

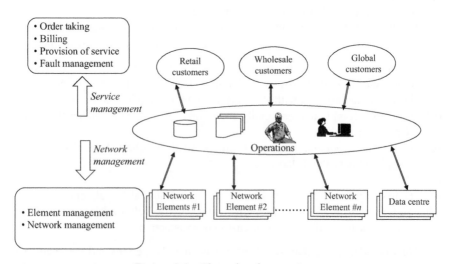

Figure 9.5　The role of operations

Figure 9.5 illustrates that the operations bifurcate into those activities facing the set of customers – known as 'service management' and those activities facing the network and data servers – known as 'network management', as discussed in section 9.4. It is generally accepted that the term network management covers not only the Telco's network of cables, exchanges, cell sites, routers, etc., but also the estate of data centres running the computer support systems for the Telco's operations and the provision of a range of data servers providing such customer services as cloud computing, web hosting and facilities management.

9.3.2   Cost drivers

We now look at the big challenge for any Telco in managing operations, namely controlling the costs. A simplified model of network operating costs is shown in Figure 9.6. The three components of network operating costs per year are: the cost associated with providing new service connections ('service provision'), maintenance costs and the running costs for the network (e.g. power and cooling). The overall total operating costs for a Telco consist of the network operating costs plus a range of other operating costs, such as the cost of maintaining the buildings, rates, etc., and payments to other licenced operators (OLO) for carrying the traffic (known as POLO). The latter can be quite a large proportion of the Telco's operational costs, which may be only partially offset by the conveyance payments for incoming traffic from other operators (receivables from other licenced operators). It is a policy decision by the Telco as to whether the cost of the data centres housing the operation's support-systems is considered part of the network or other operating costs.

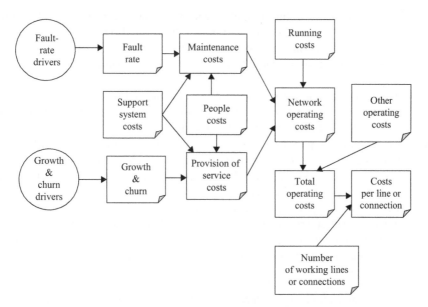

Figure 9.6   Model of a Telco's network operating costs

In order to gauge the total operating costs they are usually converted into unit costs, for example 'costs per line' in the case of a fixed network operator, and a 'cost per connection' in the case of a mobile operator. In either case, these unit cost values give a way of comparing the cost efficiencies of different network operators; it is therefore a favourite measure used in Telco benchmarking activity.

Referring to Figure 9.6, the operating costs are grouped into two main activities: provision of service and maintenance of service (sometimes also called 'assurance'). First, the costs for provision of service are influenced by both people and support-system cost drivers, with the volume driven by the growth and churn in customer demand. There are a number of factors which affect the amount of demand for the provision of service for new customers or the cessation of service for existing customers (which can be equally demanding in the amount of work required), which includes the following:

- Customer satisfaction levels.
- The effect of competition.
- Regulation – in terms of legal requirements on the Telco and any new opportunities for new services.
- The launch of new services.
- The general state of the national economy.

The maintenance costs are also driven by people cost and support-system costs, with volumes depending on the fault rate in the network. There are many factors which drive the fault rate on equipment and plant within the network, including the following:

- The overall quality of the equipment and line plant – of course, investment in upgraded and new equipment can improve this (but also see next bullet!).
- The Telco's capital investment programme. Disruption to existing parts of the network, leading to increased faults, can be incurred while new equipment is being installed.
- Network design and the resulting inherent resilience to faults.
- The maturity of the equipment being installed in the network, particularly new technologies, will influence the overall fault rate.
- The standard of repair work, both historically and currently, directly influences the incidence of repeat faults. Also, it is a well-known fact that manual interventions on network equipment during repairs invariably causes faults to occur in other equipment in the vicinity.
- The weather – especially wind and rain – can have a profound effect on the fault rate, particularly in the copper access network due to flooding of the underground ducts and joint boxes. Also, stormy weather can damage poles, overhead cables, as well as masts and mobile radio antennas.

The total operating costs for a network operator (OpEx) can be considerably greater than the capital expenditure (CapEx), primarily driven by the field staff costs – as illustrated in the example of BT in Box 9.3.

Box 9.3   Example of a Telco's operational costs

The operating costs for British Telecommunications plc for the financial year
2013/2014 are as follows [10]:

Total operating costs = £15.1 Bn, of which:

– Staff costs =	£4.7 Bn
– POLO costs =	£2.5 Bn
– Network operating & IT costs =	£0.6 Bn
– Other operating costs =	£3.7 Bn
– Depreciation of equipment and plant =	£2.1 Bn
– Amortisation of intangibles =	£0.5 Bn
– Miscellaneous costs =	£1.0 Bn

For this year the capital expenditure on the network was £2.4 Bn
(Figures have been rounded)

9.4   How operations are managed

9.4.1   The telecommunications management hierarchy

As might be expected, given the complexity of operations, there is a universally
accepted simple model that helps all in the industry – service providers, Telcos,
manufacturers and the regulators – to manage the complexity. The model, known as
the Telecommunications Management Hierarchy developed by the TM Forum
(previously known as the Telecommunications Management Forum [11]), comprises
five layers, which are often represented as a conical stack as shown in Figure 9.7.

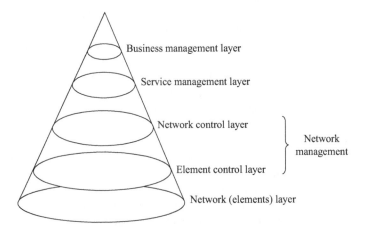

Figure 9.7   Telecommunications management hierarchy

The bottom layer represents the elements within the network, e.g. cables, transmission systems, switching systems, routers, computer data centres, etc. The second layer represents the set of systems that directly control the network elements – e.g., alarm equipment monitoring an SDH transmission system or a controller that sets up virtual paths in an ATM switch. Above this is the network control layer, which co-ordinates all the individual element controllers (in the second layer) so that a full end-to-end network view can be taken. The network-control and the element-control layers together provide, what is generally known as, 'network management'. At the fourth layer of this model resides all the activity concerned with interfacing to the customers – e.g. order taking, billing, and fault management – known collectively as 'service management'. Finally, at the top layer are all the functions associated with managing the business – e.g. setting strategy, objectives and budgets, financial tracking, budgetary control, P&L management, project management and pay roll – essentially the subject matter of much of this book!

9.4.2   *How the Telco's operations are organised*

While all models have their limitations, the TMF model of Figure 9.7 does help in organising the operations of a Telco into convenient management domains, even though there are inevitable areas of overlap in the execution of some operations. As we will see later in this chapter, the cost and quality of the operations can be significantly improved by appropriate organising of the support systems, their interconnection and information flows, as well as the procedures followed by the staff. Table 9.2 shows the top-level mapping of the range of a Telco's operational activities against the four management layers of the TMF model, together with an indication of the locations of these activities.

The execution of the network elements control layer is usually located in exchange buildings around the country so that technicians can manage equipment within the buildings, as well as remotely managing plant and equipment in their catchment area. For a fixed-network Telco there will be a considerable amount of activity in the access network, particularly on the underground and overhead copper cables and the street cabinets. In many cases, of course, a customer's premises visit is also necessary for working on the network termination equipment and internal wiring. In the case of business customers, there may be extensive telecommunications equipment owned by the customer and located on their premises which is also managed by the Telco's technicians. The network control layer additionally covers the management of the Telco's data centres, as discussed later in this chapter.

The network control layer functions are normally managed by technicians at terminals within operations and maintenance centres located in operational (i.e. having network equipment) or office buildings. These can be arranged on a regional basis – e.g. one centre controlling all types of network within its region – or on a functional basis, whereby individual centres control the full national network of one type (e.g. a national centre controlling the national IP network).

Service management is normally in customer service centres around the country, the location of which is usually away from city centres in more low-cost

Table 9.2 The Mapping of the operations to the functional layers

Layer	Activities	Location
Business (also known as 'Enterprise') management	The company's business and its governance, including: setting strategy, financial planning, budget & objective setting, product mgt., marketing, regulation, PR, investor relations, HR mgt. (payroll etc.), purchasing, R&D, standards.	Offices – HQ and regional offices.
Service management	All customer facing activities, including: customer enquiries, order taking, fault report mgt., billing mgt. & enquiries, provision of service, outgoing sales, account mgt., operators (999, assistance, directory enquiries).	Call centres, auto-manual centres (for 999, etc.), and assistance centres. (All these may be in exchange & other operational buildings or in office accommodation.)
Network management	Network control activities, including: performance mgt., real-time remedial action (congestion control), service restoration mgt., routeing table builds, control-software update mgt., interconnect mgt.	Operations and maintenance centres, regional and national control centres.
Element management	Element control activities, including: equipment configuration, service provisioning, capacity assignment, alarm mgt., equipment & line plant repair.	Network nodes – e.g. exchange buildings, transmission stations, data centres, street connection boxes, overhead plant, customers' premises.

Note: mgt. = management.

areas where there is a high availability of suitable people to staff the centres. Some network operators may off-shore this activity.

Finally, the activities of the top layer of the model – i.e. business management (also known as enterprise management) – are executed in the offices of the company. Invariably, many of the planning activities described in Chapter 5 are appropriate to this layer and are carried out in the company's offices. This is particularly appropriate for the network planning activities, since this leads to the building of the capital expenditure programme for the Telco.

Many of the operational activities (in all of the layers of the model) are supported by computer applications, often run on small local PCs. However, where large volumes of data and computation are required the applications may be run on large servers located at remote data centres. Examples of the latter include the company's payroll, customer billing, control-software management and development.

9.4.3 Example of the use of support systems

Obviously, the range of support systems and the way that they underpin the operations of a Telco is a vast subject. However, it is instructive to look at one

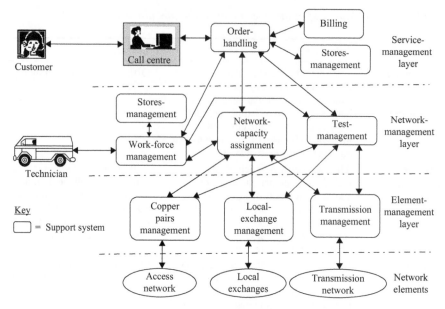

*Figure 9.8   Example of systems involved in the provision of service of a new
subscriber's line [Reproduced with permission from K.E. Ward]*

example of this relationship. Figure 9.8 shows that the support systems in the
service, network and element management layers are used in the provision of
service for a new telephone or broadband line, with a typical set of information
flows indicated by double-headed arrows. It is important to note that although the
support systems reside in their respective layers, most of the operational activities
of a Telco involve the interaction of support systems in all the layers.

With reference to Figure 9.8, we can briefly consider the flow generated by
an order for a new broadband line (ADSL over copper or VDSL over copper and
optical fibre). On receiving the customer order at the call centre, an installation
date will be given after checking that spare lines are available at that location, and
a copper pair and telephone number is allocated (network capacity assignment). If
spare line or number capacity is not available then the procedure for adding new
line plant is followed (transmission management), which may involve rearran-
gement of existing capacity or the initiation of some new cable provision in the
field. Once all capacity is allocated, the service provision work is scheduled for
the technician visiting the customer's premises, the street cabinet, overhead or
underground cable termination (work-force management and stores management). There is also the need to connect at the exchange end and test the broadband line is working to the QoS level offered for that service (local-exchange
management and test management). Finally the billing system will be advised of
the job completion.

9.5 Service management

9.5.1 Cost model

Service management covers all the activities involved in dealing with the customers. For a Telco this embraces: sales, account management, order taking, billing, dealing with complaints, fault reports, customer surveys, etc. It is generally recognised that successful customer service management is the key to achieving satisfied customers. However, the factors that influence customer's perception of the service received from a Telco are wide ranging and may include such issues as: how smart the technician looked when he installed the broadband line at a customer's house; how attractive is the set of products in the Telco's portfolio; the level of expertise exhibited by the agents in the call centres; etc. Thus, it is not surprising that service management is often described as an all-embracing wrap around the network service – sometimes referred to as the 'service surround'.

However, for the purposes of cost analysis, we consider the range of service management to comprise four main areas: service provision, service assurance (covering repair and correcting configuration), service accounting (billing) and customer-contact management. Figure 9.9 presents a simple cost model for customer service activity, showing the volume and cost drivers. In addition to the volume drivers for service provision and assurance, which were introduced in Figure 9.6, this model shows the volume drivers for service accounting and customer contact management. For example, the causes of customers contacting call centres to report faults are driven not only by the inherent reliability of the network and its rate of faults, but also by the effect of network enhancements and any new products, and also the teething problems with new products/services being launched.

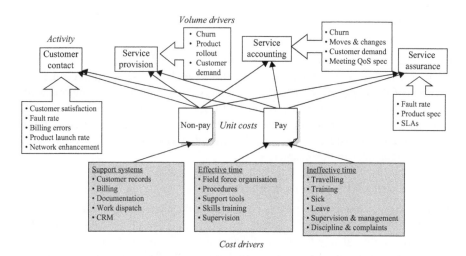

Figure 9.9   Customer service cost model

As our model (Figure 9.9) shows there are two distinct sets of unit costs affecting the customer-service functions, namely: pay and non-pay. We now briefly consider the components contributing to these unit costs.

9.5.1.1   Non-pay

This category covers the annualised current account costs of all items other than manpower costs, typically comprising:

(a)   *Computer support systems*: Customer records, billing records, technician work dispatch, customer relationship management (CRM).
(b)   Process documentation.
(c)   *Technical support*: This category covers the support to the technicians in the field, including: tools, mechanical aids, vehicles, terminals, manuals and laptops.

9.5.1.2   Pay

The pay component is not only the biggest contributor to the operational annual costs for a Telco's operations, but also offers the greatest opportunities for varying to meet cost and quality objectives for the services provided.

(i)   *Effective time*: This covers all the activity which directly contributes to providing customer service, including:
  • field force organisation and management
  • working procedures and methods

(ii)   *Ineffective time*: This category covers all the activity which does not directly contribute to the task in hand, including:
  • travelling time
  • training
  • sick absence
  • holidays
  • dealing with complaints

Clearly, all organisations try to keep the proportion of ineffective time in their workforce as low as possible, since the ratio of ineffective-to-effective time is a measure of the company's cost efficiency. This ratio is particularly important for any Telco that has to set up a service field force of technicians for customer site visits (provision of service or repair) or for running customer help desks or call centres. We will consider the cost and dimensioning aspects of a Telco's field force of technicians later in this chapter.

9.5.2   Customer-contact channels

Traditionally, customers have made contact with their telecommunications service providers (Telcos, ISP, mobile service provider, etc.) via phone calls. While this may be the preferred way for many of the customers because of the human contact received, it does incur a heavy operational cost for the Telco in having to staff the call centres with sufficient suitably qualified agents. The advent first of fax, then emails, has enabled alternative customer-contact channels, which are significantly cheaper for the Telco to provide (reduction ratios of 100 to 1 have been suggested) because the enquiries can be handled in non-real time. This enables the work load to spread over the day among a smaller set of agents. However, it is the increasing acceptance of web-based transactions that now offers the telecoms companies even greater opportunities for cost reduction resulting from the automation of the process, as well as the ability to offer extra features for customers. A good example of the latter is the ability for customers to keep track of the status of their order or fault report through the web site (similar to the order-tracking feature offered by Amazon.com). It is for this reason that it is now normal for service providers to offer discounts to customers who agree to receiving paperless bills via a website, as well as charging them for calling the service centres rather than using email or web channels. Table 9.3 summarises the set of customer contact channels and lists the pros and cons for the customers and for the Telco.

9.5.3 Customer relationship management (CRM) systems

The arrival of customer relationship management (CRM) systems has provided an important tool to assist Telcos to improve the quality of their service management, as well as providing a systematic way of using customer intelligence to improve the marketing and product management within the company. A simplified view of CRM systems and how they fit within the service management activities is shown in Figure 9.10. The role of the CRM system is to collect and correlate all the relevant customer information and present this to computer screens of the agents in the call centres and to interact with the service and business management systems within the Telco. This is achieved by assembling the data relating to individual customers (residential and business) – sales, marketing, account management, faults and billing – as shown in Figure 9.10.

For customers, CRM provides the convenience of presenting all relevant historical and status information relating to their account on the screen of the call centre agent when dealing with an enquiry. For the Telco, the CRM systems link into the set of support systems within the service management layer associated with fulfilment (i.e. service provision), assurance (i.e. fault management and QoS management) and billing. Importantly, the CRM systems also link to the systems supporting the marketing and product management activities (in the business management layer) of the company.

9.5.4 Call centre dimensioning

As described earlier, the cost of call centres is significant in terms of the annual operating costs for a Telco, due to the predominance of the staff pay costs. The ready availability of cheap, good quality international circuits and VOIP has made the possibility of outsourcing call centres to overseas English-speaking countries (such as India and South Africa), where the pay cost of agents is lower, an attractive way of minimising such costs. (However, there has been some adverse reaction from customers to this outsourcing and there can be compliance issues with regard

Table 9.3   The characteristics of customer contact channels

Contact method	Customer pros/cons	Telco pros/cons
Telephone call	• Well-established method • May be charged for or free • Human contact • Requires waiting for call to be answered	• Highly costly to provide adequate cover at call centres • Needs to be dealt with in real time • Requires high-calibre agents in call centres
Fax	• Well-established method • Provides a record of enquiry • No call waiting • Difficulty of putting problem/enquiry into writing	• Can be dealt with in non-real time • Increasingly becoming an obsolescent channel
Email	• Difficulty of putting problem/enquiry into writing • No call waiting • May not get reply	• Can be dealt with in non-real time
Via a web page	• Forces customer to navigate the website • May be difficult to convey the query • No call waiting • May not get reply • May be able to track the progress on a logged enquiry	• Can be dealt with in non-real time • Requires good, user-friendly web page design • Can provide specific information on the progress of sorting the enquiry
Via web chat	• Gives immediate response to the customer • Keeps a record of the transaction • Helps solve language and dialect issues	• Depending on the transaction, one agent can handle several web chats at a time.
Letter	• Difficulty of putting problem/enquiry into writing • No call waiting • May not get reply	• Can be dealt with in non-real time

to data-security regulations.) Call centres usually deal with a range of customer-contact activities covering receipts of incoming calls (inbound) – complaints, enquires, and orders – and the generation of outgoing (outbound) calls as part of a sales promotion campaign. Clearly, the objective of the call centres is to manage customers' satisfaction which means that they need to provide sufficient agents to be able to answer calls within the target time on one hand, while minimising the costs on the other. The main factors that determine the number of agents required in a call centre are outlined below.

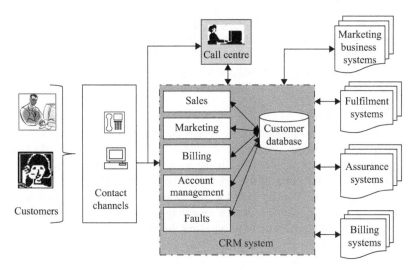

Figure 9.10   CRM system within service management

9.5.4.1   General

- If the call centre is required to be staffed 24 hours per day then each desk in the call centre needs to be allocated some 4.5–5.0 people. This is based on effective time efficiency (as described in section 9.5.1) of between 67–60%, and 3×8-hour shifts working.
- Large call centres are more efficient than small centres in terms of answering the incoming calls (due to the Erlang's traffic formula, as described in section 5.4.3). Agents can be dispersed geographically as long as they are working to one virtual queue; however, large centres also benefit from greater efficiency in the supervision of the agents.
- Good information support systems for the agents are essential (e.g. using CRM)

9.5.4.2   Inbound

- Dimensioning for inbound activity at call centres is always difficult because the Telco has to estimate the expected levels of calls throughout the day and staff the desks accordingly. This way of working is called 'reactive' because the Telco has to react to the occurrences beyond their control. If the inbound calling rate is higher than expected (or the calls are taking longer than average to be dealt with) then the customers will experience the frustration of waiting in a queue – often leading to the perception of poor customer service.
- It is difficult to staff desks optimally, particularly if agents with specific skills and knowledge (e.g. broadband or IP TV services, etc.) need to be made available throughout the day. There is consequently a balance between the costs of making a full range of expertise constantly available at desks against a deterioration in the quality of service that can be offered to the customers.

Interactive voice response (IVR) systems are often used to help steer customers to appropriately trained agents and can greatly increase efficiency for the Telco. Such systems need to be well designed so as to not irritate the customers. To avoid people waiting for a long time in a queue, some call centres will offer to call back the calling customer, which improves efficiency.

9.5.4.3   Outbound

- Dimensioning for outbound traffic is easy because this is actually generated by the agents themselves – i.e. this is 'proactive' working. Thus, the loading of the agents (in terms of number of calls made per desk) can be kept high.
- Similarly, the skills mix of the agents can be managed efficiently.
- The outgoing calls can be linked to the Telco's marketing and sales activity. However, many customers dislike receiving such calls and so operators are concentrating their sales activity on using the more effective inbound calls, as discussed in more detail in section 7.6.4.

9.6   Network and element management

9.6.1   Organisation

In Table 9.2 we saw that the two layers of network management activities were spread across the country, with the top layer (providing the network view) executed in a few regional and national centres, and the lower layer element-level activities undertaken locally and remotely at operational buildings. We can now take a closer look at the way that the network operations are organised by a Telco, referring to Figure 9.11.

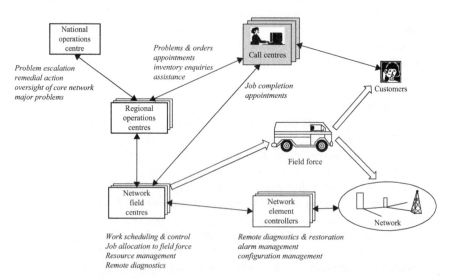

Figure 9.11   Typical Telco network operations organisation [Reproduced with permission from K.E. Ward]

The technician workforce are split into geographically separate teams and managed from network field centres (NFC), located at convenient points with good access to the road network in the service area. The NFC locations are important because, as we will see later, one of the big challenges is the minimisation of the travelling time by technicians to and from the population of subscribers or field equipment in the area. It should be noted that the technicians will be required also to visit the local and trunk exchanges, mobile network cell sites, street based cabinets and core transmission stations (most of which will be unmanned) within the area for construction, service provision and repair work. Minimisation of travel time can be achieved by giving a technician a set of jobs at the start of the day which can be visited in turn. Also some technicians will park their vans at home and pick up details of their first job at home and drive straight there, rather than spending time going to the field centre to start the day's work.

The control of the network management activity may be split between a number of separate centres spread across the area of jurisdiction – part of a country, all of the country, or even over several countries in the case of a global operator. There are two basic ways of organising the work of this set of centres – either on a geographical basis, where each centre has responsibility for all networks and services within their catchment area; alternatively, each centre has responsibility for a specific set of networks and services across all geographical areas. Irrespective of whether the regional operations centres have geographical or functional responsibilities, there is usually one national operations centre (NOC) overseeing the real-time operations across the entire national (and, if appropriate, international) network.

9.6.2   Network control layer support systems

The wide range of support systems used within the network management and control layer is shown in Figure 9.12. For convenience, the support systems and databases are shown mapped on to the set of the main activities and processes followed in network management, namely:

- *Resource management*: maintaining an inventory of the physical network plant and equipment.
- *Configuration management*: maintaining a map and inventory of the working network and spare capacity.
- *Network assignment*: allocating network elements and configurations assigned for customer (e.g. leased lines) and Telco.
- *Traffic capacity management*: network routeing and configuration.
- *Test management*: all the testing for provision, repair and restoration of service.
- *Event management*: providing real-time remedial action.
- *Performance assessment*: keeping track of the network performance achieved across the various parts of the network.
- *Security management*: network integrity and resilience.
- *Data collection*: assembly of the various management data for operations, planning and business management within the Telco.

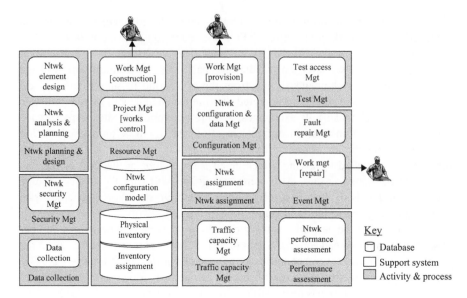

Figure 9.12   Network-control layer support systems [Reproduced with permission from K.E. Ward]

In addition, the network planning and design activity is shown in Figure 9.12, because there it is now generally recognised that there are architectural and operational efficiencies for the Telco in linking the planning of the network with the in-service utilisation of the network capacity.

You will notice that there are some support systems that interact with the field force for the following three activities:

- *Construction*: i.e. building the network – covering the installation of cables, microwave radio systems, switching systems, data switches (IP/MPLS, ATM), intelligence nodes, mobile cell sites, data centres, etc.
- *Provision*: i.e. bringing customers into service through configuring and connecting equipment (e.g. jumpering across the MDF in local exchanges, connecting up private circuits, establishing large corporate data networks).
- *Repair*: i.e. maintaining all the network elements, restoration of service (including re-routeing around network elements that can't be repaired in time).

9.6.3   Sizing the field force

For a Telco, or indeed any company having a field force as a key part of their operations, the over-arching concern is how to organise the field force, in terms of allocating responsibilities to the geographical service area, the management structure, and the location of the field control centres. The two big activities for a field force are the provision of service and the maintenance of service – both may include some travelling to customers' sites and operational buildings. There are also critical factors concerning the required expertise of the technicians and their training, together with the choice of whether to rely on a generally held comprehensive level of skill (i.e. 'multi-tasking') or a series of specialist teams. Of course, any consideration of technician skills also raises the question of whether the technicians work singly or need to work in pairs or teams, and the degree of supervision required. Other factors include the policy for holding spares and stores, since these incur costs of purchase and storage for the Telco, but they need to be readily available to the technicians across the service area.

Although the annual expenditure on the support systems estate for any network operator is significant, the large component of OpEx is invariably the cost (pay and non-pay) of the technician field force. Thus, there is continuous pressure within a Telco to minimise the size of the field force. However, of course, reducing the number of technicians available for the customer service activities of provision and maintenance is likely to have an adverse effect on the quality of service experienced by the customers. Therefore, the challenge for the Telco is to reduce the field force cost through increased efficiency – through improved organisation, use of tools and support systems and refined processes – while meeting the customer service obligations.

There are numerous ways of measuring the in-life quality of a product or service, and this is best measured with respect to customer expectations. For example:

(a) How reliable? Measures are – faults per circuit p.a.; availability or downtime p.a.
(b) Speed of repair or restoration? Measures are – mean-time to restore; percentage of faults repaired in a target time; time to respond and initiate a repair.
(c) Quality of repair? Measures might be the number of repeat faults within a given time (often 28 days).

These parameters are typically measured as either an average (i.e. arithmetic mean) or as a percentage meeting a specified target (e.g. % faults fixed within 5 hours), or as a tail measurement (e.g. the percentage of faults still outstanding after 24 hours, 48 hours or 96 hours). Senior Telco managers will normally look at three or four measures to ensure, say, repair time is not being improved at the expense of quality.

One of the key quality parameters for any network is the mean time to repair a fault – known as the MTTR [12]. This parameter will be kept constantly under review by the Telco's management team because it indicates how well the company is able to honour its commitment to repair customers fault within a specified period. (Although, care is required in interpreting results because a few bad repair times can be masked by a large number of quick repairs.) For business and corporate customers, who depend on the continuous availability of services from the Telco for running their business, the time to repair faults may be enshrined in a service-level agreement (SLA) with cost penalties for non-compliance, as described in section 9.2.2.

So, the question arises: how might an operator reduce the achieved time to repair in the field, and so improve the quality for customers? The obvious answer is to increase the number of technicians deployed to the national, regional or local service area. This will shorten the time spent waiting for a free technician to attend the fault and potentially reduce the travelling time involved in diagnosing, locating and repairing the fault. However, the increase in field staff will also reduce the efficiency of operations, in terms of number of faults cleared per technician. Therefore, the number of field staff and their deployment across the service region have to be carefully managed and included with a range of initiatives and improvements in order to obtain satisfactory MTTR at an acceptable cost.

Box 9.4 presents an analysis of the effect of the number of available technicians on the MTTR, assuming the worth case. Actually, the advent of remote diagnostics can allow a network fault to be located and its likely cause determined from the central field unit, so reducing some of the travelling time and enabling the appropriately skilled technician and test equipment to be despatched. Also, in some situations service can be restored remotely, again with consequent savings in technicians' travelling time.

Box 9.4   Analysis of mean time to repair

The MTTR is made up of the following time components:

$t_1 =$ Time taken at a reception desk to ascertain the problem in communication with the customer and logging the problem on the fault management system.

$t_2 =$ Time awaiting allocation of the repair job to an available technician.

$t_3 =$ Time taken to locate and diagnose the fault (which may be within the network or at the customer's premises) including travelling time.

$t_4 =$ Time taken to resolve the problem and restore service, including travelling time.

$t_5 =$ Time taken by the reception desk agent to get the customer to confirm the clear of the fault and to hand back service to the customer.

In general, the greater the number of technicians available within a service area the quicker a job can be allocated to a free technician (t_2). Also, increasing the number of technicians should reduce their average travelling time to get to the customer's premises or the parts of the network (e.g. street cabinet) or operational building (e.g. remote exchanges or cell sites). These improvements will only be realised if the technicians are sensibly located within the service area and – most importantly of all – they are given consecutive jobs which are in the same local area (and so avoid travelling to opposite extreme ends of the service area!). However, although increasing the technician force reduces the travelling time component of t_3 and t_4, it does not improve the time to diagnose and repair the fault. Figure 9.13 illustrates the decline in travelling time with increasing technicians, while having no impact on the net repair time (t_3 plus t_4 minus travelling time) or the reception-desk activity ($t_1 + t_5$). The graph also shows that the field-force loading – in terms of jobs done per technician per day (an important measure of cost efficiency for a Telco) – reduces as the number of technicians deployed within the area increases [12].

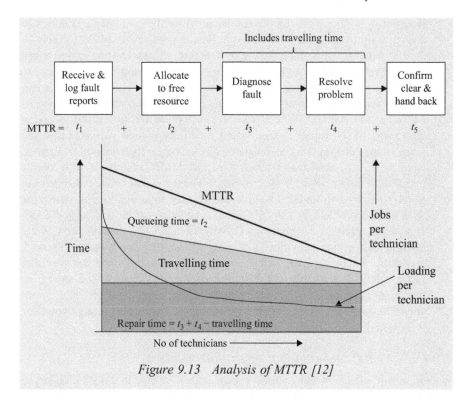

Figure 9.13   Analysis of MTTR [12]

Improvements to MTTR, therefore, also include the provision of efficient computer support systems for:

(i)   Logging and tracking faults and interacting with the customer, including flagging up to senior management any faults that take an excessively long time to clear.

(ii)   Allocating jobs to the field force, recognising the priority of the customer and the need to minimise travelling time.

(iii)   Assisting with the fault diagnosis, perhaps using expert systems and lessons from fault history.

It is now normal practice for the technicians in the field to be given laptops, tablets or specially designed work terminals and access to communication systems to provide diagnostic advice, as well as to enable the status of the job to be logged and reported back to the field control centre in real time. The whereabouts of the field force are to be constantly tracked using GPS or other technology. This approach also makes significant improvements in the efficiency of the field staff involved in provisioning of service.

Other ways of improving the efficiency factor of field staff, i.e. increasing: *Jobs per day per technician,* for both provision and maintenance operations include:

(a)   *Increased use of automation.* Examples of this are the remote testing facilities for telephone lines and broadband transmission systems (ADSL, VDSL) initiated by a phone call by the technician from the customer's premises or any location in the field. This can save time diagnosing and fault locating.

(b) *Remote working.* In general, the use of remotely activated testing of equipment saves travelling time in repair situations – e.g. application of remote loops to determine whether the fault is beyond that point or not. This technique is particularly useful for customers' data circuits (e.g. ethernet) and leased lines.

(c) *Customer involvement.* Where possible, customers are now asked to undertake some of the repair and provision functions at their premises to avoid unnecessary site visits by technicians. For example, the sending of terminal equipment, e.g. modems, through the post ('Jiffy bagging') with simple installation instructions for the customers. For maintenance the customers are usually required to undertake some preliminary investigations to determine whether the problem is with the Telco's network or the (privately-owned) customer's premises equipment. Customers may also benefit from these obvious cost saving activities by feeling involved by being better informed, especially when they take part in automatic line-testing facilities requiring just a simple call. Finally, the customer satisfaction for provision and repair can be greatly improved by making status information available on the website or through text messaging to the customer's mobile phone (e.g. 'your line is expected to be repaired by noon on Wednesday').

9.6.4   *National and regional operations centres*

We will now briefly look at the network management layer activity, which takes an overall view of the network performance and initiates or oversees any remedial actions in real time. As mentioned earlier, these functions are located at one national operations centre (NOC), usually supported by several regional operations centres (ROC), the latter either dealing with all networks/services in their catchment area, or having nation-wide responsibility for certain services/networks. All operations centres comprise large control rooms with many computer screens where engineers monitor the networks performance, and where necessary they are able to remotely interrogate the network equipment via the element controllers to determine the cause of problems.

The NOCs are often a great visitor attraction, since they usually have viewing galleries overlooking the control room with one wall having illuminated maps of the network in which hot spots due to overloads, or transmission or switching failures, are indicated in various colours. It is clear from here that remedial action can be initiated remotely to prevent the problem propagating through the network.

The roles of network management for a national Telco typically fall into the four categories described below. They all have the overall aim of managing the network resources to ensure optimum use; in the event of disruption due to overloads or breakdowns to use available resources in the most effective way.

(a) *Traffic Management*
     All exchanges are constantly monitored for levels of congestion on the switches and traffic routes. In response to overloads detected or expected, remedial actions can be initiated remotely in real time by the NOC via the

exchange element management control. For example, during the broadcast of popular TV programmes viewers voting by phone during a short period can cause call attempts in the network to increase by up to a factor of 10. An excessive volume of call attempts arriving at a destination exchange can have a serious effect on other users of the exchange, including in extreme cases, stopping the exchange from working at all, hence the importance of the NOC initiating remedial action. Such action includes:

(a) *Re-routeing* of traffic in the core network to offload congested routes (by temporarily changing the routeing tables in the exchange control systems).
(b) *Call gapping* – the temporary blocking at appropriate points in the network of a proportion of new calls directed to the specific number, as described in Box 9.5 [13].

Box 9.5   Call gapping

The most commonly used method of call gapping uses an algorithm run by each the switch control system of all participating exchanges that allows just one call attempt per time slot T to be accepted. The time slots start as soon a call has been accepted, as shown in Figure 9.14. Therefore, the time between accepted calls, t will be equal to or greater than T. Typically T is set to 0.2 s. The result of applying call gapping to a telephone network is illustrated in the lower picture in Figure 9.14.

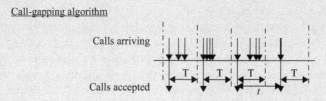

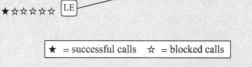

Figure 9.14   Call gapping mechanism

(b) *Signalling management*

The signalling network linking all fixed and mobile (2G and 3G) telephone exchanges using the SS7 system (see Reference 14) is managed as a separate network, and a similar set of remedial re-routeing actions as (a) above can be initiated by the NOC. It is important that overloads and breakdowns are coped with effectively since deterioration in the signalling QoS will seriously affect the service levels of the exchanges. Any remedial action is likely to require some liaison with the NOCs of other operators since Cable, PSTN and mobile operators are highly interconnected for calls and rely on the integrity of the participating SS7 networks.

(c) *Transmission management*

Much of the core transmission network has automatic restoration and protection, especially on the major routes. In some cases there may be one protection channel for a number of main channels (which are not expected to fail at the same time – '1 in N protection' [15]). Therefore, much of the NOC activity is the monitoring to ensure that these mechanisms are working appropriately, and where a restoration path has automatically been switched in, to progress the repair underlying fault. Where major breakdowns occur there may not be sufficient spare capacity in the network for full automatic protection, and the NOC will need to oversee repair activity and manage the progressive restoration of service, particularly when large customers (e.g. the military, other operators, TV broadcasters) are affected.

(d) *IP traffic management*

Whereas call gapping addresses overload in circuit-switched telephony networks, focussed overloads may also appear on IP networks. These can take the form of high traffic to a massively popular website, or malicious denial of service (DOS) attacks where traffic is artificially generated – often by virus-infected computers – in order to stop a particular website from working. In these circumstances the Telco serving the target site may have to block or restrict traffic to specific IP addresses.

(e) *Serious incident management*

Occasionally serious incidents, such as earthquakes, major fires, floods, civil unrest, storm damage, train crashes, etc., cause disruption to many services and utilities as well as the telecommunications networks. In such cases the NOC's primary aim is to take charge of the Telco's response, with a view to restoring telecoms services (fixed, mobile and Internet) – which, invariably, are in abnormally high demand by the authorities and the public due to the emergency. Even though network operational buildings normally have standby electricity generators there may be a need for additional transportable generators to be moved into the affected area where the on-site generators have been damaged (e.g. flooded) or are unserviceable. In the event of major physical damage to the exchanges the NOCs may arrange for transportable standby telephone exchanges (on lorries) to be deployed. The NOC will also have to liaise with the NOCs of other operators, and possibly provide information for the government and press.

9.6.5 Data centres

In addition to the network communications plant and equipment, Telcos also own and run one or more data centres (DC) accommodated in purpose-built facilities, usually separate buildings. These DCs typically occupy vast areas, looking like warehouses, with raised floors covering power supply cables, suits of equipment cabinets (housing IP routers, servers and mass storage systems), with overhead distribution of optical fibre and copper cables. As with telephone exchanges, the DCs have separate areas for the optical fibre transmission termination and secure power supply equipment. Large capacity cooling and ventilating systems are used to keep all the equipment operating within the permitted temperature range. Box 9.6 gives details of an example data centre.

Box 9.6   Example of a data centre

Moreland House in North London is one of over 300 data centres owned worldwide by Level 3 communications [16]. The key features are as follows:

- Seven story building, with 25,000 m^2 of floor space.
- 20 MVA power capacity, provided by dual redundant 33 kV dedicated underground feeds from two public supply sub stations.
- Connections to 13 other Telcos.
- Emergency power from four diesel driven generators, which start automatically and gain full generation capacity within 1 minute.
- Dual redundant power supply around the building serving all equipment from under floor power-track distribution system.
- Three independent optical fibre entry points.

Services provided at DCs include:

(a) range of cloud computing services: IaaS, PaaS and SaaS
(b) computer hosting services for other service providers
(c) content distribution network (CDN) services
(d) running software applications and services for the Telco's administration (e.g. pay roll, inventory)

The operation of DCs is similar to that described above for telecommunications networks, including:

- Control centre (manned continuously) to monitor the performance of the equipment. This can be particularly onerous for the DC operator because many of the customers are other service providers relying on secure wholesale delivery of video and data content, cloud computing, etc. The control centre also oversees the software upgrades on the equipment.

- Secure power supply. Maintaining a secure no-break electrical power supply is one of the most important features of the DC operations. Normally, like most telephone exchanges, electrical power is from the main supply with battery back-up to cover the breaks before the standby diesel generators get up to speed and provide standby power for three to four days operation.
- Physical security is also especially important for DCs, given the nature of the services provided. This can be challenging when areas of the DC are made available to other service providers to locate their computing equipment on a hosting basis. The security process needs to cover physical access to the equipment areas, as well as strict control on the computer terminals and the various levels of authority for systems management.
- Communications security is usually ensured by the use of three separate entry points for the communications cables (optical fibre systems and copper cable) serving the DC.

9.7   Architecture, models and processes

9.7.1   Architecture

The great challenge for the designers of a Telco's set of computer systems to support the operations is the wide range and complexity of the different functions involved and the required flows of information between them. Earlier in this chapter we looked at a simplified view of the set of support systems for the network management layer (Figure 9.12 refers) in order to appreciate the range of activities involved. However, when the necessary linkages between the systems – computer software applications and data bases – are mapped, as in Figure 9.15, the overall view becomes quite complex. The high level of interconnection between systems and data bases raises two fundamental challenges for the designers:

1. How should the data be managed? That is, where should particular information (e.g. customer name, address, service entitlements) reside, which systems should have access to it, which systems can update it, and how can data integrity be ensured?
2. What should be the structure of the set of support systems and the databases that ensure service and operational objectives are met, and that individual systems can be upgraded or new systems introduced without undue delay or cost? It is particularly important that the Telco is able to bring in new features and new services as rapidly as possible into its offerings to the market.

Both the data management and the systems structure referred to above are addressed by the support systems architecture (see more in Reference 4). Unsurprisingly, given the importance of the support systems estate in terms of the *currex* and *capex* drain on the Telcos, and the crucial role it plays in the delivery of customer service (and satisfaction) for existing and soon-to-be-launched services, there have been several major attempts at devising the optimum architecture.

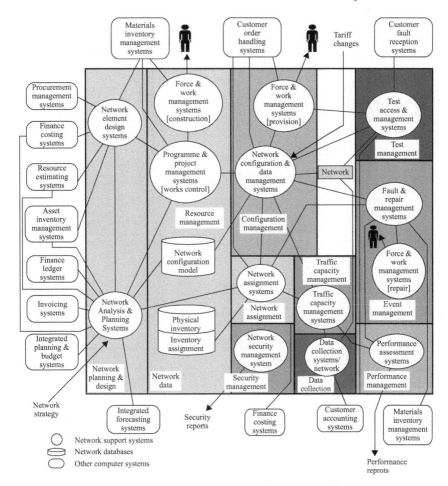

Figure 9.15   View of network control L support systems and their linkages
            [Reproduced with permission from K.E. Ward]

An indicative top-level architecture for the estate of support systems for a Telco is shown in Figure 9.16. The positioning of the systems in this architecture reflects not only the logical groupings of the various functions but also the need for data flows between them. Thus, the customer profile system is shown at the heart of the architecture to reflect the principle that all data and operational activities can easily be associated with the customer (i.e. 'customer centric'), a feature that is vital for providing good dialogue with the large business customers who may have many services spread across many sites. The various channels of communication to the customers are shown at the top, feeding into the CRM systems. These split into those dealing with volume services (relatively simple and common) for the majority of customers, and those dealing with the business customers who have complex extensive service (some might be bespoke) who warrant dedicated

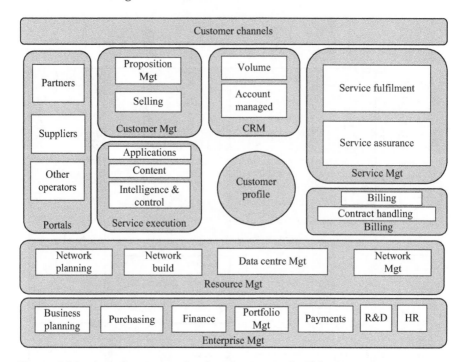

Figure 9.16   An indicative top-level architecture of a Telco's operations support systems

account managers in the Telco sales team. The propositional management systems provide support for the sales force in constructing offerings and bids for providing service to the large customers.

Figure 9.16 also shows the portals on the left cover the interactions between the Telco and their partners, other network operators (for interconnection), and equipment and service suppliers. Service management systems are grouped architecturally into those associated with providing service against customer orders ('fulfilment'), and those grouped around maintaining the service at the required QoS level ('Assurance'), and billing. In this architecture, the efficient use of resources is helped by grouping the support systems for network planning with those for network build and the real-time network (and element) management. The management of the data centres is also in this group to help optimise the planning and running of the Telco's infrastructural resources. The enterprise or business management activities are shown underpinning the architecture.

9.7.2   Models

Earlier in this Chapter, we introduced the idea of the TM Forum's simple layered model that is used by the telecommunications industry to describe the range of operations undertaken. The TM forum has introduced two important models which capture the operations centred on the three basic customer service deliverables of *fulfilment, assurance* and *billing*. We will briefly look at these two models – FAB and eTOM.

9.7.2.1 The FAB model

The FAB model takes the three customer service deliverables and decomposes the operations involved by mapping these against three sets of processes:

- customer-care processes
- service development and operations processes
- network and systems management processes

The model, although simple, does give a view of the range of activities involved in providing service to customers. This approach proves to be useful because a set of internationally agreed definitions have been proved for all the processes in the model. Thus, if a manufacturer offers a FAB-compliant order-handling system for sale, a Telco will know exactly what functions and interfaces the system will provide, and how it will interwork with other support systems [15, 17].

9.7.2.2 eTOM

The extended telecommunications operations map (eTOM) has been developed by the TMF to overcome the limitations of the simple FAB model. It covers all the functions of a Telco involved in managing the business, building and running the network, and providing customer services. It is essentially an elaboration of the FAB model, providing a set of agreed definitions of functions for all the planning and operational activities for all the layers of the TM Forum model – business (or Enterprise) management, service management, network control management and element management. The strength of the eTOM model is that for each of the functions identified there are internationally agreed detailed definitions of the characteristics, processes and interfaces [15, 17].

9.7.2.3 ITIL

In addition to using eTOM to guide the how operations of the Telco business should be undertaken, many providers of ICT services are also adopting the principles of ITIL. The Information Technology Infrastructure Library (ITIL) presents guidance on the best practice in the provision and management of IT services. In particular the library covers: service support, service delivery, ICT infrastructure management, security management and application management. For those Telcos who are in the process of changing to become providers of networked IT (i.e. ICT) as well as telecommunications the specific guidance of ITIL is clearly relevant. However, implementation of the concepts presents major challenges to the Telcos due to the changes in the processes followed and how service quality is measured. More information on ITIL can be found in Reference 18.

9.7.3 Processes

Any consideration of the operations of a Telco would not be complete without looking at the importance of processes as a driver for both costs and quality of customer service. By processes we mean the way the things are done, the sequence of actions taken and how these relate to the actions of others in the organisation. Many Telcos undertake detailed analysis of their existing processes – e.g. the provision of a new customer's line – so as to capture the current way of doing things. This analysis includes detailed flowcharts, with information flow to other functions. Ideally, the flowcharts are independent of the current organisation within the Telco, even though many of the peculiarities of a process are often due to the idiosyncrasies of the organisation! When all of the processes have been captured and placed in a hierarchical structure, the result is referred to as the 'current operational model' or sometimes the 'current mode of operation' for the company. This provides a clear *before-picture* for the Telco, against which it can compare its plans for improvement. Mapping processes is in practice a hugely complex task, which requires processes to be identified at different levels of detail – from the most specific level of working procedure (e.g. the sequence in which to use a test system) to the top-level processes of the company (e.g. 'provide a service').

Following one particular process from beginning to end results in the definition of a 'work string'. As with the processes, work strings can cross organisational boundaries and weave in and out of support systems across the company. By examining the set of work strings comprising a process, areas of improvement can be identified. Examples of problems that can be easily identified at the work string level include:

- Replication of tasks, often occurring in different parts of the Telco's organisation.
- Replication of the storage of particular information in many different databases.
- No clear method of synchronising updates between data bases.
- Circuitous sequencing of tasks.
- Unnecessarily long sequence of actions for what should be a relatively simple task.
- Insufficient information passed along the process, requiring the same information to be re-accessed several times.
- Excessive hand-offs from one group to another resulting in too many people getting involved in one overall work string, with the consequent probability of error and delay.

The Telco is well advised to investigate any areas for improvement highlighted by the work-string analysis, since this will invariably produce cost savings. For example, just improving the process for installing a new customer's line by 1% will produce significant absolute cost savings because of the many thousands of times per day that this action is executed across the country. Invariably, the process improvements are also appreciated by the customers, particularly if there is a resulting reduction in provisioning or repair times. The Telco can then either offer better provisioning times to the customers – e.g. a new line within 48 hours rather than 72 hours – or keep the existing lead times and deploy its work force more efficiently, using the time saved for other activities.

It is generally recognised that before any existing manual operation is computerised – i.e. transferred to a computer-based imitation of the manual process – it should first be studied to see if it can be improved. Many IT projects have floundered because they have been designed on the basis of just replicating basically bad manual processes.

The need for radical redesign of processes is particularly important for a Telco considering the implementation of a new technology into the network, or even an NGN. The full advantage of the CapEx savings of an NGN, as described in Chapter 6, will not be matched in the current account spent unless there is a correspondingly radical review of processes in order to exploit the full potential of the new platform. Moreover, this may necessitate a radical overhaul of the operational support systems as well.

The redesign of a company's processes is often referred to as 'process re-engineering'. Undertaking such an activity can be a complex task, involving the consultation with, and the input from, many parts of the company. Therefore, the re-engineering is normally managed as a formal project or programme (including many projects), as described in the appendix. The objective of the project is to create a new 'operational model' that offers clear improvements (in terms of costs, timescales for service delivery, or functionality) over the current operational model, or it maximises the gains for the company provided by the introduction of new technology.

9.8 Summary

In this Chapter we have addressed the wide ranging subject of a Telco's operations – i.e. all that is involved in running a day-to-day telecommunications business. We concentrated on the customer service and network management support systems. A number of simple architectures and models were introduced to help analyse the subject. Finally, the generally least glamorous part of operations, namely processes, was addressed and we recognised their contribution to the costs and quality of service offered to the customers. As a concluding thought for this chapter, we note that while it is invariably glamorous new shiny technology that captures people's interest – even those working in the Telco – it is the prosaic, but vital, areas of support systems, organisation and processes, that usually determine the success of a new network or the launch of a new service.

References

[1] Hain, J. 'The Ionica story and its relevance to day' *Knowledge Transfer Network*, www.cambridgewireless.co.uk/.../Standards%20and%20the%20New %20...

[2] ITU-T Recommendation E800.

[3] Oodan, A., Ward, K., Savolaine, C., Daneshmand, M. & Hoarth, P. *Telecommunications Quality of Service management*, IEE Telecommunications Series 48, 2003, Stevenage, UK, Chapter 4.

[4] Valdar, A. *Understanding Telecommunications Networks*, IET Telecommunications Series 52, 2006, Stevenage, UK, Chapter 11.

[5] ITU-T Recommendation G821.

[6] ITU-T Recommendation Y1540.

[7] ITU-T Recommendation G1080.

[8] ITU-T Recommendation G107.

[9] Lemke, F., Clark, M. & Wilson, H. 'What makes a Great Customer Experience?' Report by the Henley Centre for Customer Management, University of Reading, UK.

[10] BT Group plc. 'Annual Report and Form 20-F 2014'.

[11] Valdar, A. *Understanding Telecommunications Networks*, IET Telecommunications Series 52, 2006, Stevenage, UK, Chapter 7.

[12] Oodan, A., Ward, K., Savolaine, C., Daneshmand, M. & Hoarth, P. *Telecommunications Quality of Service management*, IEE Telecommunications Series 48, 2003, Stevenage, UK, Chapter 23.

[13] Whitehead, M.J. & Williams, P.M. 'Adaptive Network Overload Controls', *Telecommunications Performance Engineering* (edited by Ackerley, R.), BT Communications Technology Series 7, 2003, The IEE, Stevenage, UK, Chapter 7.

[14] TM Forum: www.tmforum.org/businessprocess framework/.

[15] Valdar, A. *Understanding Telecommunications Networks*, IET Telecommunications Series 52, 2006, Stevenage, UK, Chapter 5.

[16] 'ITP Does Data', Telcoms Professional, The Institute of Telecommunications Professionals, January–March 2015, pp. 10–11.

[17] Valdar, A. *Understanding Telecommunications Networks*, IET Telecommunications Series 52, 2006, Stevenage, UK, Chapter 11.

[18] ITIL: www.itil-officialsite.com/AboutITIL/WhatisITIL.asp.

Chapter 10

People and organisational development

10.1 Introduction

The old adage 'Our people are the company's greatest asset' may be a cliché but few managers today would disagree with the sentiment, although they might question the level of the superlative used. Most of us would recognise that the quality and behaviour of a company's employees and the way that their efforts are deployed affect the way that the company's products and services are developed and delivered, the levels of sales achieved, and the resulting customer satisfaction. However, there is also evidence that good management of people will have a direct positive effect on the financial performance of the company [1]. To a certain extent the importance of people management to a commercial company is reflected by the widespread adoption of the term 'human resource', or more usually just 'HR', to replace the old term of 'personnel' (a name much loved by the military services). Unfortunately, some of us may feel that the very term 'resource' in the context of people management leads to a certain de-humanisation of the process – employees becoming just numbers, representing costs or indications of the size (and hence importance, possibly) of a department. Perhaps to reflect such concerns the term 'human capital' is also used by many companies.

Another view is that when companies fail, e.g. due to misjudging the market, poor performance or even illegal activity, the competence of the senior management needs to be questioned. Did they – i.e. directors and general managers – have the necessary skills and experience to drive the company? Thus, it is not just the quality of the staff and how they are managed that matters, but of equal importance is the calibre of the senior managers in the company.

In this chapter we consider how a company's operations should be divided into suitable work areas, roles and necessary capabilities of managers and staff defined, how people are recruited, trained, managed, and rewarded. Clearly, this is a vast subject which we are only able to address through the key principles involved; much more detail is available in the references cited. Also, we look at this fascinating subject from the point of view of a company in the telecommunications business, that is to say: a network operator or service provider – typically having a large field force of technicians, customer service and sales people, some covering vast service areas, a central set of highly skilled middle and senior managers in headquarters functions, perhaps also with a separate R&D facility employing research and scientific teams.

We begin by looking at how peoples' behaviour and needs have been identified, and how the various types of personality are categorised with a view to forming work groups. This initial consideration of people and their skills then addresses the qualities required in leaders and managers. Next the subject of organisation design is described, again with particular reference to the telecommunications industry. This is followed by an overview of human resources management (HRM) today. We bring the subject to a close by considering the people and organisational aspects of an ICT company introducing large-scale change, resulting from M&A activity or the transformation of its network, resources and services – a common event for most companies engaged in this industry.

10.2   People at work

10.2.1   People's motivation

For any company in business there is a primary need to understand human behaviour – whether it is in order to gauge potential customer needs and expectations, or being able to manage a workforce that is motivated and efficient. There have been many studies into why humans behave the way they do, and how best to cope with or exploit their characteristics. In 1960 Abraham Maslow, who was investigating human motivation, developed his now famous model of human needs [1–3]. The premise of the model is that humans strive to satisfy a set of needs, which start with the basic physical needs of food, drink, warmth and shelter. Once these are satisfied humans strive for social needs, such as sense of belonging and relationships with others. Finally, if the physical and social needs are met people seek the personal needs of self-respect and recognition, followed by self-fulfilment and a sense of accomplishment. Figure 10.1 shows the usual representation of Maslow's hierarchy of needs.

Although Maslow's model was originally directed at scoping customer needs and motivations, it has proved to be equally applicable to the understanding of employees' behaviour. Ten Have et al. [3] have identified two groups of needs that an employer should address: those that bind a person to the company (preventing staff turnover) and those needs which motivate and define the necessary skills of the staff. These theories on human behaviour have over the last 50 years or so formed the general approach to the management of staff within a company. It is interesting to note that Maslow's hierarchy of needs has also recently been applied by Fife and Periera, of the University of Southern California, to understanding how people adapt to the demands and opportunities of 'digital life' [4].

Naturally, people within a population vary and there are important cultural differences around the world in the relative importance and priority of the various needs. For example, in Asia people tend to give greater importance to the social

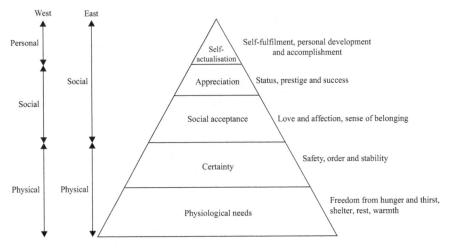

Figure 10.1   Maslow's hierarchy of needs

needs rather than personal needs, as shown in Figure 10.1. Awareness of the difference in the priorities of needs is becoming increasingly important in Europe and N. America due to the multi-national nature of today's workforces. Obviously, the cultural differences will be more pronounced when a telecoms company is addressing a market in a foreign country, or planning to set up operations (e.g. sales, customer service and engineering) there. A systematic study of cultural differences around the World was started in 1966 by Geert Hofstede, who after a number of extensive surveys over the following 15 years developed the recognised authorative model and measured values covering 66 countries [3, 5]. Hofstede identified five dimensions that characterise a national's culture, each measured by a numerical index, as follows:

- Power distance (PDI).
- Degree of individualism (IDV).
- Masculinity versus femininity (MAS).
- Uncertainty avoidance (UAI).
- Long-term orientation, which is also known as Confucian dynamism (LTO).

A description of each of these dimensions and examples of national values is given in Table 10.1. Hofstede's set of dimensions are generally recognised as giving a broad view on the nature of the cultures in different countries. They form a useful starting point for more detailed analysis of a foreign market – which is particularly important for a global industry like ICT and telecommunications. Also, taken with Maslow's hierarchy of needs model, the Hofstede cultural indices form a sound basis for understanding people's motivation and behaviour, the development of people-management practices, and organisation design.

Table 10.1   Hofstede's cultural dimensions [3, 6]

Dimension of index	Description	Dimension characteristics	
		High value	Low value
Power distance (PDI)	The PDI measures the acceptability of uneven distribution of power among individuals.	People accept the large gap between them and those in power (or with wealth), e.g. Malaysia has the highest PDI of 104, while the USA is around the middle with PDI = 40.	There is small gap between the rich and the poor and a strong belief in equality, e.g. Austria has the lowest PDI at 11, the UK and Germany both have 35 with Denmark at 18.
Degree of Individualism (IDV)	The IDV measures how people depend on groups (e.g. extended families with support) rather than act individually.	People are individualistic, ties between them loose, everyone is expected to look after themselves, e.g. USA = highest at 91, with UK with 89.	People depend on families or groups for support. E.g. India with 48, Indonesia and Pakistan with 14, with Guatemala the lowest at IDV = 6.
Masculinity versus femininity (MAS)	The MAS measures the degree of assertiveness and the acquisition of goods versus the concern for relationships and quality of life.	People show the 'masculine' characteristics of being competitive and assertive, highly acquisitive, e.g. Japan is highest with MAS of 95, UK with 66 and USA with 62.	The feminine end of the MAS dimension is modest and caring. Denmark, Netherlands, Norway and Sweden have the lowest MAS with 16, 14, 8 & 5, respectively.
Uncertainty avoidance (UAI)	The UAI measures the acceptability of living and working with uncertainty.	Dislike of dealing with uncertainty, and wants everything to be planned and controlled, e.g. Greece has the highest UAI of 112, while Germany has 65.	Comfortable in managing uncertain situations, e.g. the USA has a UAI of 46 and the UK has 35. Singapore has the lowest value with 8.
Long-Term orientation (LTO) (aka: Confucian Dynamism)	This dimension reflects the extent that the society takes a long-term rather than historic or short term point of view.	People are persistent and ordering relationships by status. They are open to new ideas, e.g. China has a LTO of 118, Hong Kong has 98 and India has 61.	People are value stability and traditions, saving face is important, e.g. West Africa (Ghana, Nigeria & Sierra Leone) have the lowest LTO of 16. UKSA and the USA have LTO of 25 and 29, respectively.

10.2.2   Teams

The idea of creating teams of people to tackle a work task has, of course, been a favoured approach for thousands of years – a classical example being the groupings of fighting men into cohorts, squadrons, platoons, etc. The team in this case is being assembled in order to get a mass of men to tackle the task of fighting. Such teams need good leaders, a common objective, and appropriate resources for the job. A particular feature of military teams is the sense of loyalty and 'team spirit' that is deliberately engendered in order to manage morale. Much is the same with the use of teams within ICT and telecommunications companies today. For example, the task of providing provision and repair to the service area of a Telco (as discussed in Chapter 9) is split into manageable chunks so that sufficient technician effort can be deployed where and when required. Again, appropriate resources are necessary, as is the management of the team against clear sets of responsibilities. We will be looking at how the latter is defined in section 10.3. The idea of morale within the teams is also applicable to the management of people in business – indeed, fostering loyalty and shared objectives across the workforce is often a high priority for senior managers anxious to meet their financial performance targets for the company.

One of the key aspects of forming teams within a business is its size. In the case of a Telco deploying technicians across a service area the main objective is the ability to be able to dispatch a suitable person to anywhere within the service area within a target time – this is best met by having a geographical spread of several small teams, all with the same roles and resources. Such a team model can be easily expanded to cover a larger service area or to cope with extra volume of work, as required. However, in the case where the team is established to tackle a single task, say within the company's head quarters (HQ), either as a permanent part of the organisation or as a special temporary task force, other factors come into play. On the one hand, the company needs sufficient people in the team to bring the right level of expertise and ideas. But large teams of people can suffer from the informal formation of 'sub groups', which can cause tensions and loss of efficiency – particularly if there is poor communication between them. On the other hand, having too small a team can lead to 'tunnel thinking' and lack of innovation. In all cases there is the need for good management, including the application of appropriate (and demonstrably fair) reward and recognition for work done by individuals within the team, as we discuss in section 10.4.

There are clearly many aspects to team management and there has been a great deal of research into techniques for business companies to follow [3, 7, 8]. One of the best known concepts for determining the necessary composition of a team was produced by Belbin in 1985 following his study into how people tackled business games. He concluded that successful teams are made up of a set of nine roles, which fall roughly into three areas – action-orientated, people-orientated, and cerebral – as shown in Figure 10.2. The nine roles are briefly summarised in Box 10.1 [3, 7].

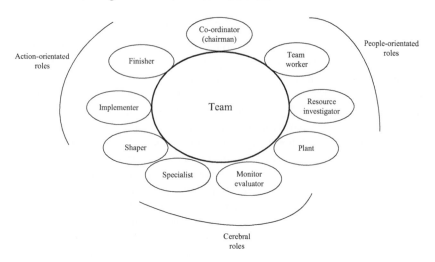

Figure 10.2   The Belbin roles within a team

Box 10.1   The set of Belbin roles within a team

The co-ordinator (also known as 'chairman') role requires a calm, mature and self-confident person who can take the chairman or leadership role within the group. They will be good at delegation, setting goals for others to achieve, and providing clarification of the task when necessary.

The team worker role relates to a person who is hardworking, co-operative and dutiful, who gets on well with the rest of the team.

The resource investigator is an outgoing, energetic person who makes contact with many others and explores new ideas. They quickly respond to new situations, but tend to have a short attention span.

The shaper is highly strung, outgoing and dynamic, apparently thriving on pressure. They may upset the conservative elements in the team as they drive ahead with a new idea.

The plant is the creator and inventor of the team, who tends to be intellectual, imaginative and at times genius. Unfortunately, their unworldliness can upset the rest of the team.

The specialist role relates to a single-minded individual who has rare knowledge or skills and so tends to be highly focussed, missing the big picture.

The monitor is a person who evaluates what the rest of the team are doing, considers the strategy and offers a judgement. Although they provide a sober view they lack drive and fail to inspire others.

The implementer is the person that efficiently turns ideas into practical actions, closely adhering to the plan without accepting suggested improvements during the project.

The completer finisher has the merits of being consciences and meticulous, anxious to be perfect and to be within the agreed timescale.

The Belbin concept is often used in practice to plan the composition of a new team, or to evaluate the make-up of an existing team within the company, recognising that any group of people will have different strengths and skills. The identification of which role each person fits can be done through the use of a questionnaire, using: self-assessment, peer-group assessment, or by reference to a third-party assessor. An ideal team will comprise individuals who fit the full range of Belbin roles. In practice, of course, the assessment of a person's Belbin role can be ambiguous, and the absence of certain traits can be acceptable, depending on the nature of the team and its task. For example, if a certain role is missing from an existing team, its absence can be partly overcome by asking members of the team to give greater emphasis to that aspect of their behaviour than they might normally do. Alternatively, systems can be used to compensate for a shortfall, e.g. a project management tool. Despite its limitations, the use of the Belbin analysis is a useful tool for organisation design in business companies today.

10.2.3 Leadership

Let's begin with a quote from one of the most famous business leaders, Sir Richard Branson:

> As much as you need a strong personality to build a business from scratch, you also must understand the art of delegation. I have to be good at helping people run the individual businesses, and I have to be willing to step back. The company must be set up so it can continue without me [9].

Of course, one of the essential requirements of a leader being able to delegate is for them to surround themselves with good people – something which Branson strongly advocates. Even a leader as charismatic as he is needs to rely on a strong supporting team beneath him in order to manage his business empire. So, what other attributes make for good leaders?

Not surprisingly, the fascinating subject of what constitutes a good leader and the different types that exist has been the subject of much research and many publications [6, 7, 10–13]. Although there are many lists of attributes and categories of the types of leader, there is a general consensus on the following ten characteristics of good leadership.

Big-picture perspective. One of the most important attribute that distinguishes a leader from others is the ability to take a strategic view, rather than be pre-occupied with the company's current problems. This does not mean that the leader is not able to tackle the details of particular issues, but rather that they are able to 'lift their heads up' to ensure the company is still heading in the right direction.

Tenacity. Whether managing the whole company or one of its business divisions is a tough job, and a leader needs both endurance and tenacity to continue to push through the inevitable difficulties encountered on the way to reaching the business objectives.

Good inter-personal skills. It is vital that a leader is able to gain the confidence of their colleagues and, ideally, earn their respect. Similarly, the leader must be able to relate well to the company's clients, customers, suppliers and the other stakeholders.

Good communicator. Most people would rate the communication skills of a manager as a primary indicator of good leadership. This means that not only does the leader have the ability to make rousing speeches to the employees or customers, but equally importantly they are able talk appropriately to individuals, whether senior or junior to them. The communication needs, of course, to be both way – so the leader should also be a good listener when necessary. Many leaders make a point of visiting all parts of their company, perhaps travelling long distances, in order to see for themselves what is happening on the 'shop floor' and talking to the staff (this is often referred to as 'Walking the job').

Hardworking. Running a big organisation is a massive job, and leaders need to be seen as hard workers prepared to put in long hours in the office or be continually on the company's intranet. They will also be expected to undertake company-related activity outside normal hours, either travelling or meeting stakeholders, as required.

Highly driven and ambitious. Getting to the top of an organisation takes much personal drive and ambition. Leaders usually have to struggle against opponents, lack of opportunity and other impediments to reach their goal. The personal drive of ambitious people is usually valued by others in the belief that a leader will apply such drive also to pushing the company (or their department) forward.

Good understanding of the company business. Whilst most of the attributes of a good leader are generic, there is a real need for such a person to have a commanding specific understanding of what the company does and how it works, as well as how it relates to the business environment of the country (or countries).

Skilled in one or more important areas. Leaders will normally be qualified in finance, accounting, law, engineering, etc., and they will have developed specialist skills during their careers. Good leaders will need to apply their expertise and knowledge to the range of issues facing the company's senior management – and where appropriate, they will need the confidence to overrule the advice from subordinates.

Good delegator. We have already seen that it is important for a leader to delegate as much of the work as possible in order to free up their time for the tasks that only the leader should do. This means that the leader has to establish effective monitoring of delegated work, since the responsibility ultimately resides with the leader.

Prepared to surround themselves with good people. Again, earlier we saw that good leaders should have the confidence to have an excellent calibre of people around them, providing the necessary expertise, without feeling their superiority is threatened in any way.

All good leaders have most, if not all, of these top ten attributes (and, indeed all managers need these qualities to some extent). There are, of course, different types of leader: 'command and control', 'charismatic', 'inspirational', 'innovative' and so on, which really reflects their personality rather than their skill set. The choice of the right type of leader, or leadership style, for a particular company usually depends on circumstances. For example, Sir Winston Churchill's inspirational personality made him the right choice to lead the United Kingdom during the Second World War, but it did not meet the needs of the nation in peace time. Whereas, a start-up company trying to establish itself in the market might benefit from an innovative leader. There is also an increasing awareness of some leaders having a darker side, either in the way that they treat people or their lack of business ethics (which we discussed later in this chapter) [14].

A form of leadership style that is gaining great credibility in the fast-moving and more flexible business world of the 21st century is *servant leadership*. Although it was first formally written about by Robert K. Greenleaf in the 1970s, it has its origins in ancient philosophy of kingship. Indeed Leo-Tzu, a scholar living in China in the Ching dynasty, wrote:

The highest type of ruler is one of whose existence the people are barely aware. Next comes one whom they love and praise. Next comes one whom they fear. Next comes one whom they despise and defy. The sage is self-effacing and scanty of words. When his task is accomplished and things have been completed, the people say 'We ourselves have achieved it!'

The servant-leader style of leadership embodies both a philosophy and set of practical applications and is as far removed from the old world 'command and control' as is possible to get. Whereas 'command and control' is characterised by hierarchy; objectives and accountability; and the exercise of power, servant leadership relies on participation, joint decision-making and the delegation of authority. It is accompanied by a desire for every employee to contribute to their maximum capability and a strong regard for personal development. The characteristics of a servant leader are therefore listening and empathy; the ability to describe a future that is good for all; persuasion; a commitment to build and contribute to teams; and a burning desire to develop those around him or her. In many ways these characteristics can be summed up as a coaching style, and often this is the practical approach that companies adopt when they wish to embed the servant leadership in their culture.

In practical terms, servant leadership is shown to improve outcomes. In particular, employees who are engaged and well respected provide excellent customer service and encourage loyalty to the company and its products. In addition, engaged employees generate greater innovation, speed of change and compliance with the company values [15].

10.2.4 The Peter principle

The popular notion that people continue to be promoted within an organisation until they reach their level of incompetence, as espoused in 1969 by Laurence

Peter and known as the 'Peter Principle' [16], is generally accepted as being true even today. This phenomenon occurs because there is an understandable tendency for organisations (particularly those with large workforces) to promote their best workers to the role of supervisor, since such people clearly have good knowledge of the intricacies of the job and will receive respect from their work colleagues. However, once the supervisors have successfully settled to their new role they are considered for even further promotion, and so on until they have reached a level beyond their competency or comfort – at which they stay. Managers can avoid the trap of the Peter principle by ensuring that they judge applicants by suitability for jobs at the higher level, rather than just considering their best performers at the current level. This approach may, of course, create resentment with the high performers who feel overlooked. A more enlightened approach, which is being followed by high-tech industries, like ICT and tele-communications, is to recognise people who perform well because of their technical expertise by appointed them as 'senior experts'. In this way they can carry on their good work (now with higher status), while not being encumbered by the additional management loads that would normally be associated with promotion up the company hierarchy.

A further factor highlighted by the Peter principle is that the nature of the job differs at the various levels of management, and each step requires an appropriate degree of leadership. Clearly, the managers at the higher levels need to be capable of overseeing bigger budgets and staff numbers, as well as having increasing responsibility for the company achieving its business objectives. It is helpful to consider how the attention of mangers at the different levels is directed when characterising the nature of the leadership required. The principle is simply illu-strated in Figure 10.3. This shows that the supervisor directs nearly all of his/her attention to the day-to-day oversight of the workforce (a). We noted earlier that supervisors tend to be recruited from the workforce, so there is a danger that they can remain one of the 'lads/girls' – resisting such a temptation requires good lea-dership skills. They are, however, normally still considered to be part of the workforce in terms of pay and conditions.

It is the supervisors' bosses who are usually considered to be the first level of a company's management. As Figure 10.3(b) shows, the lower and middle level of management split their attention between the subordinates and their line managers. Recruitment to these management levels tends to be a mix of graduate entrants and some within-company promotions.

It is at the higher levels of management that the job requires leaders with some skill in office politics and power play. Thus Figure 10.3(c) shows the attention of the managers splits essentially in three ways: up towards their bosses, down to their subordinates and sideways to their peer-group colleagues. At these levels the leadership political skills of the manager may influence their domain of control within the company as much as their technical or managerial skills. Recruitment to these management levels is predominantly through internal promotion. However, there are usually some strong personalities brought in from other parts of the

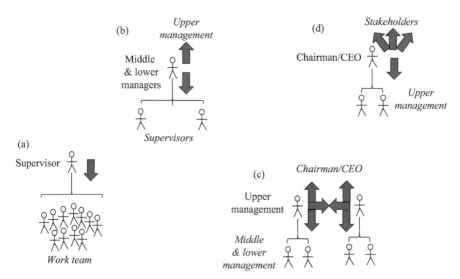

Figure 10.3 The attention directions at different levels of the management hierarchy

industry to keep the upper management on their toes, and to refresh the leadership pool within the company!

Finally, at the top of the company in the roles of chairman or CEO, leaders are required who can not only direct the company through the board and executive control, but also manage all the relationships with the company's stakeholders, as shown in Figure 10.3(d). We look at the governance structure of a company in Chapter 4 and the typical range of stakeholders for an ICT company in Chapter 1.

10.3 Organisation design

The way that a company organises its staff and managers will have a profound effect on the efficiency of its operations and business success. Bad organisation structure will result in poor communications across the company, inefficient use of its technical resources, unnecessary political in-fighting amongst its senior managers, and invariably poor customer service. In this section we look at some of the leading generic theoretical analysis of organisation design and then consider the practical aspects of organising an ICT service company.

10.3.1 Theoretical approaches

In 1981 Henry Mintzberg published his famous work on how the structure of organisations may be a key to their business success [17]. He concluded that organisational

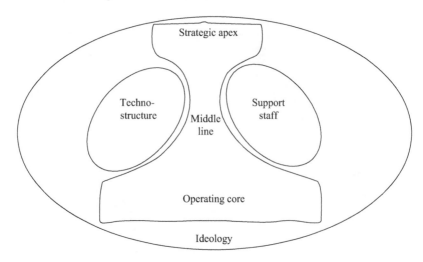

Figure 10.4   Mintzberg's organisational model [17]

structures fall into five natural configurations of a basic set of components. (This was extended in 1990 to six configurations [3].) His model is illustrated in Figure 10.4, which comprises six components of an organisation:

(i)   Operating core: the work force manufacturing or delivering the company's service
(ii)   Strategic apex: the senior management of the company
(iii) Middle line: the company's middle management
(iv)   Techno-structure: this includes strategic planning, training, production scheduling, controlling and technical administration
(v)   Support staff: this includes R&D, human resource management, legal, public relations, marketing, financial management
(vi)   Ideology: company culture covering the norms and values

Of course, the external parties – particularly customers and suppliers – also influence the organisation's effectiveness.

The application of the model to organisation design revolves around identifying which form of co-ordination mechanism is appropriate for the structure adopted and how the power is to be divided within the company. Mintzberg identifies six co-ordinating and control mechanisms, and six ways in which power is disseminated within an organisation. An appropriate structure for a business organisation depends on its role within the industry, the size and maturity of the company, etc. Mintzberg's proposal was that the organisation design should be an appropriate match of the six functional components (i to vi above, and in Figure 10.4) to the co-ordinating and control mechanism, and the way that power is disseminated. The types of organisational configurations identified by Mintzberg, their coordinating mechanisms and the methods of power dissemination are shown in Figure 10.5 and Table 10.2, respectively.

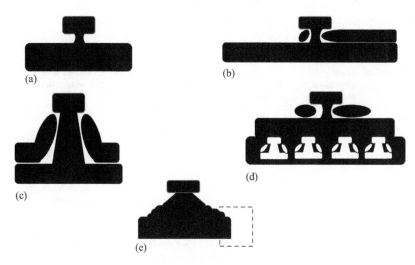

*Figure 10.5   Mintzberg's five organisational configurations [17]: (a) simple
structure; (b) professional bureaucracy; (c) machine bureaucracy;
(d) divisionalised form and (e) adhocracy*

An example of a professional organisation (shown as Figure 10.5[b]) might be
a consulting firm, in which the operating core with the body of highly qualified
people is key, whilst the know-how and client relationships they manage represent
the main asset of the company. There is strong vertical and horizontal dissemination of information and control, and the operational skills are standardised. Again,
taking the start-up ICT company with just a few employees the simple Structure
(shown as Figure 10.5[a]) applies – most of the functions of a large established
company are missing, and control is primarily via flat direct supervision.

In practice many companies may be hybrids of the basic structures identified
by Mintzberg, but the model is still considered to be useful in helping designers
understand the relationship between an organisation's structure, and its communication and power mechanisms [3, 18].

10.3.2   Organisation life cycle

Let's now look at some of the practical aspects of organisation design. The major
driver for setting up an organisation is to cope with scale and management of
expertise. Thus, a small ICT start-up company can operate efficiently as a simple
structure, since the expertise is spread among the few employees and the entrepreneurial boss can directly control all the work internally, as well as deal with
external contacts. However, as the volume of work activity increases more people
need to be taken on. Once the point is reached where there are too many employees
for the boss to manage, a supervisor needs to be brought in. Also, if more technical
know-how is required the organisation has to cope with a bigger body of experts.
Increasing size and expertise introduces the need for extra levels of supervision
and management. Normally, organisations establish a management hierarchy as

Table 10.2   Characteristics of the five Mintzberg organisation configurations [3, 18]

Configuration	Main co-ordinating mechanism	Key part of the organisation	Types of power and control	Examples and characteristics
Simple structure (a)	Direct supervision	Strategic apex	Highly centralised vertical & horizontal	– Typically start-ups & VSME. – Simple, informal with few staff & no middle managers. – Agile, but vulnerable to hostile business environment.
Professional bureaucracy (b)	Standardisation of skills	Operating core	Horizontal decentralisation (Strategic Apex shares power with operating Core)	– Typically consulting, accounting, legal firms. – Bureaucratic but decentralised teams of professionals. – Danger of professional teams forming silos.
Machine bureaucracy (c)	Standardisation of work.	Techno-structure	Limited horizontal decentralisation (Strategic Apex shares power with Techno-structure that standardises the work practice)	– Typically utilities, factories, government. – Centralised bureaucracy control through formal procedures. – Reliable & consistent, but can be preoccupied with keeping status quo.
Divisional form (d)	Standardisation of outputs	Middle management	Limited vertical decentralisation (power delegated to middle management and business divisions)	– Typically mature large conglomerates, having a number of semi-autonomous divisions separated on a market basis. – Heavily relies on effective reporting mechanisms between divisions and centre
Adhocracy (e)	Mutual adjustment	Support staff (with operating core)	Selected decentralisation (power is dispersed at various levels)	– Typically high-tech (e.g. aero industry). – Work as a set of interacting project teams. – It suffers from inefficiency due to much liaising between teams

they grow – usually following a pyramid structure, with decreasing numbers of managers at each higher level.

Organisations tend to follow a life cycle, analogous to the product life cycle described in Chapter 8. The organisation develops as the company, division or product line grows until a mature state is achieved, with the organisation operating at maximum efficiency in terms of marshalling expertise, internal communication and power flows. However, with changes of market or technology, or disruptions within the company, the organisation ceases to be optimum and it moves into the decline phase of the life cycle [1]. It is eventually replaced by another organisation design, with changes of people, reporting arrangements and responsibilities.

As indicated earlier, one of the factors driving the need for a management hierarchy is to achieve the optimum span of control for supervisors and managers. In the case of a Telco's supervisor overseeing a group of technicians in the field, all of whom are undertaking similar tasks, the main criterion is achieving a reasonable load for the supervisor. On the one hand, the company saves on numbers of super-visors required with large spans. On the other hand, small spans give the employees more time and attention from the supervisor, and a larger number of supervisors offer more promotion opportunities for technicians. Spans of between 10 and 20 are usual for supervisors of field technicians in most service industries. However, in middle management and above, particularly in the headquarters (HQ) departments, the staff are usually highly qualified experts (e.g. accountants, engineers, marketing, and finance managers) who tend to work as individuals or in small groups, so needing greater management attention. Spans of control in the HQ situation tend to be five or fewer. Having these much smaller spans in the management structure of HQ can lead to 'turf wars' between the middle and senior management if there is not a clear demarcation of roles established during the organisation design.

Figure 10.6 illustrates the range of hierarchical structures in ICT service providers or manufacturing organisations.

(a) The simple structure has already been discussed earlier.
(b) The generic line and staff topology applies throughout a hierarchical organi-sational structure, with appropriate spans of control and co-ordination. How-ever, the main organisational categories are on the right hand side of Figure 10.6, as described further.
(c) Functional. In this category there is a set of domains responsible for each of the main functional responsibilities in the company. Each of these domains, usually referred to as 'divisions', has their own separate management hierarchy and tends to be self-contained and may become isolated to some extent from the rest of the company. With this structure successful company operations and business rely on there being strong horizontal lines of communication to enable the specific expertise from each division to be made available as required, e.g. to products.
(d) Divisional. This structure takes an orthogonal approach to that of the func-tional structure by including the necessary functional resources within each product (or product group). Such an approach avoids the contention for resources typical with the functional structure, and gives a sense of belonging to the various functional people within the product group. It does, of course, mean that the product groups are multi-disciplinary (e.g. engi-neering, finance, marketing, etc.), which can be a challenging management

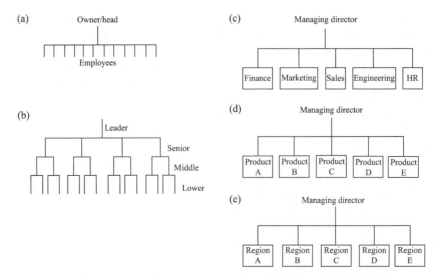

Figure 10.6   The different types of organisational structures [18]:
(a) simple; (b) line and staff; (c) functional; (d) divisional
and (e) geographical

task for the product director. Also, small groups of functional people moved into the product group can feel isolated from their functional colleagues in the centre of the company.

(e)   Geographical. In the case of service companies, with areas covering most of the country, or even many countries, the organisational split by geography is suitable. In this arrangement each of the dependent management domains is clearly defined by its geographical coverage, so there is no ambiguity of responsibility within the organisation. Service companies, such as Telcos, usually organise their provision and maintenance filed force on a geographical basis. As described in Chapter 9, one of the key issues in designing the service regions is the minimising of travelling time and grouping staff resources at local centres within the regions such that there is a critical mass of skills, and they are able to receive adequate supervision and support.

As with all organisational structures there are pros and cons with each form, and the choice of which to adopt depends on weighing the various factors. In many cases companies will swap between organisation structures every few years – typically geographical and functional configurations – as circumstances change. For example, in the case of a large Telco managing its field force providing and maintaining a network of mature technology the usual structure is geographical, as described earlier. However, when new technology is being introduced to the network the main need is for pooling the scarce technical knowledge and test equipment at a special centre – essentially following a divisional structure.

Once the technology becomes more mature and the necessary skills are more widely held by the field force the organisation reverts to a geographical structure. A further example is given in Box 10.2.

Box 10.2 Examples of product line organisation

Up until the 1980s, each of BT's products had been managed in the HQ by a small team of marketing people who reported on the product's financial performance and undertook the specification of any product upgrades, and led on producing any necessary business cases. The organisational structure was therefore functional, as in Figure 10.6(c). Then BT took the bold step of establishing a set of product lines within the HQ, as in Figure 10.6(d). Each product line contained the product management team (essentially marketing activity), as well as a small finance group and a team of engineers to oversee the planning, provisioning and maintenance procedures for the dedicated equipment used in the network to support that product. Examples of the product lines include: ISDN, telex, payphones, and the biggest of all: private circuits. The latter comprised about 100 people and included a team of 50 or so engineers responsible for planning the deployment of the special multi-plexors, digital cross-connection units and test equipment, as well as the provisioning and maintenance procedures.

Once the product line organisational structure was established it seemed to be successful, insofar as it ensured sufficient focus and resource was given to the various products. However, this very success did create tensions between the mangers in the product lines and the remainder of BT's HQ, which was still essentially a functional organisation. After about 5 years there was a major re-organisation in the HQ which, among other objectives, recreated a fully functional structure leaving just a vestige of the product lines and all functional groups returned to their original divisions.

10.3.3 Matrix management

It is now usual for high-tech industries, particularly ICT and telecommunications, to organise the groups of professionals in a divisional structure. Typically each division, section or group will exclusively have a particular expertise and role. However, the many projects and programmes being implemented within the company will either occasionally or continuously need the skills of the expert divisions. Rather than disrupt the organisational structure by re-parenting the required experts into the project teams, most companies adopt the principle of matrix working. This is a formalised method of lending the experts to project teams, such that the project manager can control their contribution while the functional-division line management retains ownership of the expert. Examples of establishing matrix-managed projects within a Telco include the launch of new network products, the development of new processes and the creation of a company-agreed response to major inquiries from the national

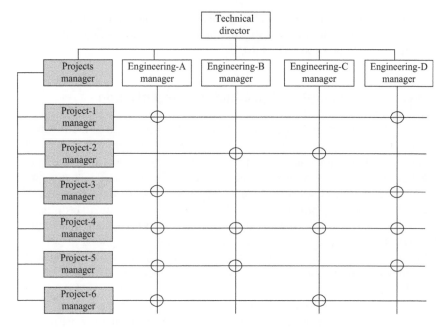

Figure 10.7   Matrix management [19]

regulator. The concept is illustrated in Figure 10.7, in which experts from the four functional engineering divisions A–D are shared across the six projects [19].

In this illustration all staff are under the same technical director, which makes overall resource management easier since just the one director's budget is being allocated. Of course, this may not always be the case and there may be tensions within the company over the allocation of resources to the various projects. The issue becomes particularly acute when projects become extended due to bad planning, poor project management, or unexpected problems, causing contention for the expert resources which have been schedule for deployment elsewhere. There are also management issues within the expert divisions if the line manager does not have good visibility of his people, potentially making reward, recognition, and career development difficult.

10.3.4   Culture and ethics

Organisations are not only defined by their structures and the quality of their management and workforce, but also by their culture and ethics. Culture, which may be defined as 'The way things are done here' [18], influences the way that people behave at work and consequently the way that the company behaves and interacts with the outside world – affecting dealings with customers and suppliers, particularly. However, the ethics followed by a company dictate how they interact with their community, and specifically how the obligations to their stakeholders are met [20]. We will briefly consider these two attributes of an organisation.

The culture of an organisation develops out of a number of influences. The biggest driver being the historical approach taken within the type of industry – e.g., the UK Civil Service has a long tradition of a formal cultural behaviour, grading structure and communication protocols which are followed by the many ministries and departments of that organisation. In fact, one of the biggest challenges facing the incumbent Telcos around the World when they change from civil-service based government organisations to commercial companies, facing competition and the move to shareholder ownership, is how to change the historically imbedded culture. As competition develops, and their dominance of the market declines, Telcos may need to change their culture again – from engineering-led network providers to customer-focused service providers.

Other influences on the culture adopted include:

- The size and age of the organisation – with the small start-up companies adopting radically different cultures to those followed by well-established and larger organisations.
- The location of the offices, operational buildings and factories, etc. – in terms of geo-type (e.g. urban, suburban, rural) and the region. Thus, companies on the West coast of California will adopt the Silicon-Valley type of culture unlike similar organisations in other regions of the United States.
- The technology used, and in particular the importance of technology to the company's business.
- The position within the company's organisation life cycle – since the culture adopted may change as the organisation develops.

The culture of a company is manifest in many ways, which may be grouped into the following three categories.

(i) *Formality.* Whilst all organisations will be governed by a set of rules and norms, there will be cultural differences in the way that they are applied and they intrude on peoples' behaviour. One of the most striking examples is that of dress codes – the wearing of ties and suits for men and modest smart wear for women is still an important norm in many organisations. In contrast, the new high-tech ICT companies tend to favour overtly casual clothes (jeans, T-shirts, etc.) whilst some companies mix the two approaches by having 'dress-down Fridays'. The other important indicator of formality is the way that people address each other: e.g., everyone by Christian names, or the use of surnames for the higher levels of management.

(ii) *Empowerment.* The degree of decision-making and the trust associated with an individual to take actions are important determinants of the type of culture being followed. Many employees become frustrated by the need for even the simplest of decisions having to be referred up the management hierarchy or taken to a committee of people. Other measures include the degree of financial authority associated with the different management levels, and to what extent individuals can use their initiative to solve problems. The empowerment attributes are dependent on the form of governance within the company and the flexibility in the structures – big companies tend to have rigid control structures, while small new companies tend to be flexible in the way control is managed. For example, one of the key characteristics of the culture of Virgin is the level of empowerment enjoyed by its employees (see Box 10.3).

(iii) *Atmosphere.* This category covers the general feel of the culture for those working in the company – and by extension this may be detected by the customers and suppliers. The attributes of atmosphere include the general buzz around the places of work, whether people seem engaged and satisfied, or disinterested and inactive. A contributor to this is the style of management within the organisation, in particular the tolerance to mistakes while people learn, the level of supervision and recognition of good work. The atmosphere of a company is also much influenced by the relationships between managers, particularly in the context of battles over demarcation of responsibilities, and the visibility of senior managers to the workforce.

Box 10.3   A classic example of employee empowerment

When Virgin Atlantic set out to challenge the incumbent BA they made a point of fully empowering their customer-facing check-in staff. Company folk-law has it that on one occasion an employee, faced with a customer who had to return home immediately to deal with a family crisis, used his/her own credit card to buy a ticket for a rival airline. It is likely that in any other airline he/she would have faced disciplinary action and never seen her money again. In Virgin he/she was immediately reimbursed and congratulated for his/her customer focus, readiness to own the problem and innovation. It doesn't really matter if this story is true; it's the telling and re-telling in Virgin which has created the culture that they wished to embed.

It should be appreciated that there is not one type of culture that is suited to all situations. Big companies with complex operations covering large terrains need a more formal type of culture to match its organisational structure. While small start-up companies tend to develop organically around a small caucus of entrepreneurs following a flexible informal culture. By the same token, we should recognise that not all cultures suit all types of people. Many employees used to working in an established organisation may feel uncomfortable in a casual highly flexible type of culture.

What is important is that the leadership of a company considers the type of culture that they ideally wish to achieve and how to achieve it. One major tool in developing and embedding the right culture is to engage the workforce in thinking about the company vision and mission (see Section 3.7.2) and, in particular, the values those employees should exhibit. Typically company values will include customer focus, team-working, being professional, promoting innovation, etc., but the most important aspects are the engagement of the employees in their creation and being seen to live by them. That is where the value lies. More and more companies are assessing employees, not only on their formal objectives but also on their behaviour and adherence to the company values.

Now we can discuss the question of ethics. All companies can be considered to have a conscience – that is that they chose how to behave within society. Many big companies now appoint a senior manager to oversee the ethics of their business, this role is often referred to as corporate responsibility (CR). The range of ethical considerations covers all of the company's operations, the main categories are listed below [20]:

(a) *Honesty in dealings with outside organisations.* It is a fundamental ethical principle that the company will be honest in all its dealings with the government, the regulator, the financial institutions and creditors, tax offices, and suppliers. In the United Kingdom the Bribery Act of 2010 lays down strict rules concerning business dealings [21].

(b) *Products and services.* It is also fundamental to the company's ethics that they treat customers with respect, and deliver the products and services in a fair manner. In addition, the needs of disabled people should also be considered, and where appropriate modified versions of products and services should be provided.

(c) *Green operations.* All businesses have an obligation to meet the government targets for reducing carbon usage and emissions of green house gasses. This represents a big challenge for the ICT industry, given the dependence on electric power to operate the services. There is also a large challenge in meeting the targets during the manufacturing of the devices and equipment used in the ICT business. Many Telcos and ISPs are investigating ways to reduce the power consumption of the telecommunications equipment, and the computer and storage energy loads in the data centres [22].

(d) *Community support.* Companies often make efforts to support the local community through activities such as donations to charities, donations of old computer equipment to schools and voluntary organisations.

10.3.5 Outsourcing

The outsourcing of activities not deemed to be part of a company's core business has become a widely adopted means of reducing costs. This results in peripheral functions, such as catering and cleaning in the company's premises, being provided by external agencies rather than a support group within the organisation. Outsourcing not only offers cost-saving advantages but it also frees up management attention within the company. In addition, it offers the chance of shifting the nature of the company's cost base cost from one which incurs capital expenditure to one which is entirely current account. Thus, many companies outsource functions such as staff and manager training, so avoiding the cost and distraction of building and maintaining training premises. The outsourcing may also cover such internal functions as the routine human resource management activities, such as salary payments.

However, poorly developed outsourcing has, on occasion, destroyed shareholder value and caused some controversy for customers. For example, the widespread use of overseas-based call centres had generated complaints about having to deal with operators who are too remote. The prevailing view is that the advantages of out-sourcing can be achieved without antagonising customers, provided the company retains strategically important functions and pro-actively manages the outsourced activity to ensure standards and quality of service are maintained [18, 23]. As we discuss in Chapters 3 and 4, ICT service providers follow a range of models covering the provision of services through, what is in effect, the outsourcing of network and support systems, becoming virtual operators.

10.4   Human resources management (HRM)

All companies need to undertake the management of their staff, through the packaging of workloads, recruitment, training and personal development, reward and recognition, and where necessary applying discipline. Because of the specialist nature of much of these tasks, the responsibility for leading and advising on the management of the people is best vested in a dedicated HR team, usually headed by a board member or senior manager. However, the application of the HRM functions is actually spread across the company with the active involvement of all managers and supervisors within the company, with the HR team acting as the drivers and advisors. (There is usually a set of internal online portals that employees and managers can access for information on the company's HR policy, as well as a system for following up enquiries.) Nonetheless, the HR team do have the responsibility for advising the senior management of the company on the need for new people and skills in the future – normally presented in the form of forward projections and plans. In this section we take a brief view of the principles of HRM as a subject, recognising that the remit of the actual HR teams (usually a Department) will vary between companies.

10.4.1   Allocating the work and recruitment

The primary HR task is the scoping of the work necessary for a company to meet its business objectives. The work needs to be partitioned into manageable chunks which can be matched to individuals or teams with specific qualifications, skills and experience. This collection of work chunks then has to be considered in the context of existing or proposed company organisation structures. Logically this leads to a set of job descriptions (JD) for each identified job, together with a specification of the qualifications, skills and experience required of the job holder. The actual writing of the JDs is the responsibility of the relevant company managers, with the HR team providing guidance. One of the main challenges for the HR team is to ensure that the emolument package (salary, bonus scheme, company car, etc.) associated with each JD reasonably matches the 'market rate' for recruitment at that level, is compatible with other jobs, and in-line with company policy.

The next big HRM area is that of recruitment of staff against the relevant JDs. There will be preferred means of seeking staff through advertising, recruitment agencies, etc., which the HR team will ensure is being followed. In many companies there is a policy for all interviews and selection procedures to have a member of the HR team present, to ensure that the national legal requirements for non-discrimination are met.

Finally, the HRM role covers the training of all employees. For new recruits some form of induction is required to acquaint them with the company objectives, policies and practices. Many companies with a large professional workforce, particularly ICT companies, provide a training programme specifically for their graduate intake. This approach usually gives the graduates a solid grounding for their management career throughout the company. Such training tends to be separate from the range of training courses provided to all employees throughout their time in the company. Normally, the training is directly provided by a specialist department within the HR team. In other cases, the actual delivery of courses may be outsourced to external agencies, colleges and universities, while under the oversight of the HR team. As discussed further, many employees feel that the opportunity to study and be trained is a reward for good performance – and hence it is generally highly valued.

10.4.2 *Managing performance*

Probably one of the most contentious aspects of HRM is that of reward and recognition of peoples' performance at work. In section 10.2.1 we discussed the various motivation drives for people, and how these vary across the population of employees. Consequently there will be differences in what features will spur people on to greater efforts. For example, it is normally expected that the sales team in the company, because of the strong competitive nature of their role, will be motivated by financial rewards linked to the level of sales – typically a bonus scheme; while other employees will respond well to verbal or written recognition of their efforts by senior managers, perhaps also with the kudos of having extra responsibilities added to their role. The appropriate reward schemes in terms of pay and bonuses need to be devised by the HR team in consultation with the line managers. Normally, the senior management of a company will place money in the budget each year for staff pay rises and bonuses, which the line managers then allocate to their staff according their performance that year – all under the watchful eye of the HR team to ensure the rules have been applied appropriately.

Clearly, any reward and recognition scheme requires an acceptable mechanism for the performance of each employee to be assessed by their supervisors and line managers. Furthermore, this scheme should ultimately ensure that the employees are contributing towards the objectives of the company. Often employees are assessed by gauging how well they have met an agreed set of specific personal objectives during the review period – sometimes referred to as 'management by objectives'. The key principle is the objectives are sufficiently well defined so that a fair and reasonable assessment of achievement can be made (see Box 10.4). However, it is important that there is scope within the assessment process to recognise extenuating circumstances and the need for the objectives to be modified. Normally, the regular (typically annual) appraisal reports produced by the staff's line mangers are monitored and stored by the HR team. It is then a matter of company policy how the reports can be used to influence future dealings with each employee, as well as acting as supporting justification. This can make the individual a candidate for any of the following:

- Award of a bonus.
- Award of pay rise.
- More training or other forms of personal development, e.g. mentoring or secondment.
- Redefinition of the job.
- Greater responsibility.
- Move to another job.
- Disciplining.
- Dismissal from the company.

Box 10.4   Defining appropriate objectives

The success of the Management by Objectives concept introduced by Peter Drucker relied on obtaining good quality definitions of objectives [24]. A simple mnemonic 'SMART' became adopted by many organisations as a way of ensuring that managers defined appropriate objectives. The generally accepted meaning of SMART is as follows:

> *Specific*: It needs to be clear exactly what the objective is addressing.
> *Measureable*: The objective must be capable of being quantified.
> *Achievable*: The importance of this feature is axiomatic.
> *Relevant*: This should ensure the objective relates clearly to the person's job.
> *Time-bound*: The timing dimension is often critical for the company, and other people's objectives may also depend on it being achieved on time.

The HR team will need to oversee the execution of the above actions to ensure that the appropriate company rules are being applied, there is fairness to the individual and the work group, and national legal requirements are being met. The areas where advice from the HR team is usually most required is in dealing with poor performers. For example, where the remedy is in redefining or change of jobs for an employee, care needs to be taken to avoid the company being open to charges of 'constructive dismissal' – i.e. creating a 'non-job' for the individual, or one which is at a demonstrably lower level than their current job. Indeed, the whole process of dealing with poor performers is problematic for all organisations: on the one hand other members of the work group will not want the management to allow people to 'get away' with not doing a proper job, while on the other hand the company is bound to follow the (usually) strict employment-protection laws of the country.

10.4.3 Legal requirements

We have seen earlier that one of the key responsibilities of the HR team is to ensure that the management is complying with the appropriate laws as well as company policy in its dealing with the employees. Overall, as a general principle all companies have a duty of care for their employees in the workplace. In addition, there is a wide range of specific employment laws administered at the national, state or federal level (or the EU in Europe). These laws cover the following areas:

(i) *Permitted hours of work.* The law sets a maximum number of hours per week that an employee should work. Of course, there are certain work situations where people are on-call and their work comes in unpredictable bursts (e.g. hospital junior doctors) where the normal limits may not be appropriate, and derogation of the law may apply.

(ii) *Employment practice.* This group of laws and codes of practice covers such issues as non-discrimination in employing people. That means a company must not apply any bias based on gender, race, religion, sexual orientation, or disability to the recruitment, promotion or selection of people for disciplinary action.

(iii) *Conditions of work.* These are usually enshrined in a set of labour laws covering terms and conditions for employees. One example is the application of the national minimum wage and where it doesn't apply.

(iv) *Health and safety at work.* There is a wide range of areas that relate that are covered by the health and safety laws. In the United Kingdom the Health and Safety Executive (HSE) provides advice to employers and employees on what the law demands and practical issues in meeting its requirements. Areas covered include:
- Work-related upper limb disorders (WRULDs) and repetitive strain injuries (RSI) which result from intensive working at machines or (more typically) computer keyboards.
- Similarly for backache.
- Health and safety in the work place: Hazard avoidance and risk assessments.
- Working at height, covering the provision of suitable safety mechanisms, as well as training.
- General working conditions, specifying amount of space, acceptable ambient temperature range, ventilation, welfare facilities and access to drinking water.

A major area of employment legislation in the United Kingdom – known as TUPE – provides for the protection of employees' rights when their company is taken over by (or merged with) another company. TUPE stands for 'Transfer of Undertakings (Protection of Employment) regulations', and is mirrored throughout Europe (Acquired Rights Directive), though not generally in North America or Asia. In the UK TUPE originally dates from 2006 and it was expanded in 2014 [25]. Essentially, TUPE ensures that when employees from any size of company are transferred to another company their terms and conditions (e.g. pay levels, allowances, holidays, pensions, hours and place of work) are also transferred and retained. Also, there are specific responsibilities identified for the donating and receiving companies. However, it is important to note that changes can be made to the employee's terms and conditions (T&C) – known as 'measures' – provided such changes are not primarily due to the transfer. That means that there must be economic, technical or organisational ('ETO') reasons for the receiving company to undertake the measures. The measures may include:

- redundancies
- workplace relocation
- changes to staff pay dates
- different work patterns
- different pension arrangements
- minor changes to T&C

Thus, e.g., the receiving company may need to rationalise its now-larger organisation to create a sustainable economic arrangement, so some measures will need to be undertaken. These may even include the introduction of redundancies. The TUPE law ensures that both donating and receiving companies provide adequate communication and negotiation with the relevant employees and their trade unions before decisions are made. It is also necessary for the donating company to make available relevant information about the employees to the potential receiving company before the transfer so that they can undertake due diligence on the workforce commitments prior to the M&A deal.

10.5   Transition management

Not surprisingly, the biggest challenges in managing people within a company arise when a major transition is being introduced. This may be in the form of a reorganisation, especially if it is company-wide. Alternatively, it may be a change in the technology being deployed in the network, computer or resource centres, with corresponding changes in the support and test systems. In either case there may also be changes in processes accompanying the new organisation design or technology. All of these changes introduce uncertainty and worry for the people concerned. Many of the staff will feel threatened by the unknown, or their ignorance of the new technology, or perhaps their loss of status in the new organisation. In this section we consider how best to manage transitions to the organisational design or operations of a company, as well as managing major change programmes to the technology used. The techniques used in managing the implementation of any type of projects – technical, process, new product introduction, organisation, etc. – are described in the appendix.

10.5.1 Organisational design changes

Big organisations in most industries suffer from the lack of efficient working internally due to the difficulties in the communication and work flows between various parts of the organisation. Informal domains tend to become established within the organisation, in which the power is achieved because of their position in the work flows. Particular individual managers and teams become 'power barons'. Invariably, the complexity tends to increase with time – new layers of management get added, and progressively more power barons get involved in each decision within the company. At some point the senior managers realise that the organisation is not fit for the purpose – this may be sparked off by a crises in customer service or financial results, by a critical review by outside consultants, or perhaps the new perspective brought by the arrival of an externally recruited person. In fact, many company re-organisations are set off by the arrival of new senior managers who, quite naturally, want to make their mark by creating a difference – i.e., the 'new broom' effect.

Initiating the trauma of a big reorganisation within a company should be done with a clear objective in mind. Ideally, it should have a strategic purpose, such as making the organisation 'customer centric', more streamlined in order to improve performance, reduce costs, etc. However, in practice the reorganisation may be just to meet the personal preferences of one or two senior managers. Nonetheless, the re-organisation will need to have a strategic reason attributed to it (ideally a credible one!) which can be communicated to the staff across the company. Furthermore, the current state of the company and the way it undertakes its main tasks (e.g. the provision of mobile service) should be measured and captured in a 'Current Operational Model' document. The improvements resulting from the change of organisation – and usually associated processes and work flows across the company – can then be assessed against the defined 'Future Operations Model'. Moving the company towards the Future Operations Model way of working will require changes to many, if not all, of the processes within the company (see Box 10.5 and Chapter 9). Whilst the HR team will take control of the 'Develop people' process rebuilding during the reorganisation, they will also need to have an oversight of the way that the changes to the other processes are introduced within the company, to ensure the people aspects are adequately addressed. It is likely that any proposed changes to roles, responsibilities or attendance patterns will need to be discussed with the relevant trade unions – also, undertaken within the oversight of the HR team.

In a well-run company, organisational change will not be a stand-alone, knee-jerk response to operational or market difficulties, much-less the brainchild of one or two senior managers. Rather, it will be a natural component of the response to a company-wide strategic review (See Chapter 3). Organisational change needs to be approached with the same rigour and discipline as any other strategic change. Companies should therefore employ a change-management technique, such as Kotter's Eight-stage Process for Strategic Change (see section 3.7.2).

Box 10.5   The set of processes within a company

Transitions within a company usually result in the need to change some or all of the processes – i.e. the way that the work flows through the organisation. Although companies will differ in the detail of their work-flows and organisational design, e.g. ICT service activities compared to manufacturing, a basic set of processes will always be required within any company. They may even spread out to suppliers and customers' organisations as well. Figure 10.8 illustrates such a generic set – with the processes grouped into managing, operating and providing support to the company's business activity.

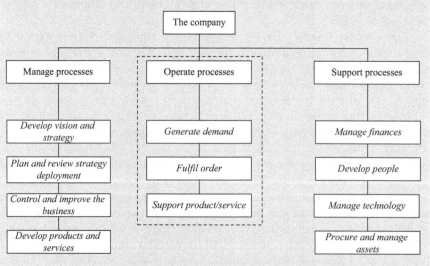

Figure 10.8   Generic process model for a company [26]

10.5.2   Changes to technology of network or resources centres

The introduction of new technology to a network, either as an enhancement to the overall network, or eventually to replace the existing network, involves a major programme of work that needs to be carefully planned. The objective of implementing the programme is to install the new technology and restructure the network, while incurring minimal adverse effect on customer service, as described in Chapter 6. Examples of this sort of network transformation include:

- Replacing the old manual exchanges by automatic exchanges.
- Replacing the electromechanical exchanges (e.g. Strowger, Rotorary) by semi-electronic exchanges.
- Replacing the semi-electronic analogue exchanges (e.g. Crossbar and Reed relay) by digital TDM exchanges.
- Replacing analogue transmission systems by digital PCM systems.
- Replacing the first-generation analogue cell mobile systems by 2G (e.g. GSM).

In all of these examples above there were consequential changes to the way that staff were deployed and the new training and skills required, changes to the work processes to be implemented, and in many cases radical changes to the accommodation and locations of work. A notable example of the latter resulted from the transformation undertaken in many countries from the analogue semi-electronic exchanges to the electronic digital TDM exchanges in the 1980s and 1990s. This shift from bulky racks of electro-magnetic switches to electronic semi-conductor-based technology resulted in significantly less floor space required, resulting in large redundant areas in the exchange buildings. More importantly, the maintenance load with the electronic equipment was less than before, and much of the fault diagnostic activity could be undertaken by remotely located technicians. The consequence of this being that fewer technicians were required to maintain the telephone service, and those that were required could be grouped in remote centres. The HR implications of this transition included:

- the need for retraining of the technician work force to cope with the new electronic equipment;
- coping with the move to remote working for the maintenance staff;
- changes to work procedures;
- reduction in the number of staff required to run the network, with the implications for redundancies and staff redeployment to other duties, as appropriate.

Of course, a major transformation for network operators is the introduction of the next generation networks (NGN) for both fixed (NGN) and mobile operators (LTE and 4G). The radical shift from the existing to the new technology handling voice calls in an NGN leads to significant changes required in many HR areas. Box 10.6 illustrates the scope of these required changes is shown as a series of new HR programmes which need careful management of their implementation.

Box 10.6   NGN human resources programme plan

The set of projects within an NGN HR programme and the objectives are listed below:

Manpower resourcing plan

- Sufficient people to plan, trial and install NGN
- Adequate resources to operate and maintain NGN post implementation
- Redundancy programme

Training

- Adequate skills to install and maintain NGN

Performance management and reward

- Set appropriate targets, objectives and pay levels for NGN staff

Employee industrial relations

- Work with trade unions to agree new attendance patterns, processes and reward

Organisational design

- Restructure work processes and management structures more appropriate to NGN operations

Employee communications

- Ensure staff are kept involved and informed in all changes and progress

Cultural change

- Appropriate changes to leadership styles, behaviours and engagement

10.6   Conclusions

In this chapter we have considered the principles of managing people in the workplace, recognising that they represent one of the most important assets for all types of companies. We then addressed the objectives of organisation design, and how the control and communication between the separate parts of the organisation will differ with the various types of structure. In particular, we considered how functional, divisional, and geographic structures apply to ICT companies. Finally, we have looked at the role of HR management within a company, noting that the HR role is actually dispersed across all managers and supervisors within the organisation, generally with the HR team taking the central drive and advisory role.

References

[1]   Lanigan, M. *Engineers in Business: The Principles of Management and Product Design*, Addison-Wesley, Wokingham, 1992, Chapter 4.

[2]   Pettinger, R. *Introduction to Management*, 3rd edition, Palgrave, 2002, Basingstoke, Chapter 20.

[3]   Ten Have, S., Ten Have, W. & Stevens, F. *Key Management Models*, FT Prentice Hall, London, 2003.

[4]   Fife, E. & Pereira, F. 'Maslow's Hierarchy of Needs and ICT: Challenges of End-User Adoption of Digital Life', *The Journal of the Institute of Telecommunications Professionals*, Vol. 2, Part 1, 2008.

[5]   http://www.clearlycultural.com/geert-hofstede-cultural-dimensions/

[6]   Ambler, T. *Marketing and the Bottom Line: The Marketing Metrics to Pump up Cash Flow*, 2nd edition, FT Prentice Hall, London, 2003, Chapter 6.

[7]   Pettinger, R. *Introduction to Management*, 3rd edition, Palgrave, 2002, Basingstoke, Chapter 19.

[8]   Pettinger, R. *Contemporary Strategic Management*, Palgrave Macmillan, Basingstoke, 2004, Chapter 17.

[9]   Virgin Entrepreneur. *Richard Branson's Top 20 Virgin Inspirational Insights*, www.virgin.com/entrepreneur.

[10] Pettinger, R. *Contemporary Strategic Management*, Palgrave Macmillan, Basingstoke, 2004, Chapter 16.

[11] Peters, T.J. & Waterman Jr., R. *In Search of Excellence – Lessons from America's Best-Run Companies*, Harper & Row, New York, 1982, Chapter 3.

[12] Burns, J.M. *Leadership*, Harper & Row, New York, 1978.

[13] Buckingham, M. 'What Great Managers Do', *Harvard Business Review*, March 2005, 70–79.

[14] Babiak, P. & Hare, R.D. 'Snakes in Suits: When Psychopaths Go to Work', Harper Press, 2006, Chapter 1.

[15] Greenleaf, R.K. 'The Servant Leader Within: A Transformative Path'. Paulist Press, Mahwah, New Jersey, November 2003.

[16] Peter, L.J. & Raymond, H. *The Peter Principle: Why Things Always go Wrong*, William Morrow & Company, New York, 1969, p. 8.

[17] Mintzberg, H. 'Organisation Design: Fashion or Fit?' *Harvard Business Review*, January–February 1981, 103–116.

[18] Pettinger, R. *Contemporary Strategic Management*, Palgrave Macmillan, Basingstoke, 2004, Chapter 12.

[19] Lanigan, M. *Engineers in Business: The Principles of Management and Product Design*, Addison-Wesley, Wokingham, 1992, Chapter 5.

[20] Pettinger, R. *Contemporary Strategic Management*, Palgrave Macmillan, Basingstoke, 2004, Chapter 14.

[21] UK Bribery Act, 2010, www.legislation.gov.uk/ukpga/2010/23.

[22] Muschamp, P. 'Energy, Carbon Emissions and ICT', *The Journal of the Institute of Telecommunications Professionals*, 2012, Vol. 6, Part 3, 2012, 8–13.

[23] Vagadia, B. 'Outsourcing in the Telco World – If it Doesn't Touch the Customer Why Can't It Be Outsourced?', *The Journal of the Institute of Telecommunications Professionals*, 2011, Vol. 5, Part 1, 9–17.

[24] Pettinger, R. *Introduction to Management*, 3rd edition, Palgrave, 2002, Basingstoke, Chapter 5.

[25] www.acas.org.uk

[26] Smart, P.A., Maull, R.S. & Childe, S.J. 'A Reference Model of 'Operate' Processes for Process-Based Change', *International Journal of Computer Integrated Manufacturing*,' 1999, Vol. 12, No. 6, 471–482.

Appendix

Project management

Introduction

A project is defined as a piece of work undertaken to produce a specified result. That means that specific resources are allocated to the project with clear objectives and a set duration. It should also have a manager responsible for orchestrating the execution of the task – i.e. the project manager (PM). Throughout this book we have identified where projects are required, e.g. to implement new products, introduce new technology to the networks or data centres, or expand the coverage area of a cellular network. Invariably, a business case is required to gain authorisation for the allocation of resources to the project. (Indeed, production of the business case itself is usually managed as a project.) This means that senior management within the company will oversee the successful completion of the project, since the result is usually a contribution to the company's potential revenue, cost base or reputation (e.g. meeting a regulatory requirement for telecommunications-service coverage). Project management is now a well-established discipline applied in most industries and administrations around the World. This appendix can aim to give only an outline of the main principles of project management and some of the tools available; more information can be found in the bibliography and references cited at the end.

The objectives of project management

The objectives of applying a systematic approach to the management of projects are to ensure delivery of the desired outcome is achieved within the planned timescale, within the cost budget and with the required level of functionality. Project management techniques enable the progress of a project during its implementation to be tracked and any deviations from the plan identified. The task of the PM is to first establish the project plan in association with the project's client or owner (i.e. the manager responsible for the successful completion of the project) and then, during the implementation, to focus on keeping track of all component project activities on behalf of the client. In the event of deviations from the plan, the PM will work with the client to agree any remedial action.

The hierarchy of projects

In practice, the implementation of a major task is managed as an overall programme which comprises a series of projects. In some cases, the projects may themselves comprise a set of sub-projects. This hierarchal assembly is shown in Figure A.1. Typically, a business case will first be presented covering the programme, giving a top-level view of the main task, its objectives and the outline financial case, and expected expenditure requirements during its life time. As explained in Chapter 4, the component projects (and sub-projects) will then have separate business cases presented during the programme – perhaps related to individual tranches of capital expenditure. This idea of splitting a large programme into the component projects enables the company to commit to relatively small levels initially, with the business cases for subsequent tranches benefitting from the successful completion of the earlier projects. Each component project should have discrete objectives and benefits so that, should the decision be made to withdraw from the overall programme, earlier expenditure is not wasted. The application of project management applies to all three levels of the hierarchy of projects, and each component will be allocated a PM.

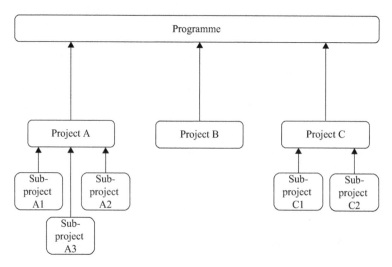

Figure A.1   The hierarchy of projects

Project scheduling and planning

The systematic approach to planning and tracking a project relies on the analysis of the task into its various activities and the identification of how these relate to each other. The most popular method of displaying the project management information is the Network technique, which dates back to the 1950s. This recognises that each task has an expected duration, a set of required resources (money, manpower, etc.), and must be completed before the next dependent task is started. For example, in building a house the first task of preparing the land must be completed before the foundations can be built, followed by the building of the walls; then the roof started once the walls are completed. The Network for this example would consist of each of these tasks with an arrowed line linking to the next dependent task. In the case where parallel activity contributes to the end result, say the laying of water pipes in the grounds leading to the house, these would be shown on the Network as a separate string of tasks with the final arrow terminating at the point in the house-building project where all utility services need to have been installed.

A convenient way of presenting the timing information associated with a network is by using the activity-on-node (AON) representation, shown in Figure A.2. The use of the AON format at each node in the Network enables all the relevant times to be easily calculated and any departures from plan to be clearly visible. It is best to explain the Network process through the use of an example, such as building a fibre access network. Table A.1 presents the schedule of the tasks A to J in our example project, showing their durations and dependencies. Task A might be to survey the estate to be fibred and Task J might be to complete customer testing. Clearly, this is a highly simplified example but it serves to explain the process. The first step is to construct the corresponding Network, as shown in Figure A.3.

The second step is to make a forward pass through all paths of the network, calculating the earliest finishing time (EFT) and earliest start time (EST) for all

Earliest start time (EST)	Duration	Earliest finish time (EFT)
Task identifier		
Latest start time (LST)	Total float	Latest finish time (LFT)

Figure A.2   Activity on node (AON) project-management networks

Table A.1   The schedule of tasks for the project

Task ID	Predecessors	Duration
A	–	2
B	A	6
C	B	5
D	A	8
E	D	10
F	A	4
G	F	7
H	G	20
I	C, H	3
J	I	9

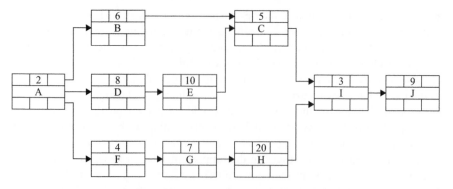

Figure A.3   Step one: construct the AON network

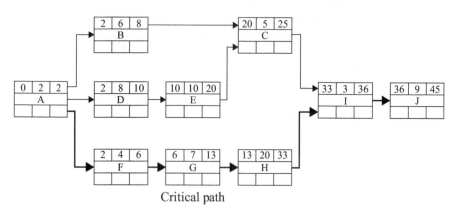

Critical path

Figure A.4   Step two: forward pass

tasks. We note that the EST of a task is equal to the EFT of the predecessor task; in the case where there are more than one predecessor, the EST is set by the predecessor with the latest EFT, as shown for our example in Figure A.4. At this stage we are able to see which of the paths through the Network will take the longest time – this is known as 'the critical path', since it sets the overall duration for the project and is therefore the most sensitive to any slippage. The non-critical paths all have some slack; the maximum amount being known as the 'total float' for that path. (The slack between two tasks is known as the 'free float'.) In our example, Task C (laying of new duct) depends on the completion of Task B (delivering plant to site) at week 8 and Task E (letting the construction contract) at week 20, so its earliest completion is 20 plus 5 weeks duration, totalling 25 weeks. Task I (installing the fibre) depends on the completion of Task C (planning fibre sizes) at week 25 and Task H (training staff) at week 33 – thus it cannot start until week 33. This leads to the completion time for the project being 45 weeks. The critical path is A-F-G-H-I-J.

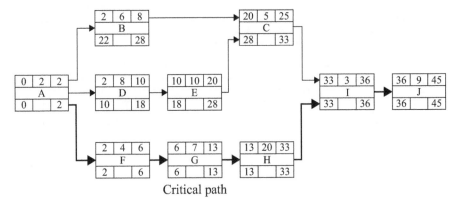

Figure A.5 Step three: backward pass

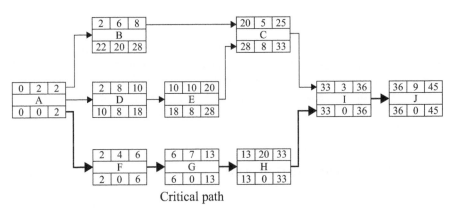

Figure A.6 Step four: calculate the floats

Step three is the backward pass through the network, in which we calculate the latest start time (LST) and the latest finish time (LFT), working from the final activity AON, as shown in Figure A.5. We note that for each node on the critical path the pairs of start times and the finish times are both equal, i.e. the EST = LST and the EFT = LFT.

Finally in step four, at the AON boxes the floats for each task (i.e. the amount of time a task can be delayed without affecting the project completion time) are calculated. The floats equal the differences between the latest and the earliest start times (i.e. LST-EST). All fields in the AON boxes are now complete (see Figure A.6). In our example, we can see that the total floats vary from 0 (on the critical path) to 20 weeks for task B.

Many people now use software tools to construct and record the project management networks – e.g. 'Microsoft Project'. There are a number of techniques associated with the use of networks, e.g. PERT (Project Evaluation and Review Technique) which applies a statistical tool to the project-management data [1].

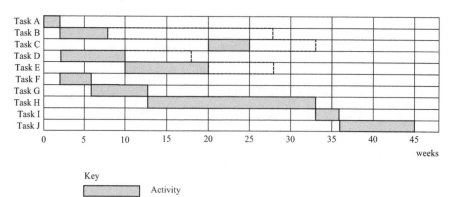

Figure A.7 Gantt chart representation of the example

A Gantt chart provides a helpful additional method of displaying the set of project tasks using calendar dates, rather than the duration in days or weeks as used in the network. Figure A.7 shows a simplified Gantt chart for our example. This gives the company's managers a convenient method of relating the project tasks to the planned dates of delivery, and so proves to be a useful project-reporting format for the PM.

The project management process

Once the project starts, the PM keeps track of its progress against the plan. The two main parameters of interest for the PM are project expenditure and time. These can be matched against the plan and the departures evaluated. The PM needs to flag up to the client when a project is in trouble due to late running or over-spend. The decision on the need for remedial action and what should be done depends on the original objectives of the project, as described below.

(a) *Cost savings.* If the objective was to create a lower cost infrastructure, the project should now be managed to minimise the cost of implementation to achieve the lower cost base, even if this means the introduction of further delays or perhaps some loss of functionality in the final delivered item.

(b) *Time critical.* If the objective was to meet a particular date – say for a product launch ahead of the competition, then remedial action should focus on doing what is necessary to meet the target date, even if this incurs extra cost or some loss of functionality.

(c) *Functionality.* If the objective was to create a platform with certain features and capabilities, then remedial action should involve extra work (e.g. design and build) in order to achieve the required functionality. This may involve extra cost and delay to the project.

On completion of the project the final task of the PM is to produce a report reviewing how well it was executed. This should highlight any serious underestimates of costs or time, and any major problems encountered – hence any lessons that might apply to future projects. In some cases it may also be appropriate to complete a further review in 12 or 24 months to ensure that the expected benefits are being achieved.

Reference

[1]   Lanigan, M. *Engineers in Business: Principles of Management and Product Design*, Addison-Wesley, Wokingham, UK, 1992, Chapter 6.

Bibliography

[1]   Meredith, J.R. & Mantel, S.J., Jr. *Project Management: A Managerial Approach*, 8th edition John Wiley & Sons Inc., New Jersey, 2012.
[2]   Microsoft Project, www.support.office.com
[3]   APMPGanttExNet.gif, www.ganttchart.com
[4]   The Association for Project Management (APM), www.apm.org.uk
[5]   The Project Management Institute (PMI), www.pmi.org

Index